T0295469

Urban Climate Change and Heat Islands

Urban Climate Change and Heat Islands

Characterization, Impacts, and Mitigation

Edited by

RICCARDO PAOLINI
School of Built Environment, Faculty of Arts, Design and Architecture, University of New South Wales (UNSW), Sydney, NSW, Australia

MATTHAIOS SANTAMOURIS
School of Built Environment, Faculty of Arts, Design and Architecture, University of New South Wales (UNSW), Sydney, NSW, Australia

ELSEVIER

Elsevier
Radarweg 29, PO Box 211, 1000 AE Amsterdam, Netherlands
The Boulevard, Langford Lane, Kidlington, Oxford OX5 1GB, United Kingdom
50 Hampshire Street, 5th Floor, Cambridge, MA 02139, United States

Notices
Knowledge and best practice in this field are constantly changing. As new research and
experience broaden our understanding, changes in research methods, professional practices,
or medical treatment may become necessary.

Practitioners and researchers must always rely on their own experience and knowledge in
evaluating and using any information, methods, compounds, or experiments described
herein. In using such information or methods they should be mindful of their own safety
and the safety of others, including parties for whom they have a professional responsibility.

To the fullest extent of the law, neither the Publisher nor the authors, contributors, or
editors, assume any liability for any injury and/or damage to persons or property as a matter
of products liability, negligence or otherwise, or from any use or operation of any methods,
products, instructions, or ideas contained in the material herein.

ISBN: 978-0-12-818977-1

For Information on all Elsevier publications
visit our website at https://www.elsevier.com/books-and-journals

Publisher: Candice Janco
Acquisitions Editor: Jessica Mack
Editorial Project Manager: Catherine Costello
Production Project Manager: Sruthi Satheesh
Cover Designer: Victoria Pearson

Typeset by MPS Limited, Chennai, India

Working together
to grow libraries in
developing countries

www.elsevier.com • www.bookaid.org

Contents

    Hassan Saeed Khan, Riccardo Paolini and Matthaios Santamouris

List of contributors

Carlos Bartesaghi-Koc
School of Architecture and Built Environment, ECMS, The University of Adelaide, Adelaide, SA, Australia

Jie Feng
School of Built Environment, Faculty of Arts, Design and Architecture, University of New South Wales (UNSW), Sydney, NSW, Australia

Kai Gao
School of Built Environment, Faculty of Arts, Design and Architecture, University of New South Wales (UNSW), Sydney, NSW, Australia

Samira Garshasbi
School of Built Environment, Faculty of Arts, Design and Architecture, University of New South Wales (UNSW), Sydney, NSW, Australia

Shamila Haddad
School of Built Environment, Faculty of Arts, Design and Architecture, University of New South Wales (UNSW), Sydney, NSW, Australia

Hassan Saeed Khan
School of Built Environment, Faculty of Arts, Design and Architecture University of New South Wales (UNSW), Sydney, NSW, Australia; Data-61, The Commonwealth Scientific and Industrial Research Organization (CSIRO), Kensington, Perth, WA, Australia

Mathew Lipson
ARC Centre of Excellence for Climate Extreme, Sydney, NSW, Australia

Negin Nazarian
School of Built Environment, Faculty of Arts, Design and Architecture, University of New South Wales (UNSW), Sydney, NSW, Australia; City Futures Research Centre, University of New South Wales (UNSW), Sydney, NSW, Australia; ARC Centre of Excellence for Climate Extreme, Sydney, NSW, Australia

Leslie K. Norford
Department of Architecture, Massachusetts Institute of Technology, MA, United States

Riccardo Paolini
School of Built Environment, Faculty of Arts, Design and Architecture, University of New South Wales (UNSW), Sydney, NSW, Australia

Gianluca Ranzi
School of Civil Engineering, The University of Sydney, Sydney, NSW, Australia

Matthaios Santamouris
School of Built Environment, Faculty of Arts, Design and Architecture, University of New South Wales (UNSW), Sydney, NSW, Australia

Giulia Ulpiani
School of Built Environment, Faculty of Arts, Design and Architecture, University of New South Wales (UNSW), Sydney, NSW, Australia

CHAPTER 1

Urban climate change: reasons, magnitude, impact, and mitigation

Matthaios Santamouris
School of Built Environment, Faculty of Arts, Design and Architecture, University of New South Wales (UNSW), Sydney, NSW, Australia

1.1 Introduction

Cities increase their boundaries and population. While the urban population in 1960 was close to 1 billion, reaching 3.33 billion in 2007, it grew to 4.1 billion by 2017, representing almost 55% of the world population (Ritchie & Roser, 2018). Future projections show that by 2050 urban population may reach 7 billion people. An increase in the urban population is associated with a spectacular growth of the size of megacities, cities hosting more than 10 million population. While in 1990, only 10 cities presented a total population above 10 million, the number increased to 28 million in 2014 and 33 million in 2017 (Young, 2019).

Apart from the urban population increase, the density of cities in the developed world has surged to unprecedented levels. Cities like Mumbai and Kolkata, India, and Karachi, Pakistan have tremendous population densities around 77,000, 62,000, and 50,000 people per square mile, respectively. Urban population densities have also increased in developed countries without reaching the aforementioned density figures. For example, population densities in Tokyo megacity, Japan and Athens, Greece, are close to 12,300 and 14,000 people per square mile, respectively, while the density in Sydney, Australia is not exceeding 1100 people per mile. Such a tremendous increase of the absolute population figures and densities are a serious challenge affecting the local climate, use of resources, disease control and health services, education and employment opportunities, networks, infrastructures, and facilities. Poverty, unemployment, and lack of proper shelters oblige almost 1 billion people to live in informal urban settlements or slums in completely unacceptable hygienic and

Urban Climate Change and Heat Islands
DOI: https://doi.org/10.1016/B978-0-12-818977-1.00002-8
1

climatic conditions (United Nations Human Settlements Programme UN-HABITAT, 2019). Unfortunately, future predictions show a dramatic growth of the slums population, and it is predicted that by 2050 it will reach about 3 billion, or more than 30% of the world population, by the middle of the current century (United Nations Department of Economic and Social Affairs, 2017).

Apart from the tremendous increase of the urban population and urban densities, several other issues influence the magnitude of urban temperatures. Design and construction practices in cities favor highly absorbing and super warm materials for buildings and open spaces like concrete and asphalt. The growth in urbanization is reflected also in the world production of cement, which has increased from almost 3700 Mt in 2010 to 4100 Mt in 2018 (International Energy Agency, 2019). In parallel, the world demand for asphalt has risen from 103.3 million metric tons in 2005 to 119.5 million metric tons by 2015 (Statista, 2016). Highly absorbing materials used in the outdoor skin of buildings and open urban places reach very high surface temperatures during the summer, transferring the absorbed and stored heat to the ambient air contributing enormously to urban overheating.

The increase of the urban population and improvement of the quality of life of the mid-class population— combined with an increasing economic capacity of households—has resulted in a dramatic growth of the number of individual cars, increasing the anthropogenic heat released in cities. It is expected that the total car sales will increase from 70 million per year in 2010 to about 125 million/year in 2025 (Bouton et al., 2015). About half of the cars are bought in cities, while future projections show that the fleet size, 1.2 billion cars, may be doubled by 2030 (Bouton et al., 2015). Cars release heat and pollution in cities, increasing the urban ambient temperature and obviously causing traffic congestion, which in several countries cost about 2%–4% of the national Gross Domestic Product (GDP) (Bouton et al., 2015).

Global climate change increases the magnitude of the ambient temperature and the frequency of extreme heat events (IPCC, 2000). Because of the important synergies between the global and local climate change, higher ambient temperatures intensify the magnitude of the regional overheating (IPCC, 2000), while increasing the demand for building cooling and raising the release of anthropogenic heat from the air conditioners. Future projections outline a considerable increase in the cooling degree days in almost every part of the planet (Warren et al., 2006). Fig. 1.1

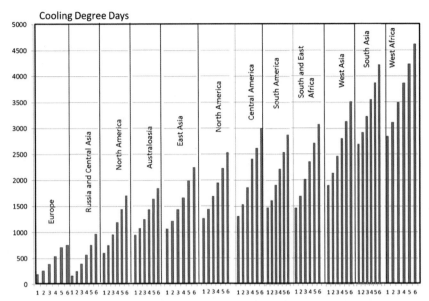

Figure 1.1 Current and future cooling degree days for the major areas of the planet and for different climatic scenarios. (1) Baseline. (2) Increase between 0K and 1K. (3) Increase between 1K and 2K. (4) Increase between 2K and 3K. (5) Increase between 3K and 4K, and 6K. (6) Increase between 4K and 5K (Santamouris, 2016a,b).

presents the predicted cooling degree days for the major parts of the planet, considering an increase in the ambient temperature between 0 and 5 K (Santamouris, 2016a,b; Warren et al., 2006). As shown, for specific regions like South-Eastern Asia, the expected increase of the cooling degree days and cooling energy demand is extremely high.

Regional climate change depends highly on the socioeconomic pathways followed in developed and developing countries. Future economic growth may define the levels of the future greenhouse emissions and thus the need for adaptation and mitigation. Predictions of the future economic growth and the world GDP depend on the specific assumptions of the models and present a high uncertainty. Existing predictions for 2050 differ substantially in terms of the predicted average GDP per capita. Fig. 1.2 presents the results of 19 published models (Santamouris, 2016a,b). As shown, the average GDP per capita may vary between 7200 and 26,400 US$ of 1990. However, a common denominator of all scenarios is the serious amplification of the economic differences between the various geographic parts of the world (Fig. 1.3). The used GDP prediction data are taken from the Massachusetts Institute of Technology emission scenario (MIT, 2021).

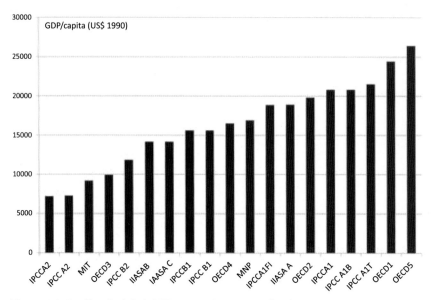

Figure 1.2 Predicted global GDP per capita in 2050 by the various emission scenarios (in US$ 1990). *From Santamouris, M., 2016a. Cooling the buildings – past, present and future. Energy and Buildings, 128, 617–638. https://doi.org/10.1016/j. enbuild.2016.07.034. Santamouris, M., 2016b. Innovating to zero the building sector in Europe: mminimising the energy consumption, eradication of the energy poverty and mitigating the local climate change. Solar Energy, 128, 61–94. https://doi.org/10.1016/j. solener.2016.01.021.*

Predictions have shown that the GDP in the less developing countries may be almost 85% lower than that of the developed ones. Given that most of the future megacities are located in the developing countries, where most of the future urban population is expected, low economic development will affect the quality of the cities, their infrastructure, the energy consumption, and the related environmental policies and very probably will result in a serious increase of the urban overheating.

Overpopulation and economic growth drive the future development of residential and commercial buildings. While in 2010, the total floor area of residential buildings in the world varied between 140 billion square meters and 190 billion square meters (Global Energy Assessment Writing Team, 2012; Urge-Vorsatz et al., 2013), it is expected to increase up to 180–290 billion square meters by 2030 and 190 – 379 billion square meters by 2050. In parallel, the total area of the commercial buildings is expected to rise by 2050 between 25 billion square meters and 30 billion square meters compared to 21–24 billion square meters in

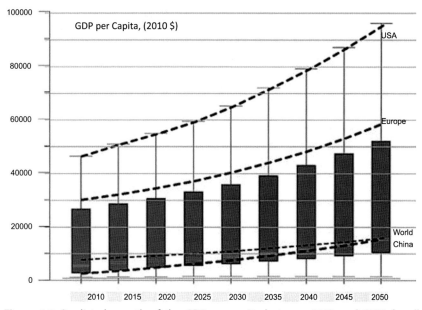

Figure 1.3 Predicted growth of the GDP per capita between 2010 and 2050 for all major areas of the world (in US$ 2010). Each boxplot comprises all data from all major zones of the world. *From Santamouris, M., 2016a. Cooling the buildings − past, present and future. Energy and Buildings, 128, 617−638. https://doi.org/10.1016/j. enbuild.2016.07.034. Santamouris, M., 2016b. Innovating to zero the building sector in Europe: mminimising the energy consumption, eradication of the energy poverty and mitigating the local climate change. Solar Energy, 128, 61−94. https://doi.org/10.1016/j. solener.2016.01.021.*

2010 (John, 2014; Urge-Vorsatz et al., 2013). Such a massive increase of the total building surface is expected to have a significant impact on their energy consumption and the released anthropogenic heat and also on the overall construction activities and material use. At the same time, it is evident that it will seriously intensify the magnitude of the local overheating.

The future energy consumption of buildings may be predicted considering all drivers affecting their energy demand at a global level. The levels of future energy consumption determine at large the magnitude of the greenhouse gas emissions and the evolution of global climate change, while the release of the additional anthropogenic heat in the urban environment may seriously intensify the levels of local overheating. Numerous prediction models considering most of the above drivers are developed around the future building energy consumption. Fig. 1.4 presents the predicted future energy demand of the residential sector as calculated by

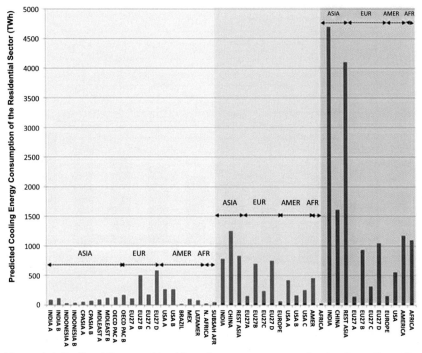

Figure 1.4 Predicted future residential cooling energy consumption by the various existing models. The blue zone (left part of the figure) is for 2030, the green (middle part of the figure) for 2050, and the red (right part), for 2100. *From Santamouris, M., 2016a. Cooling the buildings — past, present and future. Energy and Buildings, 128, 617—638. Santamouris, M., 2016b. Innovating to zero the building sector in Europe: mminimising the energy consumption, eradication of the energy poverty and mitigating the local climate change. Solar Energy, 128, 61—94. https://doi.org/10.1016/j.solener.2016.01.021.*

several models (McNeil & Letschert, 2008; Scott et al., 2008). As shown, a very substantial increase in cooling energy consumption is foreseen for the next 20 − 30 years. It is important that the highest energy consumption is expected in Asiatic countries and in particular in China and India, where the most serious urban problems are expected.

Global climate change, an increase in the urban population, higher urban densities, a significant increase in the number of buildings, and the expected tremendous energy consumption of the building sector are factors that may seriously intensify the magnitude of urban overheating.

Higher urban temperatures have a serious impact on the energy consumption of buildings, indoor and outdoor thermal comfort, the concentration of harmful pollutants, heat-related mortality and morbidity, sustainability and survivability levels of low-income households, while seriously affecting the global economy and well-being of cities (Santamouris, 2020). The precise impact of urban overheating is well quantified and is analyzed and presented in the rest of the book.

To face the problem and counterbalance the impact of local overheating, proper mitigation, and adaptation technologies, measures and policies have to be developed and implemented. In the recent years, serious research has been carried out aiming to counterbalance the impact of regional climate change while numerous large scale projects are undertaken to employ and implement advanced mitigation technologies and measures (Akbari et al., 2016). Monitoring results obtained from a high number of mitigation projects have shown that the use of the currently available mitigation technologies contributes seriously to decrease the peak ambient urban temperature up to $2.5°C-3°C$, improve comfort levels, decrease heat-related morbidity and mortality, reduce the concentration of harmful pollutants, and improve the living conditions in deprived urban areas (Santamouris et al., 2017).

This book aims to present and analyze the causes of urban overheating, the future challenges concerning the local climate change, demonstrate the latest developments of the experimental and monitoring technologies to quantify the characteristics and the magnitude of the local overheating, identify and quantify the impact of overheating on energy, health, environmental quality and economy, and finally present the recent achievements in the field of urban mitigation and adaptation technologies.

1.2 What is causing urban overheating?

The thermal balance in the built environment is defined as the sum of the heat gains, heat storage, and heat losses. As urban areas compared to the surrounding rural or suburban environment present higher thermal gains and less thermal losses, their thermal balance is more positive, and cities present a higher ambient temperature compared to their surroundings. This dynamics is well known as the urban heat island phenomenon (Fig. 1.5), which probably is the most documented phenomenon of climate change (Akbari et al., 1992).

Cities receive and absorb solar radiation. The exact amount of absorbed radiation depends on the solar absorbance of the materials and other urban structures, which is a surface property. Most materials used, like concrete and asphalt, present a high solar absorbance (i.e., the ratio of absorbed to incident solar radiation). The absorbed energy is stored into the mass of the materials,

Sketch of an Urban Heat-Island Profile

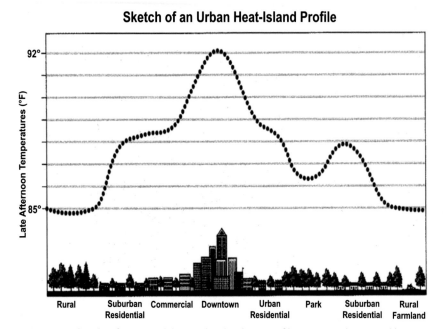

Figure 1.5 Sketch of a typical heat island urban profile. *From Taha, H., Akbari, H., Sailor, D., Ritschard, R., 1992. Causes and effects of heat islands: sensitivity to surface parameters and anthropogenic heating.*

increasing their temperature, and it is released into the atmosphere in the form of convective heat and infrared radiation. In parallel, materials absorb infrared radiation emitted by the atmosphere and the other surfaces in the built environment. Convective losses or gains between the urban surfaces and the ambient air depend mainly on the corresponding temperature difference and the wind speed and turbulence. Anthropogenic heat added to the atmosphere as released by cars, industry, power plants, and the energy systems of buildings increases the energy fluxes in the built environment. Heat transfer by advection in cities affects in a positive or negative way the energy budget as a function of the temperature difference between the ambient and the advected air. Finally, latent heat released through the evaporation of water by urban vegetation and water surfaces helps to decrease the ambient temperature.

Thus the energy balance of the surface—ambient air system can be written as:

$$Q_r + Q_T = Q_E + Q_L + Q_S + Q_A$$

where Q_r is the sum of the net radiative flux, Q_T is the released anthropogenic heat, Q_E is the sum of the sensible heat, Q_L is the latent heat, Q_s is the stored energy, while Q_A is the net energy transferred to or from the urban system through advection under the form of sensible or latent heat. The advective term can be ignored in central urban areas surrounded by an almost uniform building density. Still, it may be imported into the boundaries between the urban and the rural environment.

Usually, the absorbed solar radiation is the term presenting the highest magnitude, and that contributes more to increase the urban temperature. Thus a decrease in the solar absorbance or increase of the solar reflectance of the urban surfaces is crucial to minimize the release of the sensible heat to the atmosphere and decrease the ambient temperature. The infrared radiation emitted by the urban structures contributes highly to lower the ambient temperature, especially during the nighttime. The emissive capacity, that is, the emissivity of the urban materials and structures, highly determines the magnitude of the emitted radiation. However, as the spectral emissivity is equal to the spectral absorptivity of the materials, high emissivity values may result in increased absorption of the emitted atmospheric radiation. Especially in urban zones with a high content of water vapor or atmospheric pollution, the magnitude of the incoming atmospheric radiation may be quite high. Materials presenting a high spectral emissivity in the so-called atmospheric window, that is, between 8 and 13 micrometers, present an additional advantage as the atmospheric radiation at these wavelengths is minimum. Latent heat released by urban vegetation and water surfaces is considerably reduced in cities compared to the rural areas as a result of the limited green and water zones. An increase of the evapotranspirational flux contributes considerably to decrease the ambient temperature and rise the water content in the atmosphere.

Advection gains or losses can be a determinant of heat flux in cities. In coastal zones, the impact of sea breeze helps to reduce the levels of the ambient temperature considerably and fight overheating, especially during the afternoon hours. In parallel, urban zones located close to hot and arid zones like the desert or other heat sources, like power plants or large photovoltaic plants, may have a very negative impact as hot or warm air may be transferred to the city.

Anthropogenic heat released in the urban ambient air varies as a function of the specific characteristics of the city and the relative anthropogenic activities like transport, industry, energy systems of the buildings, etc. Although the average anthropogenic heat flux is small compared to the summertime mid-day solar radiation, waste heat from urban anthropogenic activities may play an important role in the formation and magnitude of the heat island phenomenon. Many experimental and modeling studies have documented that waste heat, mainly from urban energy, transportation systems, and power generation, contributes to increased heat island intensities (Khan & Simpson, 2001; Sailor & Lu, 2004). A methodology to estimate the magnitude of the anthropogenic heat generated in cities is proposed in Sailor and Lu (2004).

Many studies have been performed to calculate the anthropogenic heat flux in urban areas, and a value close to $100 \ W/m^2$ is suggested as an average (Grimmond, 1992; Kłysik, 1996). However, much higher values have been reported for various cities. In the past, it was estimated that the anthropogenic heat in downtown Manhattan was close to $198 \ W/m^2$ (Coutts et al., 2007), while the maximum flux in central London was close to $234 \ W/m^2$, with an average value close to $100 \ W/m^2$ (Harrison et al., 1984). An analysis of the anthropogenic heat released in US cities reports an average flux between 20 and $40 \ W/m^2$ for the summer and between 70 and $210 \ W/m^2$ for the winter period, considering the upper value as an extreme (Hosler & Landsberg, 1997). In Moscow, Budapest, Reykjavik, and Berlin, the average anthropogenic heat flux is estimated close to 127, 43, 35, and $21 \ W/m^2$, respectively, while for Montreal and Vancouver, it is 99 and $26 \ W/m^2$ (Steinecke, 1999; Taha et al., 1992). More recent studies show that the anthropogenic heat flux in the urban area in Tokyo exceeds $400 \ W/m^2$ in the daytime, while the maximum value is close to $1590 \ W/m^2$ in winter (Ichinose et al., 1999). Another analysis of the anthropogenic heat distribution for central Beijing shows that at 0800 a.m. local time, it ranges between 40 and $220 \ W/m^2$ in summer and 60 to W/m^2 in winter (Chen et al., 2007). Finally, estimations for Toulouse, France, showed that anthropogenic heat flows are around $15 \ W/m^2$ during summer and $70 \ W/m^2$ during the winter with peaks of $120 \ W/m^2$ (Pigeon et al., 2007).

Anthropogenic heat can be an important contributor to the thermal environments of cities. Numerical simulations of the urban temperature regime have shown that anthropogenic heat may increase urban temperatures by up to 3°C (Narumi et al., 2003). Using a mesoscale model calculated that the addition of anthropogenic heat in Osaka increases the urban temperature to about 1°C. Detailed simulations for the Tokyo area, reported in Kondo and Kikegawa (2003), show that anthropogenic heating in the Otemachi area resulted in a temperature increase of about 1°C.

In Ichinose et al. (1999), it is estimated that the temperature increase in the same area of Tokyo during the summer period was around 1.5°C at 10 p.m., while much higher differences have been calculated for surface energy balance components.

1.3 About the magnitude of the urban overheating

Urban overheating is experimentally documented in more than 450 large cities in the world. While the existing knowledge on the magnitude of urban overheating is quite rich, it is overshadowed by the problem's accuracy and representativeness of the results and inconsistencies of the experimental data and theoretical conclusions (Stewart & Oke, 2012). There are three main monitoring techniques employed to measure the magnitude of urban overheating: (1) Those based on the use of mobile traverses. (2) Those using standard fixed observation stations. (3) Those using nonstandard observation stations (Santamouris, 2015). In parallel, measurements vary as a function of the number of measuring stations used, the duration of data collection, the reporting format, and the criteria to select the reference station (Stewart, 2011).

Studies based on mobile traverses and nonstandard measuring stations are usually based on data collected for a relatively short period, and the reported magnitude of urban overheating is usually the maximum or the average maximum value measured during the experimental period. When standard measuring stations are used, data may be available for several years, while either the annual average, the annual average maximum, or the annual absolute maximum are reported as the magnitude of urban overheating.

An analysis of the specific levels of urban overheating in 100 Asian and Australian cities is given in Santamouris (2015). In parallel, a similar analysis for 110 European cities is reported in Santamouris (2015). As it concerns the Australian and Asian cities, when mobile traverses are used, the magnitude of the urban overheating varies between 0.4°C and 11.0°C (Santamouris, 2015), while the average intensity is close to 4.1°C. For about 23% and 58% of the examined cities, the magnitude was below 2°C and 4°C, respectively, while 27% of the cities presented an overheating intensity higher than 5°C. When nonstandard meteorological stations are used, the overheating magnitude is found to vary between 1.5°C and 10.7°C, with an average value close to 5°C. Almost 42% of the cities presented an overheating intensity higher than 5°C (Santamouris, 2015).

When measurements are based on multiyear data collected by standard meteorological stations, the annual average, annual maximum average, and annual absolute maximum intensity of the overheating is reported as the magnitude of urban overheating. Figs. 1.6 and 1.7 report the magnitude of the annual average and absolute maximum annual overheating magnitude (Santamouris, 2015). The reported average intensity of the annual mean, mean maximum, and absolute maximum overheating intensity is 1.0°C, 3.1°C, and 6.2°C, respectively. It is characteristic that for about 20% of the cities, the annual absolute maximum intensity of overheating exceeded 8°C (Santamouris, 2015).

In Europe, when standard meteorological stations are used, the corresponding average annual, average maximum, and annual absolute maximum intensity of the urban overheating is 1.1°C, 2.6°C, and 6.2°C, respectively. In parallel, the average maximum magnitude of the overheating when mobile traverses are used is found close to 6°C (Santamouris, 2016a,b).

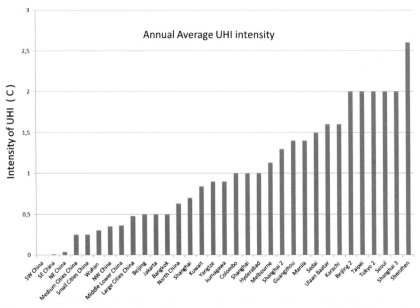

Figure 1.6 Reported intensity of the annual average urban heat island for studies based on standard measuring equipment. *From Santamouris, M., 2015. Analyzing the heat island magnitude and characteristics in one hundred Asian and Australian cities and regions. Science of the Total Environment, 512–513, 582–598. https://doi.org/ 10.1016/j.scitotenv.2015.01.060.*

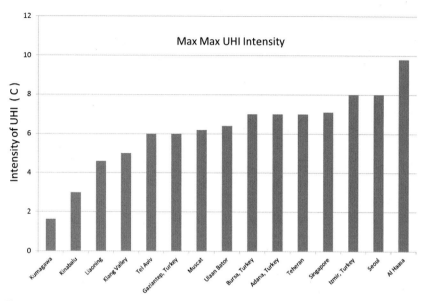

Figure 1.7 Reported intensity of the max—max urban heat island for studies based on standard measuring equipment. *From Santamouris, M., 2015. Analyzing the heat island magnitude and characteristics in one hundred Asian and Australian cities and regions. Science of the Total Environment, 512—513, 582—598. https://doi.org/10.1016/j. scitotenv.2015.01.060.*

It is evident that the intensity of urban overheating is quite high, especially during the summer period, and on average, may exceed 5°C. Such a temperature increase has a severe impact on the energy demand for cooling purposes, while it increases the concentration of harmful pollutants and raises the levels of heat-related mortality and morbidity.

1.4 About the impact of urban overheating

Important research work has been carried out in the recent years aiming to quantify the impact of urban overheating on energy, peak electricity demand, pollution levels, heat-related mortality and morbidity, as well as on urban sustainability and economy.

In particular, the impact of higher ambient temperature on the energy systems is well quantified and documented. As described in Santamouris (2020), urban overheating affects both the energy demand and supply sectors adversely. As it concerns the energy consumption of buildings, overheating raises the cooling energy consumption in cities. It is reported that the additional energy penalty induced by the urban overheating at the city level is around $0.74 \, kWh/m^2/°C$, while the Global Energy Penalty per person is close to 237 ($\pm$ 130) kWh/p (Santamouris, 2014). In parallel, urban overheating is found to cause a significant decrease in the heating demand of buildings in climatic zones with an average summer temperature below 23°C (Santamouris, 2014).

As already mentioned, climatic change is expected to cause in the next decades a very significant increase of the building cooling needs over passing the corresponding heating consumption (Phadke et al., 2014). As documented by several studies, the actual levels of the increase/decrease of the buildings' cooling and heating demand are given in Fig. 1.8 (Santamouris, 2014).

Apart from the energy demand, urban overheating is affecting the energy production sector. According to recent studies, it increases the peak electricity demand obliging utilities to built additional power plants while it raises the cost of electricity. As reported in Santamouris et al. (2015), for each degree of temperature increase, the corresponding peak electricity demand rises between 0.45% and 4.6%. This is equal to an additional peak electricity penalty of 21 ($\pm$ 10.4) W per degree of

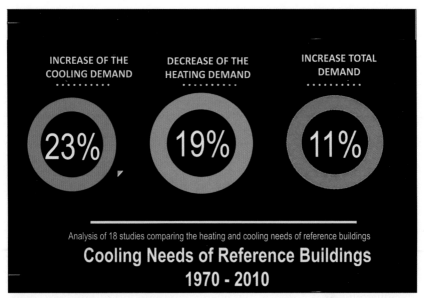

Figure 1.8 Current increase of the heating and cooling need of buildings caused by urban overheating.

temperature increase and per person (Fig. 1.9). Higher ambient temperatures affect the efficiency and the generation capacity of the nuclear and
thermal power plants, increases the power losses between substations and
transformers, and decreases the carrying capacity of the electricity transmission lines (Chandramowli & Felder, 2014; Dirks et al., 2015). Fig. 1.10
reports the main impact of urban overheating on the power production
systems.

An increase in the ambient temperature in cities affects the quality of
life seriously as well as the health of low-income households and raises the
levels of urban vulnerability (Santamouris & Kolokotsa, 2015). The vulnerable urban population used to live in low-quality houses in deprived
urban zones, and a possible increase of the ambient temperature seriously
affects indoor temperature, indoor pollution, and survivability levels
(Kolokotsa & Santamouris, 2015; Smoyer, 1998). Fig. 1.11 reports the
main impact of urban overheating on low-income and vulnerable
population.

Urban overheating increases considerably the concentration of several
harmful pollutants like the ground-level ozone and particulate matter (Lai
& Cheng, 2009). Higher ambient temperatures accelerate the

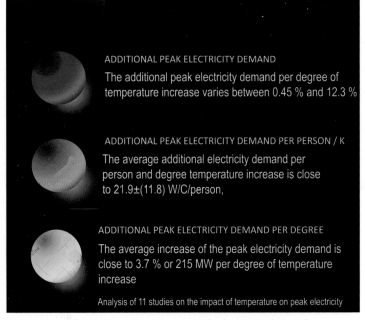

ADDITIONAL PEAK ELECTRICITY DEMAND

The additional peak electricity demand per degree of
temperature increase varies between 0.45 % and 12.3 %

ADDITIONAL PEAK ELECTRICITY DEMAND PER PERSON / K

The average additional electricity demand per
person and degree temperature increase is close
to 21.9±(11.8) W/C/person,

ADDITIONAL PEAK ELECTRICITY DEMAND PER DEGREE

The average increase of the peak electricity demand is
close to 3.7 % or 215 MW per degree of temperature
increase

Analysis of 11 studies on the impact of temperature on peak electricity

Figure 1.9 Impact of urban overheating on peak electricity demand.

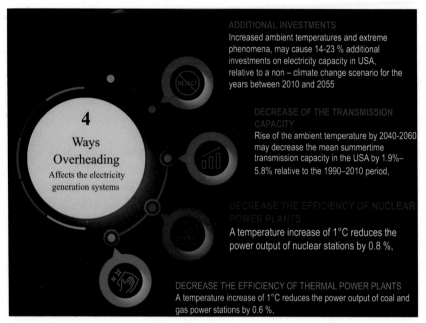

Figure 1.10 Impact of urban overheating on electricity generation systems.

photochemical reactions between pollutants resulting in higher ozone concentrations threatening the health of urban citizens (Yoshikado & Tsuchida, 1996). Fig. 1.12 reports some statistical data on the impact of urban overheating on the concentration of ground-level ozone.

The impact of urban overheating on health seems to be the more alarming one, and it is considered as a peak current and future scientific topic (Gasparrini et al., 2017). High ambient temperature is associated with increased hospital admissions and mortality as the human thermoregulation system cannot offset very high temperatures (Johnson et al., 2005). Fig. 1.13 reports the main elements associating with urban overheating and heat-related mortality. Recent research has proven that the health risk is substantially higher in urban than in rural environments (Ho et al., 2017). It is also found that the risk of heat-related mortality in warmer urban precincts is 6% higher than in cooler neighborhoods (Schinasi et al., 2018). It is characteristic that based on recent epidemiological statistics, almost 59,114 persons died between 2000 and 2007 during extreme heat events around the world (Gasparrini et al., 2017).

The threshold ambient temperature over which heat-related mortality starts to increase varies between the various parts of the planet as a function

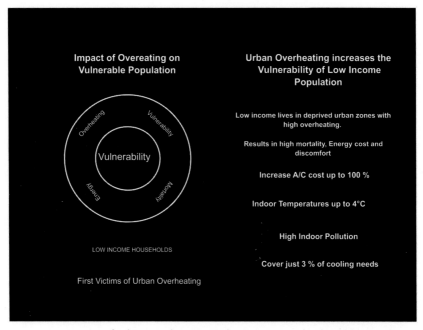

Figure 1.11 Impact of urban overheating on low-income and vulnerable population.

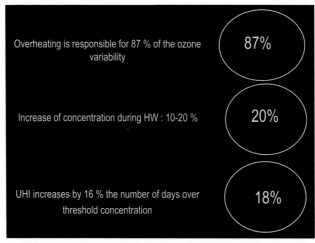

Figure 1.12 Statistical data on the impact of urban overheating on the concentration of ground ozone.

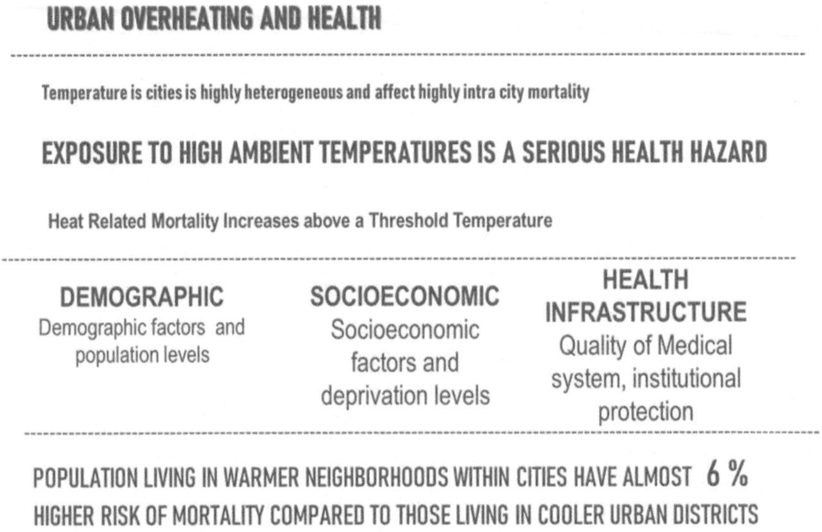

URBAN OVERHEATING AND HEALTH

Temperature is cities is highly heterogeneous and affect highly intra city mortality

EXPOSURE TO HIGH AMBIENT TEMPERATURES IS A SERIOUS HEALTH HAZARD

Heat Related Mortality Increases above a Threshold Temperature

DEMOGRAPHIC	SOCIOECONOMIC	HEALTH INFRASTRUCTURE
Demographic factors and population levels	Socioeconomic factors and deprivation levels	Quality of Medical system, institutional protection

POPULATION LIVING IN WARMER NEIGHBORHOODS WITHIN CITIES HAVE ALMOST **6 %** HIGHER RISK OF MORTALITY COMPARED TO THOSE LIVING IN COOLER URBAN DISTRICTS

Figure 1.13 Main characteristics of the association between urban overheating and heat-related mortality.

of the local adaptation (Baccini et al., 2008), and it is much higher in cooling than in heating–dominated climates. Future projections of heat-related mortality present a quite high uncertainty as issues related to human adaptation, future climate and future population, health infrastructures, and demographic evolutions still present a high uncertainty and fuzziness.

1.5 Mitigation of urban overheating

To counterbalance the serious impact of urban overheating and decrease the magnitude of the ambient temperature in cities, several mitigation technologies and strategies are developed and implemented. Mitigation technologies aim to decrease the strength of the heat sources causing urban overheating and boost the cooling potential of natural cooling sinks in cities. The main objectives, as well as the technological developments of urban mitigation engineering and science, can be concentrated in the following (Santamouris, 2018):

1.5.1 Decrease of the absorption of solar radiation in the urban fabric

The absorption of solar radiation by materials and other structures increases their surface temperature and raises the temperature of cities. To reduce the solar heat flux, shading systems and reflecting materials are mainly proposed and used. Shading devices can protect the urban fabric while reflective materials reflect the incident solar radiation back to space.

1.5.2 Increase of the emission of infrared radiation by the urban structures

Materials emit infrared radiation as a function of their temperature and the corresponding emissive capacity. High infrared losses contribute to decrease the temperature of the materials and reduce the ambient temperature. This can be achieved through the use of highly emissive materials that usually present a high reflectivity to solar radiation.

1.5.3 Increase of the ventilative cooling in cities

The ambient air around the cities is usually of lower temperature. Once it flows into the urban places, it contributes to decreasing the ambient and surface temperature of the city structure. Especially in coastal cities, the flow of the low-temperature sea breeze can cool the urban zones adjacent to the coast significantly. Proper landscape and channeling of the cool air can boost the potential of ventilative cooling techniques.

1.5.4 Decrease of the flow of advective heat

Warm aerial masses may flow into the city from the surrounding areas through advective processes. Cities in the proximity of deserts or other heat sources may suffer from warm advective fluxes increasing the ambient temperature. While it is considerably difficult to block advective fluxes in cities, proper positioning of the heat sources around the urban zones when possible, creation of peripheral buffer zones, and proper landscape design may partially block or reduce the strength of advective fluxes.

1.5.5 Increase of the evapotranspiration hear flux

Evaporation of water from the ground and water surfaces, as well as transpiration from urban greenery, are latent heat processes that contribute to decreasing the ambient temperature in cities. Increase of the urban greenery infrastructure and rise of the soil humidity using additional watering and precipitation water increase the latent heat losses and fight urban overheating, while it increases the levels of ambient humidity.

1.5.6 Decrease of the anthropogenic heat release

Anthropogenic heat released by transport, industry, and buildings increases the magnitude of the ambient temperature considerably. Minimization of

Figure 1.14 A cool pavement applied in a street in Athens. *From Kyriakodis, G.E., Santamouris, M., 2018. Using reflective pavements to mitigate urban heat island in warm climates—results from a large scale urban mitigation project. Urban Climate, 24, 326–339. https://doi.org/10.1016/j.uclim.2017.02.002.*

Figure 1.15 Visible and IR pictures of a cool roof installed on a school in Athens (Synnefa et al., 2012). (A) Roof before the installation of the cool roof. (B) Roof after the installation of the cool roof. (C) Surface temperature of the conventional roof. (D) Surface temperature of the cool roof.

the anthropogenic heat sources in the cities using nonheat emitting transport systems decreases industrial activity, and use of more efficient heating and air conditioning systems can considerably decrease ambient overheating.

1.5.7 Dissipation of the excess heat to low-temperature environmental sinks

Natural sinks like the ground in temperate climates and at a specific depth present a considerably lower temperature than the ambient one. Dissipation of the excess urban heat into the ground through the use of earth to air heat exchangers may provide cool air and fight urban overheating.

Research carried out during recent years has succeeded to advance the potential of heat mitigation technologies. More emphasis is given on the development of advanced materials for the urban fabric like reflective materials for roofs and pavements (Figs. 1.14 and 1.15) (Kyriakodis & Santamouris, 2018; Synnefa et al., 2012), thermochromic and fluorescent materials as well as innovative photonic materials able to achieve subambient surface temperatures (Santamouris & Young, 2020). The various

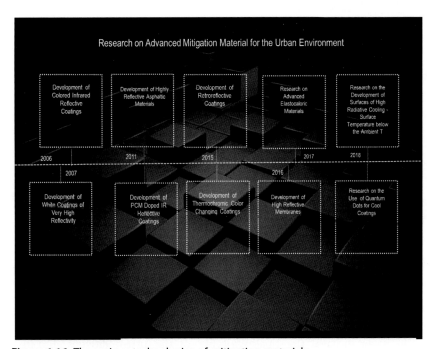

Figure 1.16 The various technologies of mitigation materials.

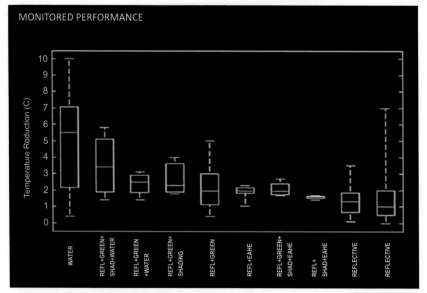

Figure 1.17 Range of the reported cooling potential of the various combinations of mitigation technologies (Santamouris, 2016a,b).

material technologies developed for mitigation purposes are shown in Fig. 1.16 and are discussed in the next chapters of the present book.

Hundreds of small or large scale urban mitigation projects are designed and implemented employing various combinations of technologies. The reported results and conclusions are summarized and analyzed in Santamouris et al. (2017). The range of the reported mitigation potential for all the considered combinations of technologies are summarized in Fig. 1.17 (Santamouris et al., 2017). However, most of the reported efficiency data refer to the cooling performance close to the mitigation source. In reality, the maximum achieved decrease of the peak ambient temperature at the city scale may never exceed 2.5°C−3.0°C, which is the upper limit of the cooling capacity of the employed mitigation technologies.

1.6 Conclusion

Local climate change and urban overheating represent a major local climatic phenomenon, increasing the temperature of urban areas considerably. This has important energy, environmental, and social consequences while it deteriorates the quality of life of the urban citizens. Existing knowledge on urban overheating is quite rich but is overshadowed by various inconsistencies related to the performed experimental and theoretical analysis. Thus there is a need for an objective experimental and communication protocol to be followed in future Urban Heat Island studies. Complete and accurate knowledge of the magnitude and the characteristics of urban overheating is a prerequisite for proper and complete planning of urban mitigation and adaptation technologies.

Research on proper mitigation technologies has permitted to develop advanced and high-quality systems and techniques that can amortize the impact of higher urban temperatures. Advanced materials for outdoor spaces and buildings are developed, allowing to alter the thermal balance of cities, decrease the energy consumption, protect human health, and improve indoor and outdoor environmental quality in the built environment. Current research efforts on the existing material technologies mainly aim to improve the thermal and optical characteristics of the components, decelerate aging effects and improve self-cleaning properties. Research on a future generation of building materials for mitigation purposes focuses on the use of advanced nanotechnology solutions to improve the performance of super cool photonic and thermoelectric materials, on new materials presenting advanced chromic optical properties, and on the development of piezoelectric components.

Research on climatic mitigation technologies should not be seen in an isolated way. It should be part of holistic research aiming to face the global challenges in the urban environment. In particular, the economic turmoil, the climatic change, the increased urbanization and the urban sprawl, the increasing age of the population, and the problem of poverty. Research on urban climatic mitigation technologies should explore interrelationships and links with advanced Information and Communication Technology technologies like Smart City Information Networks, Intelligent Urban Management, and also with Efficient Green Supply Networks, Zero Energy Settlements, Alternative Labor and Education Technologies, etc., in order to uncover new information about how our cities work and develop and provide integrated urban solutions that will improve the quality of citizen life by providing direct and personal services.

References

Akbari, H., Davis, S., Dorsano, S., Huang, J., Winnett, S., 1992. Cooling our communities. A guidebook on tree planting and light-colored surfacing. EPA. <https://escholarship.org/uc/item/98z8p10x>.

Akbari, H., Cartalis, C., Kolokotsa, D., Muscio, A., Pisello, A.L., Rossi, F., et al., 2016. Local climate change and urban heat island mitigation techniques—the state of the art. Journal of Civil Engineering and Management 22 (1), 1—16. Available from: https://doi.org/10.3846/13923730.2015.1111934.

Baccini, M., Biggeri, A., Accetta, G., Kosatsky, T., Katsouyanni, K., Analitis, A., et al., 2008. Heat effects on mortality in 15 European cities. Epidemiology (Cambridge, Mass.) 19 (5), 711—719. Available from: https://doi.org/10.1097/EDE.0b013e318176bfcd.

Bouton, S., Knupfer, S., Mihov, I., Swartz, S., 2015. Urban mobility at a tipping point. <https://www.mckinsey.com/business-functions/sustainability/our-insights/urban-mobility-at-a-tipping-point>.

Chandramowli, S.N., Felder, F.A., 2014. Impact of climate change on electricity systems and markets—a review of models and forecasts. Sustainable Energy Technologies and Assessments 5, 62—74. Available from: https://doi.org/10.1016/j.seta.2013.11.003.

Chen, F., Tewari, M., Miao, S., Liu, Y., Warner, T., Kusaka, H., 2007. Developing an integrated urban modeling system in WRF: current status and future plan. 8th WRF User's Workshop, Boulder, CO, USA.

Coutts, A.M., Beringer, J., Tapper, N.J., 2007. Impact of increasing urban density on local climate: spatial and temporal variations in the surface energy balance in Melbourne, Australia. Journal of Applied Meteorology and Climatology 46 (4), 477—493. Available from: https://doi.org/10.1175/JAM2462.1.

Dirks, J.A., Gorrissen, W.J., Hathaway, J.H., Skorski, D.C., Scott, M.J., Pulsipher, T.C., et al., 2015. Impacts of climate change on energy consumption and peak demand in buildings: a detailed regional approach. Energy 79 (C), 20—32. Available from: https://doi.org/10.1016/j.energy.2014.08.081.

Gasparrini, A., Guo, Y., Sera, F., Vicedo-Cabrera, A.M., Huber, V., Tong, S., et al., 2017. Projections of temperature-related excess mortality under climate change scenarios. The Lancet Planetary Health 1 (9), e360—e367. Available from: https://doi.org/10.1016/S2542-5196(17)30156-0.

Global Energy Assessment Writing Team, 2012. Global Energy Assessment toward a Sustainable Future. Cambridge University Press. <https://www.cambridge.org/au/academic/subjects/earth-and-environmental-science/environmental-policy-economics-and-law/global-energy-assessment-toward-sustainable-future?format = PB&isbn = 9780521182935>.

Grimmond, C.S.B., 1992. The suburban energy balance: methodological considerations and results for a mid-latitude west coast city under winter and spring conditions. International Journal of Climatology 12 (5), 481—497. Available from: https://doi.org/10.1002/joc.3370120506.

Harrison, R., McGoldrick, B., Williams, C.G.B., 1984. Artificial heat release from greater London, 1971—1976. Atmospheric Environment (1967) 18 (11), 2291—2304. Available from: https://doi.org/10.1016/0004-6981(84)90001-5.

Ho, H.C., Knudby, A., Walker, B.B., Henderson, S.B., 2017. Delineation of spatial variability in the temperature — mortality relationship on extremely hot days in greater Vancouver, Canada. Environmental Health Perspectives 125 (1), 66—75. Available from: https://doi.org/10.1289/EHP224.

Hosler, C.L., Landsberg, H.E., 1997. The effect of localised man-made heat and moisture sources in mesoscale weather modification. Energy and Climate. National Academy of Sciences, New York (1977).

Ichinose, T., Shimodozono, K., Hanaki, K., 1999. Impact of anthropogenic heat on urban climate in Tokyo. Atmospheric Environment 33 (24—25), 3897—3909. Available from: https://doi.org/10.1016/S1352-2310(99)00132-6.

International Energy Agency, 2019. Cement Production 2010—2018. <https://www.iea.org/data-and-statistics/charts/cement-production-2010-2018>.

IPCC, 2000. In: Nakicenovic, N., Swart, R. (Eds.), Emissions scenarios. Cambridge University Press., p. 570. Available from: https://www.ipcc.ch/report/emissions-scenarios/.

John, D., 2014. Building drivers and metrics. Household, floor area and energy performance forecasts. John Dulac, Energy Analyst. In collaboration with Tsinghua University.

Johnson, H., Kovats, R.S., McGregor, G., Stedman, J., Gibbs, M., Walton, H., et al., 2005. The impact of the 2003 heat wave on mortality and hospital admissions in England. Health Statistics Quarterly/Office for National Statistics 25, 6—11.

Khan, S.M., Simpson, R.W., 2001. Effect of heat island on the meteorology of a complex urban airshed. Boundary-Layer Meteorology 100 (3), 487—506. Available from: https://doi.org/10.1023/A:1019284332306.

Kłysik, K., 1996. Spatial and seasonal distribution of anthropogenic heat emissions in Lodz, Poland. Atmospheric Environment 30 (20), 3397—3404. Available from: https://doi.org/10.1016/1352-2310(96)00043-X.

Kolokotsa, D., Santamouris, M., 2015. Review of the indoor environmental quality and energy consumption studies for low income households in Europe. Science of the Total Environment 536, 316—330. Available from: https://doi.org/10.1016/j.scitotenv.2015.07.073.

Kondo, H., Kikegawa, Y., 2003. Temperature variation in the urban canopy with anthropogenic energy use. Pure and Applied Geophysics 160 (1—2), 317—324. Available from: https://doi.org/10.1007/s00024-003-8780-9.

Kyriakodis, G.E., Santamouris, M., 2018. Using reflective pavements to mitigate urban heat island in warm climates—results from a large scale urban mitigation project. Urban Climate 24, 326—339. Available from: https://doi.org/10.1016/j.uclim.2017.02.002.

Lai, L.W., Cheng, W.L., 2009. Air quality influenced by urban heat island coupled with synoptic weather patterns. Science of the Total Environment 407 (8), 2724—2733. Available from: https://doi.org/10.1016/j.scitotenv.2008.12.002.

McNeil, M., Letschert, V., 2008. Future air conditioning energy consumption in developing countries and what can be done about it: the potential of efficiency in the residential sector. Lawrence Berkeley National Laboratory. Available from: https://escholarship.org/uc/item/64f9r6wr.

MIT, 2021. 2021 Global Change Outlook. MIT. <https://globalchange.mit.edu/sites/default/files/newsletters/files/2021-JP-Outlook.pdf>.

Narumi, S, Kondo, Y., Minoru, M., 2003. Effect of anthropogenic waste heat upon urban thermal environment using mesoscale meteorological model. In Fifth International Conference on Urban Climate.

Phadke, A., Abhyankar, N., Shah, N., 2014. Avoiding 100 new power plants by increasing efficiency of room air conditioners in India: opportunities and challenges. Lawrence Berkeley National Laboratory. Available from: https://www.osti.gov/servlets/purl/1136779.

Pigeon, G., Legain, D., Durand, P., Masson, V., 2007. Anthropogenic heat release in an old European agglomeration (Toulouse, France). International Journal of Climatology 27 (14), 1969—1981. Available from: https://doi.org/10.1002/joc.1530.

Ritchie, H., Roser, M., 2018. Urbanization. <https://ourworldindata.org/urbanization>.

Sailor, D.J., Lu, L., 2004. A top-down methodology for developing diurnal and seasonal anthropogenic heating profiles for urban areas. Atmospheric Environment 38 (17), 2737—2748. Available from: https://doi.org/10.1016/j.atmosenv.2004.01.034.

Santamouris, M., 2014. On the energy impact of urban heat island and global warming on buildings. Energy and Buildings 82, 100—113. Available from: https://doi.org/10.1016/j.enbuild.2014.07.022.

Santamouris, M., 2015. Analyzing the heat island magnitude and characteristics in one hundred Asian and Australian cities and regions. Science of the Total Environment 512−513, 582−598. Available from: https://doi.org/10.1016/j.scitotenv.2015.01.060.

Santamouris, M., 2016a. Cooling the buildings—past, present and future. Energy and Buildings 128, 617−638. Available from: https://doi.org/10.1016/j. enbuild.2016.07.034.

Santamouris, M., 2016b. Innovating to zero the building sector in Europe: minimising the energy consumption, eradication of the energy poverty and mitigating the local climate change. Solar Energy 128, 61−94. Available from: https://doi.org/10.1016/j. solener.2016.01.021.

Santamouris, M., 2018. Minimizing Energy Consumption, Energy Poverty and Global and Local Climate Change in the Built Environment: Innovating to Zero. Elsevier.

Santamouris, Mattheos, 2020. Recent progress on urban overheating and heat island research. Integrated assessment of the energy, environmental, vulnerability and health impact. Synergies with the global climate change. Energy and Buildings 207, 109482. Available from: https://doi.org/10.1016/j.enbuild.2019.109482.

Santamouris, M., Kolokotsa, D., 2015. On the impact of urban overheating and extreme climatic conditions on housing, energy, comfort and environmental quality of vulnerable population in Europe. Energy and Buildings 98, 125−133. Available from: https://doi.org/10.1016/j.enbuild.2014.08.050.

Santamouris, M., Cartalis, C., Synnefa, A., Kolokotsa, D., 2015. On the impact of urban heat island and global warming on the power demand and electricity consumption of buildings—a review. Energy and Buildings 98, 119−124. Available from: https://doi. org/10.1016/j.enbuild.2014.09.052.

Santamouris, M., Ding, L., Fiorito, F., Oldfield, P., Osmond, P., Paolini, R., et al., 2017. Passive and active cooling for the outdoor built environment—analysis and assessment of the cooling potential of mitigation technologies using performance data from 220 large scale projects. Solar Energy 154, 14−33. Available from: https://doi.org/ 10.1016/j.solener.2016.12.006.

Santamouris, M., Young, Y.G., 2020. Recent development and research priorities on cool and super cool materials to mitigate urban heat island. Renewable Energy 792−807. Available from: https://doi.org/10.1016/j.renene.2020.07.109.

Schinasi, L.H., Benmarhnia, T., De Roos, A.J., 2018. Modification of the association between high ambient temperature and health by urban microclimate indicators: a systematic review and meta-analysis. Environmental Research 161, 168−180. Available from: https://doi.org/10.1016/j.envres.2017.11.004.

Scott, M.J., Dirks, J.A., Cort, K.A., 2008. The value of energy efficiency programs for US residential and commercial buildings in a warmer world. Mitigation and Adaptation Strategies for Global Change 13 (4), 307−339. Available from: https://doi.org/ 10.1007/s11027-007-9115-4.

Smoyer, K.E., 1998. Putting risk in its place: methodological considerations for investigating extreme event health risk. Social Science and Medicine 47 (11), 1809−1824. Available from: https://doi.org/10.1016/S0277-9536(98)00237-8.

Statista, 2016. Global asphalt demand from 2005 to 2015. <https://www.statista.com/statistics/264581/asphalt-demand-worldwide/>.

Steinecke, K., 1999. Urban climatological studies in the Reykjavik subarctic environment, Iceland. Atmospheric Environment 33 (24−25), 4157−4162. Available from: https:// doi.org/10.1016/S1352-2310(99)00158-2. Elsevier Science Ltd.

Stewart, I.D., 2011. A systematic review and scientific critique of methodology in modern urban heat island literature. International Journal of Climatology 31 (2), 200−217. Available from: https://doi.org/10.1002/joc.2141.

Stewart, I.D., Oke, T.R., 2012. Local climate zones for urban temperature studies. Bulletin of the American Meteorological Society 93 (12), 1879−1900. Available from: https://doi.org/10.1175/BAMS-D-11-00019.1.

Synnefa, A., Saliari, M., Santamouris, M., 2012. Experimental and numerical assessment of the impact of increased roof reflectance on a school building in Athens. Energy and Buildings 55, 7−15. Available from: https://doi.org/10.1016/j.enbuild.2012.01.044.

Taha, H., Akbari, H., Sailor, D., Ritschard, R., 1992. Causes and effects of heat islands: sensitivity to surface parameters and anthropogenic heating. Lawrence Berkeley Laboratory Report No. 29864.

United Nations Department of Economic and Social Affairs, 2017. World Population Prospects 2017. United Nations; Data Booklet (ST/ESA/SER.A/401).

United Nations Human Settlements Programme (UN-HABITAT), 2019. Population living in slums (% of urban population). <http://data.worldbank.org/indicator/EN.POP.SLUM.UR.ZS>.

Urge-Vorsatz, D., Petrichenko, K., Staniec, M., Eom, J., 2013. Energy use in buildings in a long-term perspective. Current Opinion in Environmental Sustainability 5 (2), 141−151. Available from: https://doi.org/10.1016/j.cosust.2013.05.004.

Warren, R., Arnell, N., Nicholls, R., Levy, R., Price, J., 2006. Understanding the regional impacts of climate change. Research report prepared for the Stern Review on the Economics of Climate Change. Working 90.

Yoshikado, H., Tsuchida, M., 1996. High levels of winter air pollution under the influence of the urban heat island along the shore of Tokyo Bay. Journal of Applied Meteorology 35 (10), 1804−1813. https://doi.org/10.1175/1520-0450(1996) 035 < 1804:HLOWAP > 2.0.CO;2.

Young, A., 2019. The world 33 megacities. <https://www.msn.com/en-us/money/realestate/the-worlds-33-megacities/ar-BBUaR3v>.

CHAPTER 2

Experimental and monitoring techniques to map and document urban climate change

Riccardo Paolini
School of Built Environment, Faculty of Arts, Design and Architecture, University of New South Wales (UNSW), Sydney, NSW, Australia

2.1 Introduction

Urban climate change, in contrast with global climate change, which is a background phenomenon, can be defined as the complex of phenomena that influence local urban climates. These phenomena include urbanization and the generation of heat islands, as well as the interaction with mesoscale conditions, especially with regional winds and advective fluxes, which without the presence of the city with its increased aerodynamic roughness, temperature, and pressure distribution, would have been different.

Probably the most common concept in the description of urban climates is that of the urban heat island (UHI), which is normally defined as an urban area where the ambient temperature is significantly higher than in a reference, nonurban adjacent area (Oke, 1987). This way of looking at the characterization of the urban climate originates in the first experiments. It is generally accepted that the very first set of measurements that documented the magnitude of the UHI was performed by Luke Howard in London and published first in 1811 and then in 1833 in *The Climate of London* (Howard, 1833, 1818). Luke Howard first recorded air temperatures at a rural site and at the Somerset House in the center of London. Since then, a multitude of similar campaigns and comparisons have been performed over the years, especially during the last three decades, contributing to the burgeoning literature on the magnitude of the UHIs. Most of these studies compared the ambient temperature at one or more urban sites with the records at one or more reference (nonurban) sites by means of automobile transects or comparisons of datasets collected by different weather stations (Rizwan et al., 2008; Santamouris, 2015).

Indeed, historically many climate observations started in urban areas, often at astronomical observatories, such as at the Clementinum in Prague, CZ, in 1752 (Clementinum, 2022) or in at the Osservatorio di Brera in Milan, Italy (INAF, 2022), with the first continuous records only related to air temperature. Then, weather observations have been chiefly performed at airports or racecourses to enable synoptic observations in undisturbed turbulence conditions and weather forecasting, also in support to aviation, more than to collect climatological records. For this reason, the almost entirety of typical weather years used for building energy simulations are developed on data collected at airports, with urban observations used only in research studies (Santamouris, 2014).

After the introduction of dataloggers, weather observations could be performed continuously without a human operator, by means of automatic weather stations (AWS), which became progressively cheaper, allowing for growth in the number of stations in networks, and a proliferation in the number of independent networks, either run by different agencies—often competing for funding—or more recently by amateur meteorologists. At the same time, this increase in the diffusion of sensors has also led to inconsistency in the installation practices and metadata collection.

Following the increase in the availability of long-term measurements, the literature on point-to-point comparisons (urban vs rural) snowballed, focusing on the magnitude of UHIs, as mentioned earlier. However, the very definition of the UHI highlights some challenges in its quantification (Martilli et al., 2020), as it is normally difficult to identify one or more reference conditions. Therefore while the body of literature on urban climates is immense, it is worth pondering whether the measurement systems currently available were designed to answer the research questions related to urban overheating or were designed for other purposes. Further, it is also worth reconsidering which are the research questions to be addressed and if they are all assessed with existing approaches.

In fact, most of the literature on urban climates focused on the quantification of the magnitude of the UHI, often referred to as "urban heat island intensity," which is defined as the urban-reference temperature difference. However, these measurements do not directly respond to the research questions on urban overheating (Martilli et al., 2020), which are instead if urban overheating has negative impacts on human life and how these can be mitigated. Indeed, in some cities, a cool island develops, with nonurban surroundings showing higher ambient temperatures than the inner city, which can still require mitigation. This is the case of many desert cities (Chow et al., 2011; Haddad et al., 2019; Middel et al., 2014) or some tropical urban areas (Johansson and Emmanuel, 2006; Yang et al., 2017). Winter heat islands, for instance, can be of the same magnitude as summer heat islands (Giridharan and Kolokotroni, 2009) and decrease building energy needs for heating considerably (Paolini et al., 2017) as well as cold-related mortality (Macintyre et al., 2021). Therefore reducing the magnitude of UHIs per se is not an objective to pursue, and it is not investigated.

Most papers addressing the mitigation of urban overheating do not use the "urban heat island intensity" as an indicator of the performance of mitigation interventions but a comparison of mitigated and unmitigated (simulation) scenarios (Martilli et al., 2020; Santamouris et al., 2017). The only category of papers that rightfully uses urban and rural datasets is that of studies focusing on the urban versus nonurban building performance simulation and the appropriate dataset to be used as an input. In fact, while this topic has been thoroughly investigated and it is documented that, on average, urban buildings show cooling energy needs 13% higher than nonurban buildings (Santamouris, 2014), practitioners still use airport datasets to design buildings.

Therefore the research questions that a sensing network should help to address in the context of urban overheating are as follows:

- What is the magnitude of urban overheating beyond conditions acceptable to human life (rather than a nonurban reference)?
- What is urban overheating caused by?
- If heat mitigation has been implemented, are the implemented technologies effective and what is their performance?

These are the questions addressed by modeling studies comparing heat mitigation technologies (Santamouris et al., 2017), and thus the questions that a network of real sensors should address, too. Obviously, the answer to the first question identifies the performance target for urban heat mitigation, while the answer to the second question allows researchers and decision-makers to identify the adequate countermeasures, which cannot be the same in all scenarios. Finally, monitoring the performance of implemented heat mitigation allows local governments to allocate incentives and avoid a performance gap.

Consequently, the objective of this chapter is to respond to the following questions:

- Are the existing and commonly used sensing approaches adequate to address the aforementioned research questions?
- If not, what should a sensing network look like, and who should use the data?
- Is a sensing network sufficient at all?

Hence, in this chapter, the purpose of different measurement systems and approaches is addressed, their design and the different types of experimental and monitoring techniques to map urban climate change are introduced. Further, the different scopes and limitations are contrasted, and the need for different systems and approaches is discussed in detail. The objective of this chapter is thus to take a pause and consider where we are now with measurement systems and consider if what we are doing—as a scientific community—is fully serving the scope.

2.2 Measurement approaches in urban climatology

Urban climatology benefits from a wide range of measurement approaches, ranging between weather stations within and out of urban areas, transects and remote sensing, which in turn comprises satellites, flights with drones [Unmanned Aerial Vehicle (UAV)] or aeroplanes, typically equipped with infrared cameras or other sensors. Instead, outdoor thermal comfort measurement approaches and, in particular, specific measurement methods of mean radiant temperature (Thorsson et al., 2007) are beyond the scope of this chapter. In this section, an overview of the different techniques is offered, discussing the advantages and limitations of each technique.

2.2.1 Networks of weather stations—continuous monitoring
2.2.1.1 Sensing networks managed by agencies and research institutions
2.2.1.1.1 Weather stations

The most common approach used in urban climatology to describe the urban climate relies on existing networks of weather stations, which were initially established either by the airforce or the Met Office. As mentioned earlier, these stations were not designed to observe the urban climate. The vast majority of stations is located at airports, racecourses, or other unobstructed sites far from buildings, aiming to avoid microclimatic influences from nearby objects (World Meteorological Organization, 2018). Met offices typically follow the World Meteorological Organization (WMO) Guide to Meteorological Instruments and Methods of Observation (World Meteorological Organization, 2018), or at least its inspiring principles. These design principles prescribe to avoid local influences that could make the observations of micro or local significance rather than representative for a large area (i.e., from 100 km^2 to 1000 km^2). In brief, the main requirements for a temperature (and humidity) station are: at >1.25 m above the ground, distant from obstacles that reduce air circulation, not on steep slopes or hollows, away from trees and objects casting shades (for what is possible). An example of a class 1 site is offered in Fig. 2.1. To discern whether a site can be considered unaffected by microclimatic influences, the WMO devised a classification system (Table 2.1). Urban environments present many constraints, and therefore finding an ideal site is seldom possible, especially in high-density metropolitan areas, where the cost of land is high, and the closest "class 1" site may be kilometers away from the city, thus not representative of where most of the population resides or works.

Detailed: artificial heat sources and reflective surfaces or expanse of water, occupying:

(i) less than 50% of the surface within a 10-m radius around the screen;

(ii) less than 30% of the surface within a 3-m radius around the screen.

Figure 2.1 An automated weather station at an unobstructed site within an urban area after its setup in Penrith, New South Wales.

Table 2.1 Adapted and commented summary table of the siting classification for surface observing stations on land provided by CIMO Guide (World Meteorological Organization, 2018). Commonly accepted compromises in urban climatology (adaptation of the original WMO interpretations to urban areas) are clearly identified in the table. For the detailed version, please refer to the original document, which is openly accessible online.

Feature	Reference	Ideal	Good	Acceptable compromise	No class
Site added uncertainty WMO station class* for temperature and humidity	— Class 1	— Class 2	<1°C Class 3	<2°C Class 4	< 5°C Class 5
Sensors height above the ground	1.25 – 2 m	1.25 – 2 m	Compromise: street signs or light poles at 3 m, to prevent vandalization	Compromise: between 3 m and 4 m, to prevent vandalization	Site not meeting the requirements of Class 4
Distance from artificial heat sources or reflective surfaces (buildings, concrete surfaces, car parks, etc.), and from an expanse of water (unless significant of the region)	>100 m	>30 m	>10 m	Easy requirement: 5 m from buildings*	
Slope	Slope inclination less than 1/3 (19°)		NA	NA	
Ground cover	Natural and low vegetation (<10 cm) representative of the region		Ground covered with natural and low vegetation (<25 cm) representative of the region		
Cast shades	Away from all projected shade when the sun is higher than 5°	Away from all projected shade when the sun is higher than 7°	Away from all projected shade when the sun is higher than 7°	Away from all projected shade when the sun is higher than 20°	

As the affordability of AWSs, the availability of resources, and the possibility of managing complex networks of stations increases, it is progressively more common to organize networks where not all the nodes have the same hierarchy. This approach allows the nesting of information, with stations (sites) with lower uncertainty serving as a pivot for stations with higher uncertainty (Fig. 2.2). Even if the multiple climatic stations (sites) of classes 2 and 3 (or even 4) are affected by higher site-related uncertainty, multiple points allow for averaging, and thus the aggregated information offers representativity at the urban scale. An example of commonly accepted different hierarchies is that of networks of air quality stations (commonly deployed and managed by environmental protection agencies).

Often these networks include (Demerjian, 2000; Munir et al., 2019; Righini et al., 2014) the following:

- Reference station(s), one or more stations out of the urban boundaries, which is exposed to low pollution levels, unless the nonurban area.
- Urban background station(s). One or more stations are typically located within (large) urban parks, which measure the background pollution for the area.
- Urban or suburban stations. These are (e.g., small green spot or local situation) local air quality conditions.
- Street stations. These are typically the most numerous stations within the network, normally located along main roads or squares, where the maximum pollution is expected and that represent points of concern for public health.

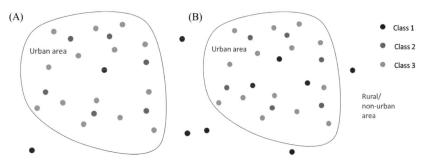

Figure 2.2 Different configurations of networks of urban and nonurban weather stations: (A) only one class 1 urban station with multiple stations of classes 2 and 3, and (B) multiple class 1 urban stations surrounded by clusters of class 2 stations, which are consequently surrounded by class 3 stations.

Different air quality stations serve different purposes. For instance, the main driver for a network of air quality stations is public health and management of limitations on traffic and planning for mobility investments. As air quality has been and—in many areas of the world—still is a major concern for public health due to emissions from buildings and vehicular traffic, this hierarchy has been adopted in multiple cities, including Milan and Rome, Italy (Maranzano, 2022) and Riyadh, Saudi Arabia (Bian et al., 2018).

Further, building rooftops are locations that the WMO recommends to avoid as a site for weather stations, as rooftops display their own microclimate and are exposed to higher wind speeds than within the urban canopy layer (UCL), and often they are covered with high albedo materials, which increase the local air temperature in proximity of the roof surface. This is particularly true for large-scale solar absorbing roofs (such as shopping centers). Green et al. (2020) measured the air temperatures above uncoated metal roofing (solar absorbance = 0.28 and thermal emittance = 0.55) at different heights, and even at 1.5 m above the surface, the air temperature was up to 4°C−5°C higher (2.53°C on average) than the undisturbed temperature measured on a mast at 8 m height over the rooftop, with similar numbers reported in the literature for solar absorptive roofs. Despite these challenges, some factors still make rooftop stations useful for specific uses:

- Collected data serve as input for building performance simulations or continuous commissioning of buildings (i.e., monitoring their performance against climate data). In this case, data collected at the top of the UCL and above the surface boundary layer of the building envelope are representative of the actual forcing more than data from a station within an urban park, for instance. While, in theory, the boundary conditions should be provided separately for roof and facades, this differentiation is not possible in most building energy simulation environments (e.g., all the software tools using the standard version of EnergyPlus as the simulation engine). This is probably the most relevant use for data collected by rooftop stations.
- If the purpose of the station is to collect data in that position in support of calibration of a microclimate simulation model, recording data at a node in the proximity of the rooftop or the facade.
- If weather data are in support of other measurements such as solar or infrared incoming radiation, for which an unshaded rooftop is the ideal location.

Moreover, dense built environments might not present opportunities for other locations, and if rooftop stations are part of a system collecting information at different levels within and above the UCL, then they have a defined role and offer information to validate or calibrate multilayer urban climate simulations.

The uncertainty in the location of weather stations has been assessed by the WMO for ground stations (Table 2.1). On the other hand, the siting-related uncertainty for rooftop stations has been evaluated in the literature to be contained below the upper limit of 1°C in the most unfavorable situation and 0.7°C in most conditions (Curci et al., 2017), which would still be considered a site for a good climatic station (between class 2 and class 3, as in Table 2.1). This relates to the uncertainty associated with different rooftop sites within an urban environment and does not compare data from rooftop stations (i.e., at the top of the UCL) with data from stations at class 1 or class 2 sites at ground level, which would be illogical. In fact, the ambient temperature is typically 1°C−3°C higher within than above or at the top of the UCL (Niachou et al., 2008; Offerle et al., 2006; Pearlmutter et al., 1999), and the wind speed typically ranges between 25% and 30% of the wind speed above the rooftops (Bottillo et al., 2014; Eliasson et al., 2006; Gülten et al., 2016; Lemonsu et al., 2004), with the variability influenced by the urban geometry (predominantly the height to width ratio of the urban canyon), or the presence of street trees (Gülten et al., 2016; Sabatino et al., 2008), among other factors.

One of the key aspects is the collection and record-keeping of metadata[1] related to the sites where the stations are located. This information typically includes the following:
- Purpose of the station and design intent,
- Instrumentation and metrology metadata (Guo et al., 2016),
- Station location (coordinates),
- Siting descriptors,
- Picture of the station,
- Maintenance information, and
- Data completeness.

---

[1] Metadata are data about the data, and it is basic information that describes what the dataset is about. Without metadata, a dataset is almost useless. More about metadata can be found online at: https://www.ands.org.au/working-with-data/metadata

The lack of a standardized and broadly accepted protocol to collect and report the metadata, also in a quasi-quantitative fashion, constitutes a hurdle toward the analysis of urban weather data from multiple stations. This is of particular concern when multiple networks, even designed for different purposes, are used within the same study. The use of data mining techniques has been proven helpful to identify redundancies or substantial deviations (Wang et al., 2018), but the collection of metadata still constitutes the backbone that allows the use of recorded measurements.

2.2.1.1.2 Street stations

The measurement of ambient temperature within the UCL is instead subject to a substantial range of compromises, including installations on lamp posts at 2–3 m or more to prevent vandalization (Fig. 2.3). These constraints introduce uncertainty in their representativity of the thermal conditions experienced by pedestrians because over 1.5 m the air temperature

Figure 2.3 Example of a temperature sensor installed on a lamp post in the city center of Parramatta, New South Wales, Australia, at 2.9 m. The sensor is at the edge of a bracket extending 0.50 m from the pole to prevent reduced airflow through the naturally ventilated radiation shield and limit the impact of reflected shortwave and longwave radiation from the pole to the radiation shield.

decreases by approximately 0.25°C/m (World Meteorological Organization, 2018). However, this is often the only option to monitor street-level thermal conditions, as the possibility of installing fenced-off stations at 1.25–2 m in dense urban areas are extremely limited. While this sensing approach is not a standard, it is a commonly accepted compromised in urban climatology and used in several research studies such as the Escompte campaign (Mestayer et al., 2005).

Despite the site-related uncertainty, information from local temperature stations may offer a reliable appraisal of the magnitude of urban overheating at street level once data from multiple stations are averaged (Fig. 2.4). For example, in a campaign conducted in Parramatta, Australia, 20 temperature sensors were deployed at 2.9 m within the local government area, and data from these sensors have shown a ± 2°C site-related variability due to the local conditions (Santamouris et al., 2020). Data from the stations within the same area were subject to a reduced variability instead. The advantage of such an approach is the possibility of

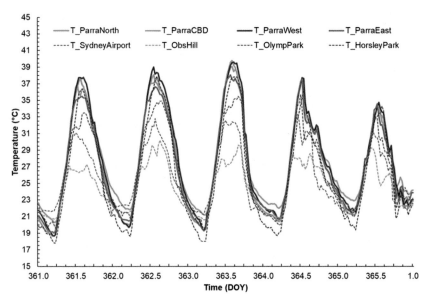

Figure 2.4 Data from Parramatta weather network. Parramatta North, CBD, West and East are the average of the stations in the relative area. Range: ± 2°C within the network of 20 sensors spread across the local government area. The solid lines are data from the sensors on lamp posts within the urban canopy layer (at 2.9 m), while the dashed lines are nonurban stations surrounding Parramatta and managed by the Australian Bureau of Meteorology.

collecting information in dense urban areas and controlling the uncertainty by averaging multiple data sources. This way, it is possible to capture urban overheating patterns, which would be lost by interpolating data between nonurban stations.

2.2.1.1.3 Stations above the urban canopy layer

The last type of continuous monitoring is implemented above the UCL and is recommended by the WMO for urban observations (World Meteorological Organization, 2018), following the measurement philosophy initially by Oke (Morrison et al., 2021; Oke, 1982; Stewart et al., 2014). This approach consists of positioning an eddy covariance station, including a 3-axis sonic anemometer and a gas analyzer, and a net radiometer on a mast on a rooftop or other structure (e.g., a trellis or radio tower) at 2—3 times the average height of buildings in the area. This measurement approach delivers measurements above the UCL and at the boundary between the top of the urban boundary layer and roughness sublayer, thus at the interface between the first atmospheric layer connecting the urban tile and the mesoscale model in climate models (Chen et al., 2011; Loridan et al., 2010). In addition to temperature and humidity, the measured quantities are the turbulent sensible and later heat fluxes and incoming and outgoing shortwave and longwave radiation (Kotthaus and Grimmond, 2014a, 2014b). While this approach is undoubtedly the most scientifically robust and the most appropriate for the calibration and validation of surface energy balance models used as urban tiles in climate modeling, such as urban canopy parametrizations, it is also pursued only for research purposes in a limited set of cities, due to its complexity and high costs. A research network of such setups comprises less than 30 installations worldwide (Urban Flux Network—IAUC, 2022), which are used to train and improve local climate models. In fact, these observations are representative of only a portion of the city and only the local climate zone, which is within the field of view of the net radiometer and fetch of the eddy covariance sensors. Furthermore, these approaches are sensitive to local influences such as reflections from buildings and spatial heterogeneity of the urban surface within the field of view (Kotthaus and Grimmond, 2014a).

2.2.1.2 Amateur networks and citizen science approaches

Since the advent of affordable weather stations and the possibility of sharing data online, amateurial networks of weather stations have become popular in many cities worldwide. For a relatively small cost, weather enthusiasts can procure a weather station (e.g., Vantage stations by Davies have been among the most popular for a long time) and connect their personal weather station among peers and any user via a web portal such as Weather Underground or regional equivalents (Weather Underground, 2022). Personal weather stations carry two main potential advantages: a distributed sensing network contributed by hundreds of participants in each big city and the removal of many of the site access issues faced by agencies because the participants install their stations on their properties. However, soon the main disadvantage became clear: the lack of standardized metadata collection and reporting methods makes data from these stations extremely fuzzy and difficult to use.

An intercomparison conducted in the United Kingdom reported significant discrepancies between observations from the Met Office and amateurial weather stations (Fig. 2.5), often exceeding 2°C−3°C (Bell et al., 2015). The authors of the study also concluded that insufficient metadata make interpretation difficult. Amateurial weather stations could be representative of local overheating conditions as much as they could be simply affected by bias due to installation conditions. In fact, personal weather stations are often installed close to the edge of the roof, within its boundary layer and thus highly influenced by the rooftop temperature and turbulence, or even on balconies (Figs. 2.6 and 2.7).

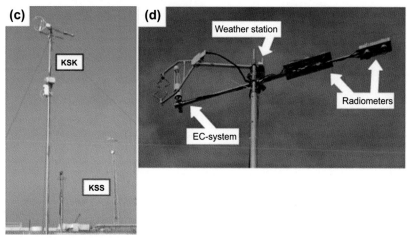

Figure 2.5 Eddy covariance systems and net-radiometers on masts at the Kings College London. *From Kotthaus, S., Grimmond, C.S.B., 2014b. Energy exchange in a dense urban environment—Part I: temporal variability of long-term observations in central London. Urban Climate 10, 261−280. <https://doi.org/10.1016/j.uclim.2013.10.002>.*

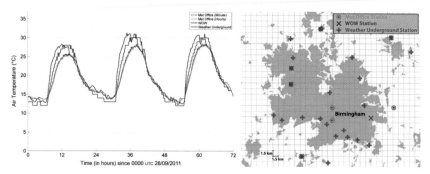

Figure 2.6 Comparison of data from Met Office and amateurial weather stations. *From Bell, S., Cornford, D., Bastin, L. 2013. The state of automated amateur weather observations. Weather, 68, 36−41. https://doi.org/10.1002/wea.1980.*

Figure 2.7 Examples of personal weather stations improperly installed too close to the building surface.

2.2.1.3 Validation of data from weather stations

In addition to the challenges in the interpretation of data, a further layer of complexity is represented by weather data validation and gap infilling. All historical series of AWSs are affected by gaps, unrealistic values or realistic but incorrect records due to instrumentation or logging bias or failure, maintenance issues (e.g., a malfunctioning instrument is not replaced for days or weeks), or systematic equipment problems. Therefore to avoid the use of unchecked data series, weather data validation procedures have been developed (Allen, 1996; Estévez et al., 2016, 2011; Meek and Hatfield, 1994; Robinson, 1998) and are progressively more and more adopted (Yun et al., 2020), also to perform quality control on data from personal weather stations (de Vos et al., 2019). Tests conducted on weather datasets typically include (Table 2.2):

- Range tests (e.g., relative humidity comprised between 0.8% and 100%).
- Step tests (e.g., air temperature change of less than 7°C in 1 hour, or relative humidity variation not exceeding 45% in 1 hour).
- Persistence tests (e.g., steady values for long periods are filtered as equipment failure).
- Relational tests (i.e., if the wind speed is null, then the wind direction cannot be different from 0 degree).
- Spatial consistency tests, which are based on a comparison of data from a target station with data collected by surrounding stations. Obviously, if the target station is within the city center and the surrounding stations are out of the urban area, for instance, a comparison with spatially interpolated values would be misleading.

The specific thresholds largely depend on the local climate conditions. For instance, a sudden variation exceeding 10°C within an hour would be anomalous in an inland city in a temperate climate. Instead, it does occur in desert or coastal cities where a subject change in wind direction or a summer storm may produce an almost instantaneous variation in the ambient conditions. Therefore when a record exceeds a threshold, it should not be automatically invalidated but flagged and analyzed. Sudden variations that can be explained by wind direction changes or a thunderstorm should be retained, filtering only records that are due to data transmission or other issues, and have no relation to actual climate events. For this reason, the availability of a reference weather station under strict calibration and quality control in climatic chambers can prove crucial in determining the acceptable ranges for the area and thus allow the validation of the entire dataset.

Initially, validation was performed by human operators on datasets of reduced size, followed by a second stage when range, step, persistence, and relational tests were performed simply considering the relations, and then simple spatial interpolations were considered to perform spatial consistency tests. In the last decades, artificial intelligence has increased the capabilities of performing both spatial consistency tests and infill gaps in the data series (i.e., missing values).

Table 2.2 Examples of the structure of tests used to invalidate unrealistic readings for temperature, relative humidity, wind speed and direction, air pressure, and solar radiation.

Quantity	T (°C)	RH (%)	WS (m/s)	WD (°)	P (hPa)	SR (W/m²)
Range test	$-20 \leq T \leq 55$	$0.8 \leq RH \leq 100$	$0 \leq T \leq 90$	$0 \leq WD \leq 360$	$750 \leq P \leq 1200$	$0 \leq SR \leq 1200$ $SR = 0$ when solar elevation < 0
Step test	$\lvert T_t - T_{h+1}\rvert \leq 7$	$\lvert RH_h, RH_{h+1}\rvert \leq 45$	$\lvert WS_h - WS_{h+1}\rvert \leq 10$	—	$\lvert P_h - P_{h+1}\rvert \leq 15$	—
Persistence test	No steady records for ≥ 3 h	No steady records for ≥ 3 h	No steady records for ≥ 3 h	No steady records for ≥ 3 h	No steady records for ≥ 3 h	—
Relational test	Sudden temperature variations exceeding the step test threshold only with rain, thunderstorms, or wind change	Sudden humidity variations exceeding the step test threshold only with rain, thunderstorms, or wind change	WS	If WS = 0, WD must be 0	—	—

Short-term discontinuities—in the range of a few hours, depending on the weather conditions and time of the day—can be handled with simple linear interpolations, spatially weighted averages, or other simple approaches. However, more significant gaps and spatial patterns when the nodes of a network are not all within the same local climate zone require either climate modeling—which is computationally very intensive—or machine learning approaches (Coulibaly and Evora, 2007; Lee et al., 2014). A variety of techniques has been proposed and applied to different datasets (MeshkinKiya and Paolini, 2020), ranging between Decision Tree Learning (Lee et al., 2014), Artificial (Back-propagated) Neural Networks (Paolini et al., 2017), Support Vector Regressions (Lee et al., 2018), or Genetic Algorithms with symbolic regression (Santamouris et al., 2020).

All these techniques search for correlations that minimize the error function given by the discrepancy between the data and the generated model or network and capture trends and relations between different stations, which simple interpolations (linear, polynomial, or spatial) cannot represent. The validation and gap-filling procedures have to be applied iteratively. As yet, advanced methods for validation and gap infilling do not routinely consider the stations' metadata, including metrological metadata (Guo et al., 2016).

2.2.1.4 Validity and representativity of networks over time

An apparently trivial but important issue concerns the use, validation, and interpretation of collected data: metadata and boundary conditions are not constant over time. As land-use changes, a weather station site that could have been considered a class 1 or class 2 site for temperature measurement can be subject to microclimatic influences. However, this specific issue might affect some stations and not others within the network, while some land-use changes may affect an entire network of stations.

As discussed earlier, among the most frequent investigations in urban climatology is quantifying the magnitude of urban overheating (or UHIs, in the scenario where no significant advective flows occur). One of the issues related to the representativity of historical series is connected to increasing urbanization. For instance, historical observatories that in the 1800s were 0.5−1 km from the city boundary (Bacci and Maugeri, 1992) are now several kilometers far from the periphery of the city and thus the first nonurban station, such as in Milan, Italy, where the minimum radius of the densely built urban area has now exceeded 10 km. Conversely, nonurban observations, including sites used to assess global warming, are affected by convective flows from cities and land-use change in the surrounding (Parker, 2010).

2.2.2 Short-term terrestrial campaigns

Short-term terrestrial campaigns are often performed almost only for research purposes and include a variety of techniques such as:
- Weather stations or other sensors deployed on weather poles or connected to lamp posts for a short duration (from hours to a few weeks).
- Transects with vehicles or carts mounting sensors or weather stations or wearable sensors.
- Wearable or other compact sensors used by the general population (assisted or not).

These options for short-term terrestrial campaigns are not alternatives and are often performed complementarily, also in conjunction with data collection from fixed-position long-term weather stations. In the following subsections, these different measurement techniques are explored.

2.2.2.1 Temporary weather stations or other fixed sensing elements

Field campaigns are often supported by deploying temporary fixed-position weather stations or other sensing elements such as temperature and humidity sensors. These are installed for a few hours or weeks on weather poles, lamp posts, or other similar infrastructure (Pfautsch and Rouillard, 2019), as well as on tripods (Piselli et al., 2018). In addition to the measurements performed with weather stations or sensors similar to those used in the medium and long term, short-term campaigns use vehicles equipped with masts (Niachou et al., 2008, 2005; Santamouris et al., 2020) or tripods to deploy weather stations and other sensors, including net-radiometry (Fig. 2.8).

The advantage of this approach is to achieve continuous synchronous monitoring of the desired quantities with a dense network for the duration of the field campaign. At the same time, transects with moving sensors do not allow for synchronous measurements at all points of the measurement grid. In brief, a dense local network is built temporarily, which is then typically used for either only thermal mapping or also for the calibration or validation of a climate simulation model. These campaigns usually are conducted at a small scale (e.g., 1−5 km × 1−5 km). On some occasions, multiple nested measurement domains are set up, with a coarse grid of sensing elements out of the primary measurement

Figure 2.8 A temporary station used in the Parramatta Urban Overheating project (Santamouris et al., 2020). A weather station and a net-radiometer are installed on a mast elevated to 10 m.

area and finer points with various measurement techniques within the inner domain, in a wide range of combinations.

In addition to the advantage of densification in thermal mapping, the primary purpose of this approach is to measure quantities that are not routinely monitored within the built environment. Traditional weather stations, such as those typically managed by met offices, usually measure the main weather parameters: dry-bulb air temperature, dew point temperature from which relative humidity is derived, wind speed and direction, air pressure, and rainfall (Australian Bureau of Meteorology, 2017). In particular, field campaigns designed to calibrate or validate climate (Eliasson et al., 2006;

Grimmond et al., 1996; Offerle et al., 2006) or computational fluid dynamic models (García-Sánchez et al., 2014) include eddy covariance techniques and net-radiometry installed on masts. For instance, the setup used by Eliasson et al. (2006) comprised a mast with multiple sonic anemometers installed within an urban canyon and a taller mast on the top of one of the buildings delimiting the street canyon. Such setup was allowed by the municipality for a limited period, as was the case for many field campaigns with major equipment deployed within pedestrian areas or car parks.

Moreover, short- and mid-term campaigns with fixed sensors deployed in the built environment often employ custom-made assemblies pursued to reduce costs and allow for more points in data collection. A review performed on 100 studies in Asian and Australian cities highlighted a high rate of mobile transverses using nonstandard equipment (Santamouris, 2015). Nonstandard equipment does not always effectively shield against reflected solar radiation (Tarara and Hoheisel, 2007), which is particularly important for measurements carried out in the built environment. In several studies, the accuracy and calibration of nonstandard equipment are not documented. Even commercial radiation shields that do not protect against reflected radiation may lead to an overestimation of the ambient temperature by up to 1.5°C (Green et al., 2020).

2.2.2.2 Transects with vehicles, carts, or wearable equipment

In alternative to or conjunction with data collection from temporarily installed fixed sensing elements, one of the traditional techniques used in urban climatology is the measurement of ambient temperature and other quantities in continuous or discontinuous mode at different predetermined positions within the area of interest. These transects are mobile measurements performed by mounting weather stations or other sensing elements on motor vehicles (Busato et al., 2014; Kousis et al., 2021), bicycles, or carts (Fig. 2.9), or even on wearable setups (Pigliautile and Pisello, 2018), for monitoring in dense pedestrian areas or complex terrain in historical city centers (Fig. 2.10).

The main purpose and product of transects is the creation of climate maps, which were first drawn in Germany and later largely used for urban design and planning in more than 15 countries worldwide (Ren et al., 2011). These climate maps (or climate atlas as originally referred to in Germany) include different layers of information, integrating both measurements and geographic information (Fig. 2.11). When informed by local-scale measurements (Fig. 2.12), climate maps identify the hot spots

Figure 2.9 Cart equipped (A) with a weather station during a measurement in an urban park and (B) in a different setup with the addition of a net-radiometer (Santamouris et al., 2020).

Figure 2.10 Different wearable systems for monitoring in pedestrian areas. (A) A helmet equipped with temperature, humidity, solar radiation, illuminance sensors, and a thermal camera (Pigliautile and Pisello, 2018). (B) Backpack equipped with sensors (Cureau et al., 2022).

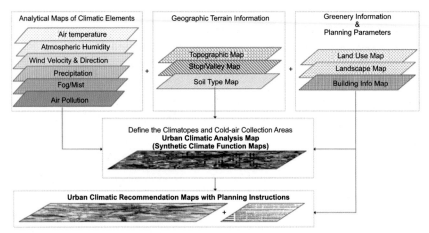

Figure 2.11 The structure of urban climate maps (Ren et al., 2012).

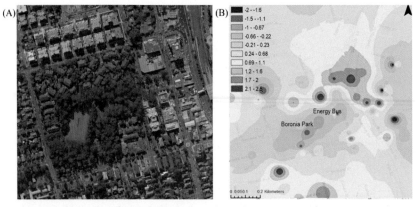

Figure 2.12 Example of a suburb-scale urban climate map after transverses performed in Epping, NSW, Australia, during the Parramatta Overheating Study (Santamouris et al., 2020). (A) Visible satellite image (MapSix NSW) and (B) climate map of the area, with the temperature scale in °C identifying the differences with respect to a reference location.

within an area that require heat mitigation and the extent of the mitigation delivered by urban parks, for instance (Fig. 2.12). Studies on the cool islands generated by urban parks and the extent of the mitigation effect—namely investigating the temperature reduction as a function of the distance from the park—are typically performed with transects (Motazedian et al., 2020; Skoulika et al., 2014; Yan et al., 2018).

A substantial advantage of this approach is that transects can be performed even within pedestrian areas or where the deployment of temporary stations would not be possible because of lack of infrastructure to connect the sensors to (e.g., no lamp posts in narrow streets) or high risk of vandalization. However, some metrologic challenges are intrinsic in mobile measurements performing transects:

1. *Metadata.* Location metadata do not relate to a point but a path, making the interpretation and validation of data orders of magnitude more complex. The insufficient or lacking metadata collection during transects might explain some of the observed inconsistencies in reported temperature differences measured across transects. Transects metadata are seldom reported or discussed in the literature (Fig. 2.13).

2. *Location uncertainty.* Before the advent of global positioning system (GPS)-equipped sensors, the position of data collection was subject to uncertainty. Still, for campaigns performing multiple measurement sessions at the same location, different teams might perform measurements at slightly different locations altering the boundary conditions (e.g., measurement in the sunlit or shaded portion of a street). Within

Location		Team	
Mount Druitt NSW 2770		A	

#	Time start hh:mm	Time end hh:mm	Impervious Sun/shade	Impervious Light/dark mat.	grass/soil Sun/shade	grass/soil Dry/wet	Notes(Thermal camera image number)
1	11:26	11:36	x	x	x		No. 155-160
2	11:41	11:50	x	x	x		No. 161-172
3	11:54	12:02	x	x	x		No. 173-186
4	12:05	12:13	x	x	x		No. 187-200
5	12:18	12:30	x	x	x		No.201-218
6	12:34	12:42	x	x	x		No.219-230
7	12:45	12:53	x	x	x		No.231-240
8	12:55	13:03	x	x	x		No.241-258

Figure 2.13 Example of metadata collection sheet for measurements performed with a weather station mounted on a cart.

a dense urban environment, an uncertainty of $1-2$ m in the data point may be associated with different boundary conditions and thus thermal influence (e.g., proximity to a source of heat, producing only microscale disturbance).

3. *Influence of movement and response time.* Several measurements, such as wind speed, and temperature, are affected by movement. The apparent wind speed can be GPS-corrected in some systems for mobile transverses (Pigliautile and Pisello, 2020; Xia et al., 2021). However, other aspects such as the response time of the radiation shield of the temperature and humidity sensor (Burt and de Podesta, 2020) cannot be overcome if not allowing for long-enough stops at measurement locations. Most temperature sensors combined with naturally ventilated radiation shields have a response time in the order of magnitude of approximately $5-10$ minutes (90% of change), with the temperature sensor already having a 63% response time of the order of magnitude of 2 minutes in most cases (Burt and de Podesta, 2020). For instance, the MX2301A with an RS1 radiation shield by Hobo has a response time of slightly less than 8 minutes with a wind speed of 1 m/s. Often, this quantity is not declared by the manufacturer, and it appears to be overlooked by many campaigns documented in the literature, which report continuous measurements without a stop. Moreover, as temporarily installed sensors, mobile transects often employ custom-made assemblies pursued to reduce costs or integrate with vehicles or wearable setups.

To date, there is no systematic metrological study on the measurement uncertainty of mobile approaches in urban climatology, given the many ways these are implemented, all influenced by the project-specific context. However, there are multiple indications of inconsistencies in the implementation of measurement protocols in mobile measurements and transects, as the reported magnitude of the heat island for the same city broadly differs when information is derived from transects or stationary weather stations or different transects (Rizwan et al., 2008; Santamouris, 2015).

2.2.2.3 Citizen science climate mapping and ubiquitous sensing

On the opposite side of the spectrum of urban heat measurements, with respect to the starting point of professionally managed long-term stationary weather stations, are short-term measurements performed by volunteers. In this case, urban heat is characterized in a discontinuous mode,

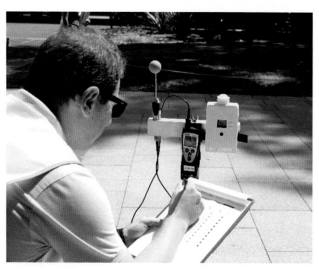

Figure 2.14 Citizen science urban heat measurements with a thermohygrometer and a globe thermometer mounted on a tripod 1.1 m from the ground (Rajagopalan et al., 2018).

either with instruments provided by researchers or institutions as part of dissemination activities and citizen science projects or with wearable sensors.

2.2.2.3.1 Climate mapping by citizen scientists

Climate mapping performed by citizen scientists may be performed with instruments supplied by researchers or institutions or environmental community groups, as supported by local government. In a project on urban microclimate in Australian cities[2], for instance, community members were provided with thermohygrometers, globe thermometers, anemometers, and infrared guns (Figs. 2.14 and 2.15) to map the local conditions in their suburb (Rajagopalan et al., 2018). In principle, these measurements are the same as those performed by professional researchers, with the accuracy of the equipment limited by cost and ease of use. Further, community members performing measurements for the first time might not be trained to collect metadata and follow a strict protocol or perform long measurement sessions necessary to map a sufficiently large area.

However, despite these apparent limitations and skepticism within the scientific community, there is a growing body of literature demonstrating

[2] Urban Microclimate Citizen Science Project <https://citizenscienceproject.org.au/>

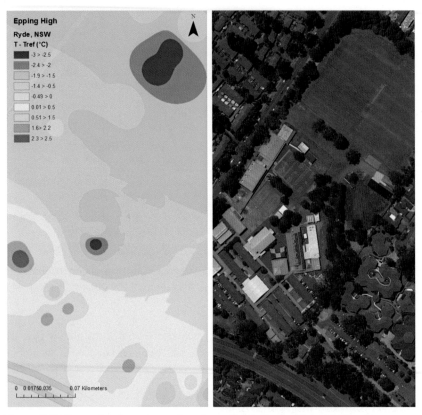

Figure 2.15 Example of a microscale urban climate map after citizen science measurements performed in Ryde, NSW, Australia (Rajagopalan et al., 2018).

good data quality in environmental citizen science projects. A critical review by Kosmala et al. concluded that volunteers can collect data with accuracy and quality similar to or superior to that of professional researchers and that the determinant factors are the design of the experiment and data quality and validation procedures, rather than who performs the measurements (Kosmala et al., 2016).

In particular, Kosmala et al. highlight the importance of iterative project development, training of volunteers, use of standardized and calibrated equipment, data validation, replication, and statistical analysis. The same methodological infrastructure is identified by other authors too (Crowston and Prestopnik, 2013; Wiggins et al., 2011). Findings of good data quality and accuracy are also reported concerning urban tree inventories (Roman et al., 2017) and similar projects that deliver important metadata for urban heat mitigation. As citizen science projects gain momentum, other aspects are identified as critical, such as protecting sensitive information and privacy when community members collect data (Anhalt-Depies et al., 2019).

2.2.2.3.2 Wearable sensors

The most granular level of information on urban heat exposure is the individual one, achieved by citizens wearing or carrying sensors (e.g., on their purse or backpack). A review by Nazarian and Lee analyzed the personal assessment of urban heat, also reporting on a shift toward a "human-centric" approach in sensing (Nazarian and Lee, 2021). These options include wearable sensors worn on the wrist (e.g., watches, bracelets), chest, head or shoes, or embedded in smart garments (Nazarian and Lee, 2021).

Such human-centric data collection systems still require a focus on accuracy and standardization beyond the intrinsic issues related to privacy and human ethics. However, wearable sensors already offer a layer of previously unavailable information. In particular, the opportunities concern the direct measurement of thermal discomfort and strain experienced by urban dwellers (Kuras et al., 2017). Epidemiological analyses of heat and cold-related mortality and morbidity almost exclusively rely on data from weather stations (Gasparrini et al., 2015; Macintyre et al., 2018; Vardoulakis et al., 2014), while most heat-related deaths occur indoors, affecting the elderly and people with preexisting medical conditions (Fouillet et al., 2006; Michelozzi et al., 2005). Therefore wearable or local (indoor) sensing may overcome the information gap.

Further, one of the parameters having the greatest influence on the estimation of thermal comfort with physically based models is skin temperature, which also is affected by great uncertainty due to assumptions in calculation (Metzmacher et al., 2018; Wu and Cao, 2022). Wearable sensors offer the opportunity to directly measure skin temperature (Nazarian et al., 2021), thus overcoming one of the many aspects that determine the poor performance of existing thermal comfort models. In fact, conventional thermal comfort models display a prediction accuracy not exceeding 75% and with the interquartile range comprised between 50% and 55% (Kim et al., 2018). Another relevant parameter often collected by wearable sensors relates to noise levels or acoustic pollution (Marquart et al., 2021), which are not routinely mapped across urban areas. These advantages potentially delivered by a distributed network of wearable sensors do not make them a substitute but rather a complement to all other traditional measurement approaches previously discussed.

2.2.3 Remote sensing

In addition to terrestrial measurements, urban climates have been largely characterized by remote sensing techniques, including terrestrial-level remote sensing (e.g., using thermal cameras), aerial, and satellite sensing.

The number and range of satellites orbiting around the Earth that detect Land Surface Temperature (LST) have constantly been increasing, progressively enhancing the resolution and widening the scope of remote sensing (Stathopoulou and Cartalis, 2011). This increased sensing capacity is also reflected in many publications on surface urban heat islands (SUHI). A recent review analyzed 579 studies published between 2000 and 2020, with almost 400 published just in the last five years of the analysis period (de Almeida et al., 2021). Most studies (68%) focused on LST followed (56%) by Land Use Land Cover estimates, principally concerning temperate climates in the northern hemisphere. Most LST studies used Landsat, with a spatial resolution of 100 m resampled at 30 m and orbital frequency of 16 days (de Almeida et al., 2021). Further, visible and near infrared data are used to compute the normalized difference vegetation index, which identifies whether the target area contains live vegetation or not (Stathopoulou and Cartalis, 2011).

The use of aeroplanes (Ban-Weiss et al., 2015) or drones (Haddad et al., 2019) solves some issues connected to satellite imagery, such as the need for atmospheric correction, while retaining the intrinsic uncertainty associated with the identification of the emissivity of surfaces, and facing further challenges related to flight restrictions over urban areas. UAVs or drones are an affordable alternative to flight with traditional aeroplanes and suited for microscale urban overheating projects (Figs. 2.16 and 2.17), also considering that most UAV platforms now deliver ultrahigh resolution in remote sensing, with less than 10 cm of ground sampling distance (Yao et al., 2019). In addition to the rapidly changing airspace regulations and consequent restrictions, another operational issue faced by UAV is the flight time, limited by payload and battery duration (Mohd Noor et al., 2018). Finally, in addition to satellite or aerial platforms for remote sensing, traditional terrestrial thermography is still performed. It is the only method that can accurately capture vertical surfaces or local situations not accessible from the sky (Fig. 2.18).

Figure 2.16 A drone (DJI600 hexacopter) equipped with a high-resolution thermal camera used in the Parramatta Overheating Project (Santamouris et al., 2020).

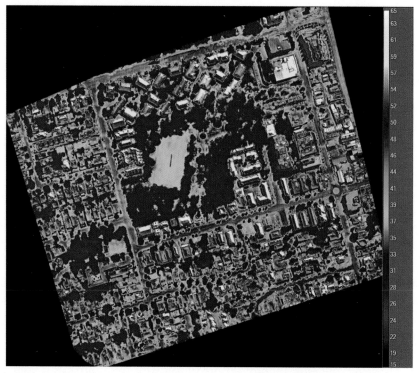

Figure 2.17 A drone-captured infrared ortomap, after a flight over Epping, NSW, Australia. The temperature scale is in °C (Santamouris et al., 2020).

Figure 2.18 Ground-based remote sensing with a hand-held thermal camera The temperature scale is in °C.

The main limitations of satellite imagery (and in part aerial remote sensing) relate to:

- Spatial, temporal, and spectral resolution (Stathopoulou and Cartalis, 2011).
- Validation and calibration with ground-based temperature measurements (Yu et al., 2017).
- Uncertainty in the atmospheric (Rasul et al., 2017; Schaepman-Strub et al., 2006) and emissivity correction (Coutts et al., 2007).

In particular, estimating the emissivity of metal roofs is difficult without ground validation, resulting in the underestimation of LST for metal roofs by even more than 2°C (Coutts et al., 2016). Several approaches have been developed, including temperature-emissivity separation methods, which require multispectral information (Gillespie et al., 1998; Wang et al., 2011). Nonetheless, in most cases, the emissivity values assigned to urban surfaces largely depend on the land cover classification and rely on information from databases (Coutts et al., 2016; Kotthaus et al., 2014).

However, the estimation of the actual value for metal surfaces is highly uncertain and influenced by factors such as the thickness of the factory-applied coating applied onto the metal sheeting (if present), rusting, smoothness of the surface, roof geometry, and soiling, which does not alter the emissivity of nonmetallic surfaces instead (Paolini et al., 2020; Sleiman et al., 2011). Also, the correction for emissivity is complicated by the high variability of radiative properties of urban surfaces. Green et al. measured large-scale roofs of shopping center images (Green et al., 2020). Their campaign included the ground validation of emissivity with a portable emissometer (ASTM, 2015) and contact surface temperature measurements, used to assign different emissivities to the drone-captured infrared images (Green et al., 2020).

Several findings of LST investigations are local and cannot be extrapolated to other contexts. For instance, Stathopoulou et al. found that high surface temperatures are associated with industrial, commercial, densely built, and port areas in several Greek cities (Stathopoulou et al., 2010). In many suburban residential areas in Australia, instead, the prevalence of an architectural fashion led to widespread use of solar absorptive roofing with higher surface temperatures than commercial areas (Green et al., 2020; Jamei et al., 2019). In fact, the magnitude of the daytime SUHI reported in the literature varies between $-6°C$ and $+10°C$ (Rasul et al., 2017).

Further, the magnitude of the UHI is weakly correlated to the SUHI spatial distribution (Streutker, 2010). Thus the purpose of remote sensing should be only to allow for the identification of land portions with a high surface temperature that thus transfer a high amount of heat by convection (increased turbulent sensible heat flux), increasing the ambient air temperature. Unfortunately, satellite imagery has been often used without the support of ground-truthing and not as part of a holistic sensing strategy. Still, remote sensing approaches, especially with multispectral data from satellites, are a useful tool to derive urban characteristics such as land use/cover and vegetation cover (Stathopoulou and Cartalis, 2011).

2.3 Climate and nonclimate data to support urban heat mitigation: challenges and prospects

The data collection platforms discussed so far have been widely used in a multiplicity of studies and reached technology maturity. With the availability of affordable sensors, data collection is going toward more instruments deployed within the urban environment, managed by various actors, from researchers, to environmental protection agencies, met offices, and weather enthusiasts or citizen scientists. Consequently, the main challenges relate to handling, processing big data, and integrating multiple layers of information from a multiplicity of sources (Chen et al., 2020; Middel et al., 2019).

Beyond the availability, integration, and validation of "more of the same" data, the main challenge is to ascertain whether existing measurement methods are suited to assist urban heat mitigation. This question cascades into identifying the research questions that a sensing network assisting urban heat mitigation should address. The three main questions relate to:

- The identification of the causes of urban overheating, which are not the same in all cities.
- The assessment of the performance of urban heat mitigation.
- The measurement of parameters that have an indirect influence on countermeasures to urban overheating or its effects on human health.

2.3.1 Measurement of advective flows and causes of urban overheating

The first research question relates to the identification of the root causes of urban overheating in each scenario, even in complex scenarios where advective flows determine a strong contribution to the local climate (Khan et al., 2021; Yun et al., 2020). This would require strengthening measurements assessing the surface energy balance (net-radiometry and eddy covariance) and the establishment of spatially resolved networks (Bassett et al., 2016). Only some networks of weather stations may be suited for the scope. Another quantity that is not measured directly is anthropogenic heat flux (Allen et al., 2011; Dong et al., 2017).

2.3.2 Measurement of parameters that influence the performance of urban heat mitigation technologies

The performance of urban heat mitigation technologies is not constant, but it is affected by the boundary conditions. Therefore several parameters offer insight into the following:

- *Black carbon measurements.* Cool materials suffer a performance loss when weathering, soiling, and biological growth decrease the albedo of the surface (Paolini et al., 2020; Sleiman et al., 2011). The main parameter that causes solar reflectance loss is black carbon (Berdahl et al., 2002). This quantity is not necessarily correlated to fine particulate matter, and it is seldom measured routinely. Ultraviolet radiation and other quantities that produce aging of materials are not commonly measured, but they are not the main determinant of performance loss and are instead screened during product development.

- *Soil moisture and evaporation potential.* While evaporation potential is more commonly measured by several agencies, soil moisture and other parameters that affect the performance of vegetation are less commonly monitored, even in a representative reduced set of locations. In fact, at high temperatures and low soil moisture levels, evapotranspiration, and therefore the cooling capacity, drastically decreases (Gräf et al., 2021).
- *Health status of trees and foliage.* The drought-induced mortality risk of plants can be measured in association with stomatal conductance (Marchin et al., 2022), preventing damage to foliage during extreme heat events (Teskey et al., 2015) by early intervention before a safety threshold is trespassed. The maintenance of healthy greenery and the measurement of the actual performance of greenery is fundamental to understanding the real mitigation achievable. Further, the actual tree coverage and shading is another descriptor of the response of urban greenery. However, while there are consolidated research approaches to quantify the effectiveness of shade (Middel et al., 2021), these cannot be implemented at the city scale.
- *Boundary layer height.* The height of the planetary boundary layer and the mixing layer determines the interaction between mesoscale advective flows and the urban boundary layer. The boundary layer height also predicts the dispersion of urban pollutants (Davies et al., 2007). Automatic lidars and ceilometers are becoming more widespread given their low maintenance, therefore enabling the possibility of routine measurements (Kotthaus and Grimmond, 2018).

2.3.3 Mapping of urban pollution and noise levels

While urban air quality is routinely measured, mapping of noise levels is limited to research projects (Nourmohammadi et al., 2021). The latter indicates the possibility of enabling natural ventilation, which is one of the main passive cooling techniques to reduce building overheating and can assist designers in identifying site-specific options to counteract indoor overheating (Fig. 2.19).

2.4 Conclusion

In this chapter, the main methods to characterize urban climates have been presented, analyzed, and discussed. These include networks of

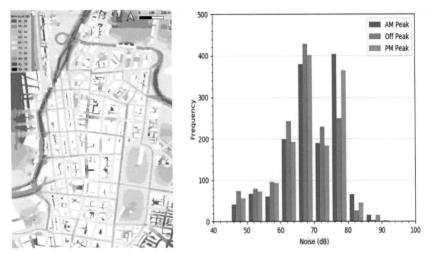

Figure 2.19 Noise levels in the CBD of Sydney. Mapping urban environmental performance with (Nourmohammadi et al., 2021).

weather stations, surface energy balance measurements (with net-radiometry and eddy covariance), and temporary measurements, comprising transects or mobile measurements with instruments on vehicles or carts or sensors worn by humans. In addition to terrestrial measurements, satellite and aerial remote sensing measurements provide information on the LST. These techniques have been widely used in urban climatology over the past four decades, in different combinations and to varying levels of integration and maturity. However, while no single framework is predominantly acknowledged in the literature to monitor the urban climate, there is consensus on the need for a multiplicity of techniques at the temporal and spatial scales. Therefore all the presented methods are used, and their interaction is progressively reaching maturity. Perhaps the area of sensing showing the most significant development in recent years is personal heat exposure with wearable sensors.

Nonetheless, all these methods have been designed and implemented to observe the urban climate. A significant challenge for the future is to rethink urban climate monitoring networks so they can address the research questions around urban overheating: what is causing urban overheating and what is affecting the performance of urban heat mitigation technologies? The main aspects that should be more commonly measured are the advective flows, in addition to the surface energy balance, and the height of the atmospheric boundary layer, for what concerns the climate side. Then, parameters such as black carbon, which affects the reflectance of surfaces, and the health of vegetation would monitor the performance of heat mitigation technologies. The integrated system would enable researchers and environmental agencies to monitor the effectiveness and efficiency of heat mitigation and the evolution of local climate change.

References

Allen, R.G., 1996. Assessing integrity of weather data for reference evapotranspiration estimation. Journal of Irrigation and Drainage Engineering 122, 97–106. Available from: https://doi.org/10.1061/(ASCE)0733-9437(1996)122:2(97).

Allen, L., Lindberg, F., Grimmond, C.S.B., 2011. Global to city scale urban anthropogenic heat flux: model and variability. International Journal of Climatology 31, 1990–2005. Available from: https://doi.org/10.1002/joc.2210.

Anhalt-Depies, C., Stenglein, J.L., Zuckerberg, B., Townsend, P.M., Rissman, A.R., 2019. Tradeoffs and tools for data quality, privacy, transparency, and trust in citizen science. Biological Conservation 238, 108195. Available from: https://doi.org/10.1016/J.BIOCON.2019.108195.

ASTM, 2015. ASTM C 1371: Standard test method for determination of emittance of materials near room temperature using portable emissometers.

Australian Bureau of Meteorology, 2017. Climate Data [WWW Document]. URL <http://www.bom.gov.au/climate/data/> (accessed 24.08.17).

Bacci, P., Maugeri, M., 1992. The urban heat island of Milan. Il Nuovo Cimento C 15, 417–424. Available from: https://doi.org/10.1007/BF02511742.

Ban-Weiss, G.A., Woods, J., Levinson, R., 2015. Using remote sensing to quantify albedo of roofs in seven California cities, Part 1: methods. Solar Energy 115, 777–790. Available from: https://doi.org/10.1016/J.SOLENER.2014.10.022.

Bassett, R., Cai, X., Chapman, L., Heaviside, C., Thornes, J.E., Muller, C.L., et al., 2016. Observations of urban heat island advection from a high-density monitoring network. Quarterly Journal of the Royal Meteorological Society 142, 2434–2441. Available from: https://doi.org/10.1002/QJ.2836.

Bell, S., Cornford, D., Bastin, L., 2015. How good are citizen weather stations? Addressing a biased opinion. Weather 70, 75–84. Available from: https://doi.org/10.1002/WEA.2316.

Berdahl, P., Akbari, H., Rose, L.S., 2002. Aging of reflective roofs: soot deposition. Applied Optics 41, 2355. Available from: https://doi.org/10.1364/ao.41.002355.

Bian, Q., Alharbi, B., Shareef, M.M., Husain, T., Pasha, M.J., Atwood, S.A., et al., 2018. Sources of PM2.5 carbonaceous aerosol in Riyadh, Saudi Arabia. Atmospheric Chemistry and Physics 18, 3969–3985. Available from: https://doi.org/10.5194/ACP-18-3969-2018.

Bottillo, S., de Lieto Vollaro, A., Galli, G., Vallati, A., 2014. Fluid dynamic and heat transfer parameters in an urban canyon. Solar Energy 99, 1–10. Available from: https://doi.org/10.1016/j.solener.2013.10.031.

Burt, S., de Podesta, M., 2020. Response times of meteorological air temperature sensors. Quarterly Journal of the Royal Meteorological Society 146, 2789–2800. Available from: https://doi.org/10.1002/QJ.3817.

Busato, F., Lazzarin, R.M., Noro, M., 2014. Three years of study of the urban heat island in Padua: experimental results. Sustainable Cities and Society 10, 251–258. Available from: https://doi.org/10.1016/J.SCS.2013.05.001.

Chen, Y., Yue, W., la Rosa, D., 2020. Which communities have better accessibility to green space? An investigation into environmental inequality using big data. Landscape and Urban Planning 204, 103919. Available from: https://doi.org/10.1016/J.LANDURBPLAN.2020.103919.

Chen, F., Kusaka, H., Bornstein, R., Ching, J., Grimmond, C.S.B., Grossman-Clarke, S., et al., 2011. The integrated WRF/urban modelling system: development, evaluation, and applications to urban environmental problems. International Journal of Climatology 31, 273–288. Available from: https://doi.org/10.1002/joc.2158.

Chow, W.T.L., Pope, R.L., Martin, C.A., Brazel, A.J., 2011. Observing and modeling the nocturnal park cool island of an arid city: horizontal and vertical impacts. Theoretical and Applied Climatology 103, 197–211. Available from: https://doi.org/10.1007/S00704-010-0293-8/FIGURES/7.

Clementinum, 2022. CHMI portal : Historical data : Weather : Prague Clementinum [WWW Document]. URL <https://www.chmi.cz/historicka-data/pocasi/praha-klementinum?l = en> (accessed 17.02.22).

Coulibaly, P., Evora, N.D., 2007. Comparison of neural network methods for infilling missing daily weather records. Journal of Hydrology 341, 27–41. Available from: https://doi.org/10.1016/J.JHYDROL.2007.04.020.

Coutts, A.M., Beringer, J., Tapper, N.J., Coutts, A.M., Beringer, J., Tapper, N.J., 2007. Impact of increasing urban density on local climate: spatial and temporal variations in the surface energy balance in Melbourne, Australia. Journal of Applied Meteorology and Climatology 46, 477–493. Available from: https://doi.org/10.1175/JAM2462.1.

Coutts, A.M., Harris, R.J., Phan, T., Livesley, S.J., Williams, N.S.G., Tapper, N.J., 2016. Thermal infrared remote sensing of urban heat: hotspots, vegetation, and an assessment of techniques for use in urban planning. Remote Sensing of Environment 186, 637–651. Available from: https://doi.org/10.1016/J.RSE.2016.09.007.

Crowston, K., Prestopnik, N.R., 2013. Motivation and data quality in a citizen science game: a design science evaluation. Proceedings of the Annual Hawaii International Conference on System Sciences, pp. 450–459. <https://doi.org/10.1109/HICSS.2013.413>.

Curci, S., Lavecchia, C., Frustaci, G., Paolini, R., Pilati, S., Paganelli, C., 2017. Assessing measurement uncertainty in meteorology in urban environments. Measurement Science and Technology 28. Available from: https://doi.org/10.1088/1361-6501/aa7ec1.

Cureau, R.J., Pigliautile, I., Pisello, A.L., 2022. A new wearable system for sensing outdoor environmental conditions for monitoring hyper-microclimate. Sensors 22, 502. Available from: https://doi.org/10.3390/S22020502.

Davies, F., Middleton, D.R., Bozier, K.E., 2007. Urban air pollution modelling and measurements of boundary layer height. Atmospheric Environment 41, 4040–4049. Available from: https://doi.org/10.1016/J.ATMOSENV.2007.01.015.

de Almeida, C.R., Teodoro, A.C., Gonçalves, A., 2021. Study of the urban heat island (UHI) using remote sensing data/techniques: a systematic review. Environments 8, 105. Available from: https://doi.org/10.3390/ENVIRONMENTS8100105.

de Vos, L.W., Leijnse, H., Overeem, A., Uijlenhoet, R., 2019. Quality control for crowdsourced personal weather stations to enable operational rainfall monitoring. Geophysical Research Letters 46, 8820–8829. Available from: https://doi.org/10.1029/2019GL083731.

Demerjian, K.L., 2000. A review of national monitoring networks in North America. Atmospheric Environment 34, 1861–1884. Available from: https://doi.org/10.1016/S1352-2310(99)00452-5.

Dong, Y., Varquez, A.C.G., Kanda, M., 2017. Global anthropogenic heat flux database with high spatial resolution. Atmospheric Environment 150, 276–294. Available from: https://doi.org/10.1016/j.atmosenv.2016.11.040.

Eliasson, I., Offerle, B., Grimmond, C.S.B., Lindqvist, S., 2006. Wind fields and turbulence statistics in an urban street canyon. Atmospheric Environment 40, 1–16. Available from: https://doi.org/10.1016/j.atmosenv.2005.03.031.

Estévez, J., Gavilán, P., Giráldez, J.V., 2011. Guidelines on validation procedures for meteorological data from automatic weather stations. Journal of Hydrology 402, 144–154. Available from: https://doi.org/10.1016/j.jhydrol.2011.02.031.

Estévez, J., García-Marín, A.P., Morábito, J.A., Cavagnaro, M., 2016. Quality assurance procedures for validating meteorological input variables of reference evapotranspiration in Mendoza province (Argentina). Agricultural Water Management 172, 96–109. Available from: https://doi.org/10.1016/J.AGWAT.2016.04.019.

Fouillet, A., Rey, G., Laurent, F., Pavillon, G., Bellec, S., Guihenneuc-Jouyaux, C., et al., 2006. Excess mortality related to the August 2003 heat wave in France. International Archives of Occupational and Environmental Health 80, 16–24. Available from: https://doi.org/10.1007/s00420-006-0089-4.

García-Sánchez, C., Philips, D.A., Gorlé, C., 2014. Quantifying inflow uncertainties for CFD simulations of the flow in downtown Oklahoma City. Building and Environment 78, 118–129. Available from: https://doi.org/10.1016/j.buildenv.2014.04.013.

Gasparrini, A., Guo, Y.-L.L.Y.Y.L.L., Hashizume, M., Lavigne, E., Zanobetti, A., Schwartz, J., et al., 2015. Mortality risk attributable to high and low ambient temperature: a multicountry observational study. The Lancet 386, 369–375. Available from: https://doi.org/10.1016/S0140-6736(14)62114-0.

Gillespie, A., Rokugawa, S., Matsunaga, T., Steven Cothern, J., Hook, S., Kahle, A.B., 1998. A temperature and emissivity separation algorithm for advanced spaceborne thermal emission and reflection radiometer (ASTER) images. IEEE Transactions on Geoscience and Remote Sensing 36, 1113–1126. Available from: https://doi.org/10.1109/36.700995.

Giridharan, R., Kolokotroni, M., 2009. Urban heat island characteristics in London during winter. Solar Energy 83, 1668–1682. Available from: https://doi.org/10.1016/J.SOLENER.2009.06.007.

Gräf, M., Immitzer, M., Hietz, P., Stangl, R., 2021. Water-stressed plants do not cool: leaf surface temperature of living wall plants under drought stress. Sustainability 2021 13, Page 3910 13, 3910. Available from: https://doi.org/10.3390/SU13073910.

Green, A., Ledo Gomis, L., Paolini, R., Haddad, S., Kokogiannakis, G., Cooper, P., et al., 2020. Above-roof air temperature effects on HVAC and cool roof performance: experiments and development of a predictive model. Energy and Buildings 222, 110071. Available from: https://doi.org/10.1016/j.enbuild.2020.110071.

Grimmond, C.S.B., Souch, C., Hubble, M.D., 1996. Influence of tree cover on summertime surface energy balance fluxes, San Gabriel Valley, Los Angeles. Climate Research . Available from: https://doi.org/10.3354/cr006045.

Gülten, A., Aksoy, U.T., Öztop, H.F., 2016. Influence of trees on heat island potential in an urban canyon. Sustainable Cities and Society 26, 407–418. Available from: https://doi.org/10.1016/J.SCS.2016.04.006.

Guo, Y., Yu, Z., Men, Y., -, al, Leach, R., Haitjema, H., Su, R., et al., 2016. The importance of metrological metadata in the environmental monitoring. Journal of Physics: Conference Series 733, 012033. Available from: https://doi.org/10.1088/1742-6596/733/1/012033.

Haddad, S., Ulpiani, G., Paolini, R., Synnefa, A., Santamouris, M., 2019. Experimental and theoretical analysis of the urban overheating and its mitigation potential in a hot arid city–Alice Springs. Architectural Science Review . Available from: https://doi.org/10.1080/00038628.2019.1674128.

Howard, L., 1818. The Climate of London, 1st (ed.) W. Phillips, London, UK.

Howard, L., 1833. The Climate of London, 2nd (ed.) Harvey and Darton, London, UK.

INAF, 2022. Museo astronomico di brera | Lo sapevi che... le osservazioni meteorolo-giche sono state per molti secoli compito degli osservatori astronomici? [WWW Document]. URL <http://museoastronomico.brera.inaf.it/osservazioni-meteorolo-giche/> (accessed 17.02.22).

Jamei, Y., Rajagopalan, P., Sun, Q.(Chayn), 2019. Spatial structure of surface urban heat island and its relationship with vegetation and built-up areas in Melbourne, Australia. Science of The Total Environment 659, 1335—1351. Available from: https://doi.org/10.1016/J.SCITOTENV.2018.12.308.

Johansson, E., Emmanuel, R., 2006. The influence of urban design on outdoor thermal comfort in the hot, humid city of Colombo, Sri Lanka. International Journal of Biometeorology 51, 119—133. Available from: https://doi.org/10.1007/s00484-006-0047-6.

Khan, H.S., Santamouris, M., Paolini, R., Caccetta, P., Kassomenos, P., 2021. Analysing the local and climatic conditions affecting the urban overheating magnitude during the Heatwaves (HWs) in a coastal city: a case study of the greater Sydney region. Science of the Total Environment 755, 142515. Available from: https://doi.org/10.1016/j.scitotenv.2020.142515.

Kim, J., Schiavon, S., Brager, G., 2018. Personal comfort models—A new paradigm in ther-mal comfort for occupant-centric environmental control. Building and Environment 132, 114—124. Available from: https://doi.org/10.1016/J.BUILDENV.2018.01.023.

Kosmala, M., Wiggins, A., Swanson, A., Simmons, B., 2016. Assessing data quality in citi-zen science. Frontiers in Ecology and the Environment 14, 551—560. Available from: https://doi.org/10.1002/FEE.1436.

Kotthaus, S., Grimmond, C.S.B., 2014a. Energy exchange in a dense urban environ-ment—Part II: impact of spatial heterogeneity of the surface. Urban Climate 10, 281—307. Available from: https://doi.org/10.1016/j.uclim.2013.10.001.

Kotthaus, S., Grimmond, C.S.B., 2014b. Energy exchange in a dense urban environment—Part I: temporal variability of long-term observations in central London. Urban Climate 10, 261—280. Available from: https://doi.org/10.1016/j.uclim.2013.10.002.

Kotthaus, S., Grimmond, C.S.B., 2018. Atmospheric boundary-layer characteristics from ceilometer measurements. Part 1: a new method to track mixed layer height and clas-sify clouds. Quarterly Journal of the Royal Meteorological Society 144, 1525—1538. Available from: https://doi.org/10.1002/QJ.3299.

Kotthaus, S., Smith, T.E.L., Wooster, M.J., Grimmond, C.S.B., 2014. Derivation of an urban materials spectral library through emittance and reflectance spectroscopy. ISPRS Journal of Photogrammetry and Remote Sensing 94, 194—212. Available from: https://doi.org/10.1016/j.isprsjprs.2014.05.005.

Kousis, I., Pigliautile, I., Pisello, A.L., 2021. Intra-urban microclimate investigation in urban heat island through a novel mobile monitoring system. Scientific Reports 11 (1 11), 1—17. Available from: https://doi.org/10.1038/s41598-021-88344-y.

Kuras, E.R., Richardson, M.B., Calkins, M.M., Ebi, K.L., Hess, J.J., Kintziger, K.W., et al., 2017. Opportunities and challenges for personal heat exposure research. Environmental Health Perspectives 125. Available from: https://doi.org/10.1289/EHP556.

Lee, M.K., Moon, S.H., Kim, Y.H., Moon, B.R., 2014. Correcting abnormalities in meteorological data by machine learning. Conference Proceedings—IEEE International Conference on Systems, Man and Cybernetics 2014-January, 888—893. <https://doi.org/10.1109/SMC.2014.6974024>.

Lee, M.K., Moon, S.H., Yoon, Y., Kim, Y.H., Moon, B.R., 2018. Detecting anomalies in meteorological data using support vector regression. Advances in Meteorology 2018. Available from: https://doi.org/10.1155/2018/5439256.

Lemonsu, A., Grimmond, C.S.B., Masson, V., 2004. Modeling the surface energy balance of the core of an old Mediterranean city: Marseille. Journal of Applied Meteorology .

Loridan, T., Grimmond, C.S.B., Grossman-Clarke, S., Chen, F., Tewari, M., Manning, K., et al., 2010. Trade-offs and responsiveness of the single-layer urban canopy parametrisation in WRF: an offline evaluation using the MOSCEM optimisation algorithm and field observations. Quarterly Journal of the Royal Meteorological Society 136, 997−1019. Available from: https://doi.org/10.1002/qj.614.

Macintyre, H.L., Heaviside, C., Cai, X., Phalkey, R., 2021. The winter urban heat island: impacts on cold-related mortality in a highly urbanised European region for present and future climate. Environment International 154, 106530. Available from: https:// doi.org/10.1016/J.ENVINT.2021.106530.

Macintyre, H.L., Heaviside, C., Taylor, J., Picetti, R., Symonds, P., Cai, X.-M., et al., 2018. Assessing urban population vulnerability and environmental risks across an urban area during heatwaves—Implications for health protection. Science of The Total Environment 610−611, 678−690. Available from: https://doi.org/10.1016/J. SCITOTENV.2017.08.062.

Maranzano, P., 2022. Air quality in Lombardy, Italy: an overview of the environmental monitoring system of ARPA Lombardia. Earth 3, 172−203. Available from: https:// doi.org/10.3390/EARTH3010013.

Marchin, R.M., Backes, D., Ossola, A., Leishman, M.R., Tjoelker, M.G., Ellsworth, D.S., 2022. Extreme heat increases stomatal conductance and drought-induced mortality risk in vulnerable plant species. Global Change Biology 28, 1133−1146. Available from: https://doi.org/10.1111/GCB.15976.

Marquart, H., Ueberham, M., Schlink, U., 2021. Extending the dimensions of personal exposure assessment: a methodological discussion on perceived and measured noise and air pollution in traffic. Journal of Transport Geography 93, 103085. Available from: https://doi.org/10.1016/J.JTRANGEO.2021.103085.

Martilli, A., Krayenhoff, E.S., Nazarian, N., 2020. Is the urban heat island intensity relevant for heat mitigation studies? Urban Climate 31. Available from: https://doi.org/ 10.1016/j.uclim.2019.100541.

Meek, D.W., Hatfield, J.L., 1994. Data quality checking for single station meteorological databases. Agricultural and Forest Meteorology 69, 85−109. Available from: https:// doi.org/10.1016/0168-1923(94)90083-3.

MeshkinKiya, M., Paolini, R., 2020. Preparing weather data for real-time building energy simulation. Proceedings of eSim 2021, 14-16 June 2021, Vancouver, Canada. http:// www.ibpsa.org/proceedings/eSimPapers/2021/Contribution_1161_final_a.pdf.

Mestayer, P.G., Durand, P., Augustin, P., Bastin, S., Bonnefond, J.-M., Benech, B., et al., 2005. The urban boundary-layer field campaign in marseille (ubl/clu-escompte): setup and first results. Boundary-Layer Meteorology 114, 315−365. Available from: https:// doi.org/10.1007/s10546-004-9241-4.

Metzmacher, H., Wölki, D., Schmidt, C., Frisch, J., van Treeck, C., 2018. Real-time human skin temperature analysis using thermal image recognition for thermal comfort assessment. Energy and Buildings 158, 1063−1078. Available from: https://doi.org/ 10.1016/J.ENBUILD.2017.09.032.

Michelozzi, P., de Donato, F., Bisanti, L., Russo, A., Cadum, E., de Maria, M., et al.,Team, E.C. for D.P. and C. (ECDC)-H.C.U.-E. editorial 2005. The impact of the summer 2003 heat waves on mortality in four Italian cities. Eurosurveillance: bulletin Europeen sur les maladies transmissibles = European Communicable Disease Bulletin 10, 161−165. Available from: https://doi.org/ 10.2807/esm.10.07.00556-en.

Middel, A., Häb, K., Brazel, A.J., Martin, C.A., Guhathakurta, S., 2014. Impact of urban form and design on mid-afternoon microclimate in Phoenix Local Climate Zones.

Landscape and Urban Planning 122, 16−28. Available from: https://doi.org/10.1016/J.LANDURBPLAN.2013.11.004.

Middel, A., AlKhaled, S., Schneider, F.A., Hagen, B., Coseo, P., 2021. 50 grades of shade. Bulletin of the American Meteorological Society 102, E1805−E1820. Available from: https://doi.org/10.1175/BAMS-D-20-0193.1.

Middel, A., Lukasczyk, J., Zakrzewski, S., Arnold, M., Maciejewski, R., 2019. Urban form and composition of street canyons: a human-centric big data and deep learning approach. Landscape and Urban Planning 183, 122−132. Available from: https://doi.org/10.1016/J.LANDURBPLAN.2018.12.001.

Mohd Noor, N., Abdullah, A., Hashim, M., 2018. Remote sensing UAV/drones and its applications for urban areas: a review. IOP Conference Series: Earth and Environmental Science 169, 012003. Available from: https://doi.org/10.1088/1755-1315/169/1/012003.

Morrison, W., Kotthaus, S., Grimmond, S., 2021. Urban surface temperature observations from ground-based thermography: intra- and inter-facet variability. Urban Climate 35, 100748. Available from: https://doi.org/10.1016/j.uclim.2020.100748.

Motazedian, A., Coutts, A.M., Tapper, N.J., 2020. The microclimatic interaction of a small urban park in central Melbourne with its surrounding urban environment during heat events. Urban Forestry & Urban Greening 52, 126688. Available from: https://doi.org/10.1016/J.UFUG.2020.126688.

Munir, S., Mayfield, M., Coca, D., Jubb, S.A., 2019. Structuring an integrated air quality monitoring network in large urban areas—Discussing the purpose, criteria and deployment strategy. Atmospheric Environment: X 2, 100027. Available from: https://doi.org/10.1016/J.AEAOA.2019.100027.

Nazarian, N., Lee, J.K.W., 2021. Personal assessment of urban heat exposure: a systematic review. Environmental Research Letters 16, 033005. Available from: https://doi.org/10.1088/1748-9326/ABD350.

Nazarian, N., Liu, S., Kohler, M., Lee, J.K.W., Miller, C., Chow, W.T.L., et al., 2021. Project Coolbit: can your watch predict heat stress and thermal comfort sensation? Environmental Research Letters 16, 034031. Available from: https://doi.org/10.1088/1748-9326/ABD130.

Niachou, K., Hassid, S., Santamouris, M., Livada, I., 2005. Comparative monitoring of natural, hybrid and mechanical ventilation systems in urban canyons. Energy and Buildings 37, 503−513. Available from: https://doi.org/10.1016/j.enbuild.2004.09.016.

Niachou, K., Livada, I., Santamouris, M., 2008. Experimental study of temperature and airflow distribution inside an urban street canyon during hot summer weather conditions. Part II: airflow analysis. Building and Environment 43, 1393−1403. Available from: https://doi.org/10.1016/j.buildenv.2007.01.040.

Nourmohammadi, Z., Lilasathapornkit, T., Ashfaq, M., Gu, Z., Saberi, M., 2021. Mapping urban environmental performance with emerging data sources: a case of urban greenery and traffic noise in Sydney, Australia. Sustainability 2021 13, 605. Available from: https://doi.org/10.3390/SU13020605.

Offerle, B., Eliasson, I., Grimmond, C.S.B., Holmer, B., 2006. Surface heating in relation to air temperature, wind and turbulence in an urban street canyon. Boundary-Layer Meteorology 122, 273−292. Available from: https://doi.org/10.1007/s10546-006-9099-8.

Oke, T.R., 1982. The energetic basis of the urban heat island. Quarterly Journal of the Royal Meteorological Society 108, 1−24. Available from: https://doi.org/10.1002/qj.49710845502.

Oke, T.R., 1987. Boundary Layer Climates. Methuen and Co., New York.

Paolini, R., Terraneo, G., Ferrari, C., Sleiman, M., Muscio, A., Metrangolo, P., et al., 2020. Effects of soiling and weathering on the albedo of building envelope materials:

lessons learned from natural exposure in two European cities and tuning of a laboratory simulation practice. Solar Energy Materials and Solar Cells 205, 110264. Available from: https://doi.org/10.1016/j.solmat.2019.110264.

Paolini, R., Zani, A., MeshkinKiya, M., Castaldo, V.L.V.L.V.L., Pisello, A.L.A.L., Antretter, F., et al., 2017. The hygrothermal performance of residential buildings at urban and rural sites: sensible and latent energy loads and indoor environmental conditions. Energy and Buildings 152, 792−803. Available from: https://doi.org/10.1016/j.enbuild.2016.11.018.

Parker, D.E., 2010. Urban heat island effects on estimates of observed climate change. Wiley Interdisciplinary Reviews: Climate Change 1, 123−133. Available from: https://doi.org/10.1002/WCC.21.

Pearlmutter, D., Bitan, A., Berliner, P., 1999. Microclimatic analysis of "compact" urban canyons in an arid zone. Atmospheric Environment 33, 4143−4150. Available from: https://doi.org/10.1016/S1352-2310(99)00156-9.

Pfautsch, S. Rouillard. Benchmarking heat in Parramatta, Sydney's Central River City, 2019. Available from: https://doi.org/10.26183/5D4B69D465DD6.

Pigliautile, I., Pisello, A.L., 2018. A new wearable monitoring system for investigating pedestrians' environmental conditions: development of the experimental tool and start-up findings. Science of the Total Environment 630, 690−706. Available from: https://doi.org/10.1016/J.SCITOTENV.2018.02.208.

Pigliautile, I., Pisello, A.L., 2020. Environmental data clustering analysis through wearable sensing techniques: new bottom-up process aimed to identify intra-urban granular morphologies from pedestrian transects. Building and Environment 171, 106641. Available from: https://doi.org/10.1016/J.BUILDENV.2019.106641.

Piselli, C., Castaldo, V.L., Pigliautile, I., Pisello, A.L., Cotana, F., 2018. Outdoor comfort conditions in urban areas: on citizens' perspective about microclimate mitigation of urban transit areas. Sustainable Cities and Society 39, 16−36. Available from: https://doi.org/10.1016/J.SCS.2018.02.004.

PWS Network Overview | Weather Underground [WWW Document], 2022 URL <https://www.wunderground.com/pws/overview> (accessed 28.02.22).

Rajagopalan, P., Andamon, M.M., Paolini, R., Santamouris. M., 2018. Developing experimental protocol for collecting large scale urban microclimate data through community participation. In P. Rajagopalan & M.M. Andamon (Eds.), Engaging Architectural Science: Meeting the Challenges of Higher Density: 52nd International Conference of the Architectural Science Association 2018. Melbourne. ISBN: 978-0-9923835-5-8. Online at: https://anzasca.net/wp-content/uploads/2019/01/66-Developing-experimental-protocol-for-collecting-large-scale-urban-microclimate-data-through-community-participation.pdf.

Rasul, A., Balzter, H., Smith, C., Remedios, J., Adamu, B., Sobrino, J.A., et al., 2017. A review on remote sensing of urban heat and cool islands. Land 6, 38. Available from: https://doi.org/10.3390/LAND6020038.

Ren, C., Ng, E.Y.Y., Katzschner, L., 2011. Urban climatic map studies: a review. International Journal of Climatology 31, 2213−2233. Available from: https://doi.org/10.1002/JOC.2237.

Ren, C., Spit, T., Lenzholzer, S., Yim, H.L.S., van Hove, B.H., Chen, L., et al., 2012. Urban climate map system for Dutch spatial planning. International Journal of Applied Earth Observation and Geoinformation 18, 207−221. Available from: https://doi.org/10.1016/J.JAG.2012.01.026.

Righini, G., Cappelletti, A., Ciucci, A., Cremona, G., Piersanti, A., Vitali, L., et al., 2014. GIS based assessment of the spatial representativeness of air quality monitoring stations using pollutant emissions data. Atmospheric Environment 97, 121−129. Available from: https://doi.org/10.1016/J.ATMOSENV.2014.08.015.

Rizwan, A.M., Dennis, L.Y.C., Liu, C., 2008. A review on the generation, determination and mitigation of urban heat island. Journal of Environmental Sciences 20, 120–128. Available from: https://doi.org/10.1016/S1001-0742(08)60019-4.

Robinson, P.J., 1998. Monthly variations of dew point temperature in the coterminous united states. International Journal of Climatology 18, 1539–1556. Available from: https://doi.org/10.1002/(SICI)1097-0088(19981130)18:14.

Roman, L.A., Scharenbroch, B.C., Östberg, J.P.A., Mueller, L.S., Henning, J.G., Koeser, A.K., et al., 2017. Data quality in citizen science urban tree inventories. Urban Forestry & Urban Greening 22, 124–135. Available from: https://doi.org/10.1016/J.UFUG.2017.02.001.

Sabatino, S., di, Buccolieri, R., Pulvirenti, B., Britter, R.E., 2008. Flow and pollutant dispersion in street canyons using FLUENT and ADMS-urban. Environmental Modeling & Assessment 13, 369–381. Available from: https://doi.org/10.1007/s10666-007-9106-6.

Santamouris, M., 2014. On the energy impact of urban heat island and global warming on buildings. Energy and Buildings 82, 100–113. Available from: https://doi.org/10.1016/j.enbuild.2014.07.022.

Santamouris, M., 2015. Analysing the heat island magnitude and characteristics in one hundred Asian and Australian cities and regions. Science of The Total Environment 512, 582–598. Available from: https://doi.org/10.1016/j.scitotenv.2015.01.060.

Santamouris, M., Ding, L., Fiorito, F., Oldfield, P., Osmond, P., Paolini, R., et al., 2017. Passive and active cooling for the outdoor built environment—Analysis and assessment of the cooling potential of mitigation technologies using performance data from 220 large scale projects. Solar Energy 154, 14–33. Available from: https://doi.org/10.1016/j.solener.2016.12.006.

Santamouris, M., Paolini, R., Haddad, S., Synnefa, A., Garshasbi, S., Hatvani-Kovacs, G., et al., 2020. Heat mitigation technologies can improve sustainability in cities. An holistic experimental and numerical impact assessment of urban overheating and related heat mitigation strategies on energy consumption, indoor comfort, vulnerability and heat-related mortality and morbidity in cities. Energy and Buildings 217, 110002. Available from: https://doi.org/10.1016/j.enbuild.2020.110002.

Schaepman-Strub, G., Schaepman, M.E., Painter, T.H., Dangel, S., Martonchik, J.V., 2006. Reflectance quantities in optical remote sensing—definitions and case studies. Remote Sensing of Environment 103, 27–42. Available from: https://doi.org/10.1016/J.RSE.2006.03.002.

Skoulika, F., Santamouris, M., Kolokotsa, D., Boemi, N., 2014. On the thermal characteristics and the mitigation potential of a medium size urban park in Athens, Greece. Landscape and Urban Planning 123, 73–86. Available from: https://doi.org/10.1016/J.LANDURBPLAN.2013.11.002.

Sleiman, M., Ban-Weiss, G., Gilbert, H.E., François, D., Berdahl, P., Kirchstetter, T.W., et al., 2011. Soiling of building envelope surfaces and its effect on solar reflectance—Part I: analysis of roofing product databases. Solar Energy Materials and Solar Cells 95, 3385–3399. Available from: https://doi.org/10.1016/j.solmat.2011.08.002.

Stathopoulou, M., Cartalis, C., 2011. Use of satellite remote sensing in support of urban heat island studies. Advances in Building Energy Research 1, 203–212. Available from: https://doi.org/10.1080/17512549.2007.9687275.

Stathopoulou, M., Cartalis, C., Keramitsoglou, I., 2010. Mapping micro-urban heat islands using NOAA/AVHRR images and CORINE Land Cover: an application to coastal cities of Greece. International Journal of Remote Sensing 25, 2301–2316. Available from: https://doi.org/10.1080/01431160310001618725.

Stewart, I.D., Oke, T.R., Krayenhoff, E.S., 2014. Evaluation of the 'local climate zone' scheme using temperature observations and model simulations. International Journal of Climatology 34, 1062−1080. Available from: https://doi.org/10.1002/joc.3746.

Streutker, D.R., 2010. A remote sensing study of the urban heat island of Houston, Texas. Available from: https://doi.org/10.1080/01431160110115023.

Tarara, J.M., Hoheisel, G.A., 2007. Low-cost shielding to minimise radiation errors of temperature sensors in the field. HortScience: A Publication of the American Society for Horticultural Science 42, 1372−1379. Available from: https://doi.org/10.21273/HORTSCI.42.6.1372.

Teskey, R., Wertin, T., Bauweraerts, I., Ameye, M., Mcguire, M.A., Steppe, K., 2015. Responses of tree species to heat waves and extreme heat events. Plant, Cell & Environment 38, 1699−1712. Available from: https://doi.org/10.1111/pce.12417.

Thorsson, S., Lindberg, F., Eliasson, I., Holmer, B., 2007. Different methods for estimating the mean radiant temperature in an outdoor urban setting. International Journal of Climatology 27, 1983−1993. Available from: https://doi.org/10.1002/joc.1537.

Urban Flux Network - IAUC [WWW Document], 2022 URL <https://ibis.geog.ubc.ca/urbanflux/> (accessed 28.02.22).

Vardoulakis, S., Dear, K., Hajat, S., Heaviside, C., Eggen, B., McMichael, A.J., 2014. Comparative assessment of the effects of climate change on heat- and cold-related mortality in the United Kingdom and Australia. Environmental Health Perspectives 122, 1285−1292. Available from: https://doi.org/10.1289/ehp.1307524.

Wang, H., Xiao, Q., Li, H., Zhong, B., 2011. Temperature and emissivity separation algorithm for TASI airborne thermal hyperspectral data. 2011 International Conference on Electronics, Communications and Control, ICECC 2011— Proceedings, pp. 1075−1078. <https://doi.org/10.1109/ICECC.2011.6066288>.

Wang, C., Zhao, L., Sun, W., Xue, J., Xie, Y., 2018. Identifying redundant monitoring stations in an air quality monitoring network. Atmospheric Environment 190, 256−268. Available from: https://doi.org/10.1016/J.ATMOSENV.2018.07.040.

Wiggins, A., Newman, G., Stevenson, R.D., Crowston, K., 2011. Mechanisms for data quality and validation in citizen science. Proceedings—7th IEEE International Conference on e-Science Workshops, eScienceW 2011, pp. 14−19. <https://doi.org/10.1109/ESCIENCEW.2011.27>.

World Meteorological Organization, 2018. Preliminary edition of the CIMO GUIDE (WMO No. 8).

Wu, Y., Cao, B., 2022. Recognition and prediction of individual thermal comfort requirement based on local skin temperature. Journal of Building Engineering 49, 104025. Available from: https://doi.org/10.1016/J.JOBE.2022.104025.

Xia, T., Catalan, J., Hu, C., Batterman, S., 2021. Development of a mobile platform for monitoring gaseous, particulate, and greenhouse gas (GHG) pollutants. Environmental Monitoring and Assessment 193, 1−22. Available from: https://doi.org/10.1007/S10661-020-08769-2/FIGURES/5.

Yan, H., Wu, F., Dong, L., 2018. Influence of a large urban park on the local urban thermal environment. Science of The Total Environment 622−623, 882−891. Available from: https://doi.org/10.1016/J.SCITOTENV.2017.11.327.

Yang, X., Li, Y., Luo, Z., Chan, P.W., 2017. The urban cool island phenomenon in a high-rise high-density city and its mechanisms. International Journal of Climatology 37, 890−904. Available from: https://doi.org/10.1002/JOC.4747.

Yao, H., Qin, R., Chen, X., 2019. Unmanned aerial vehicle for remote sensing applications—A review. Remote Sensing 11, 1443. Available from: https://doi.org/10.3390/RS11121443.

Yu, W., Ma, M., Li, Z., Tan, J., Wu, A., 2017. New scheme for validating remote-sensing land surface temperature products with station observations. Remote Sensing 9, 1210. Available from: https://doi.org/10.3390/RS9121210.

Yun, G.Y., Ngarambe, J., Duhirwe, P.N., Ulpiani, G., Paolini, R., Haddad, S., et al., 2020. Predicting the magnitude and the characteristics of the urban heat island in coastal cities in the proximity of desert landforms. The case of Sydney. Science of the Total Environment 709. Available from: https://doi.org/10.1016/j.scitotenv.2019.136068.

CHAPTER 3

Synergies and exacerbations—effects of warmer weather and climate change

Hassan Saeed Khan[1,2], Riccardo Paolini[1] and Matthaios Santamouris[1]
[1]School of Built Environment, Faculty of Arts, Design and Architecture, University of New South Wales (UNSW), Sydney, NSW, Australia
[2]Data-61, The Commonwealth Scientific and Industrial Research Organization (CSIRO), Kensington, Perth, WA, Australia

3.1 Urban heat islands and urban overheating

An urban heat island (UHI) is a local-scale phenomenon that exhibits higher temperatures in urban areas compared to adjacent nonurban areas (Santamouris et al., 2017). The UHI is largely employed to quantify the degree to which urbanization modifies land-atmosphere interactions (Cui and Shi, 2012). Urban overheating (UO) is a more comprehensive term, considering detrimentally high urban temperatures (for human life) as a result of a complex of phenomena including the UHI but also synoptic conditions and advective flows, which are typical of coastal areas. In fact, urbanization modifies the sensible, latent, radiative, and aerodynamic responses of the surfaces and results in differential heating and cooling rates between urban and rural areas. The distinctive features of the urban fabric (e.g., less vegetated surfaces, surfaces retaining more heat, having lower albedo, and reduced permeability), the city's configuration (e.g., geometry, size, topography, population density (PD), industrial development), the pronounced anthropogenic effect in the cities, higher pollutant levels, and synoptic-scale weather conditions are primarily responsible for the urban − rural thermal gradient (Oke et al., 2017; Santamouris, 2015a). Cities consist of various land-covers and land-use classes (LCLU), and every LCLU exhibits its unique thermal characteristics due to different surface properties (Wu, 2008). Further, according to 2018 statistics, 55% of the global population lives in the cities, and the percentage is expected to escalate to 68% by 2050 (World Urbanization Prospects, 2019). A further change in land surface characteristics will have a severe impact on local temperatures and affect the global climate by altering regional climate patterns.

Urban Climate Change and Heat Islands
DOI: https://doi.org/10.1016/B978-0-12-818977-1.00005-3

3.1.1 Urban overheating causes

Modified urban surfaces are the major contributor to thermal imbalance between urban and peripheral areas. Thermal emissivity, solar reflectance, and heat storage capacity are the prominent characteristics of the materials that define the urbanization impact. Construction materials such as concrete and asphalt generally have higher thermal storage capacity than the rural land cover due to high thermal inertia (for instance, than tilled soil). Therefore construction materials absorb more heat during the day and release it at nighttime (Elsayed, 2012). Thermal emissivity is the metric to measure the emission of longwave radiation from the surfaces. Surfaces with higher thermal emissivity emit the heat quickly to stay cooler. Similarly, the solar reflectance, also referred to as "albedo," is used to define the percentage of the solar radiation reflected by a surface. Dark surfaces in the urban fabric reflect fewer radiations and have lower solar reflectance. Lack of vegetation and pervious surfaces in the urban fabric reduces evaporation and evapotranspiration rates. Through evaporation and evapotranspiration, the moisture released in the air by the plants or the pervious surfaces in the form of latent heat flux assists in dissipating the ambient heat (Elsayed, 2012; Khan and Asif, 2017). Reduced tree canopy cover in the urban fabric is also reported as another prominent factor. Typically, higher tree canopy cover not only enhances the evapotranspiration rates but also provides shading that further alleviates the ambient temperatures (Stone and Norman, 2006). Further, the evapotranspiration rate regulated by the trees is comparatively higher than grassland.

Urban geometry also plays an important role in influencing the UO magnitude. Building height, the aspect ratio (building height to street width), street orientation, and urban roughness affect the wind pattern and the emission of the longwave radiations at nighttime (Akbari et al., 2008). Absorption of shortwave radiation and reabsorption of longwave radiation are enhanced in the complex urban geometry due to multiple reflections or radiation entrapment (Elsayed, 2012). Anthropogenic heat released in the urban spaces due to human activities, including building heating and cooling, transportation and manufacturing (exhaust from vehicles and industrial plants), not only intensifies the UO (Hu et al., 2012; Khan et al., 2017) but is also a source of pollutant formation (Sailor, 2013). The air pollutants affect the incoming shortwave radiations and increase the absorption of reflected longwave radiations that consequently amplifies the ambient temperatures (Rizwan et al., 2008). The topography of the city (e.g., mountainous or plain region), the elevation of the city (Vahmani and Ban-Weiss, 2016), location of the city (e.g., coastal or non-coastal), size of the city (e.g., compacted or sprawl city) (Li et al., 2020), and PD (Ward et al., 2016) also influence the UO magnitude.

3.1.2 Urban overheating quantification methods

Typically, UO is estimated as the temperature difference between urban and rural/suburban areas (UO = $T_{urb} - T_{rural}$). The near-surface temperatures or the land surface temperatures both are employed to compute the UO. Generally, surface UO is comparatively higher than the UO computed at 2-m height (Nichol et al., 2009). However, UO at 2-m height is considered more influential as it directly affects thermal comfort. Several expressions of temperature are used to estimate the UO. For instance, hourly temperature (Founda and Santamouris, 2017; Pyrgou et al., 2020), daily maximum or minimum temperature (Kumar and Mishra, 2019; Scott et al., 2018), daily average temperature (Tewari et al., 2019; Fenner et al., 2019), daytime and nighttime average temperatures (Li and Bou-Zeid, 2013), and monthly average temperature (Ward et al., 2016) have been used to measure the UO magnitude.

The urban increment method is also deployed to estimate UO (Chew et al., 2021). In the urban increment method, the whole spatial extension of the city is considered, and the temperature difference is taken between the controlled and the experimental cases. In the experimental case, all urban surfaces are replaced by rural surfaces, while the original city case is considered in a controlled case (Bohnenstengel et al., 2011). To take into account the spatial extension of the city, the composite urban temperatures (mean temperature of all urban stations) and the composite rural temperatures (mean temperature of all rural stations) are also utilized to compute the UO (Basara et al., 2010; Morris and Simmonds, 2000).

3.2 Heatwaves

A heatwave is a regional-scale phenomenon exhibiting amplified air temperatures in the whole region for a prolonged period, and it is typically influenced by synoptic-scale weather conditions, soil moisture, and land– atmosphere interactions (Perkins, 2015). Heatwaves are largely associated with clear sky conditions, intense solar radiation, and low wind speed (Meehl and Tebaldi, 2004; Liu et al., 2018). Globally, an increase in extreme heat events, while a decrease in the extreme cold events is recorded (Sheridan and Lee, 2018). Many studies have also concluded a projected increase in the frequency, duration, and severity of extreme heat events in the 21st century (Meehl and Tebaldi, 2004; Seneviratne et al., 2014; Horton et al., 2016).

3.2.1 Heatwaves identification methods

People respond to extreme heat events differently, and their response varies according to geographical locations. Due to the varying thermal adaptability at disparate locations and varying heatwave impacts in different sectors, including health, energy, bushfire management, etc., there is no consensus on the definition of heatwaves (Meehl and Tebaldi, 2004). Further, there is also no consensus if heatwaves are only a summertime phenomenon or may occur in other seasons as well, and accordingly, various indices exist (Perkins and Alexander, 2013). However, three criteria are considered essential to elucidate the phenomenon, which are (1) the threshold temperature, (2) the duration of the extreme heat events, and (3) the geographical extension of the city where the regional-scale phenomenon occurs (Perkins and Alexander, 2013; Chen et al., 2015; Tong et al., 2015).

In literature, the threshold temperature is defined with different metrics, including an absolute temperature threshold (ATT) (Ao et al., 2019; Schatz and Kucharik, 2015) and a relative temperature threshold (RTT) (Pyrgou et al., 2020; Rogers et al., 2019). In the ATT metric, a fixed temperature, whereas in the RTT metric, percentile-based temperatures are calculated to define the threshold temperature. In contrast to the ATT metric, the RTT metric may consider the seasonal variability as the calculations are based on the available dataset for a particular period (Perkins and Alexander, 2013). In addition to the various metrics, temperature expressions such as daily maximum temperature (T_{max}) (Founda and Santamouris, 2017; Pyrgou et al., 2020) and daily minimum temperature (T_{min}) both have been deployed to define the threshold temperature (Kumar and Mishra, 2019; Scott et al., 2018). For instance, while utilizing the ATT metric, T_{max} greater than or equal to 32.2°C (Ramamurthy and Bou-Zeid, 2017), 35°C (Ao et al., 2019), or 37°C was defined as threshold temperature, which primarily represents the thermal stress at daytime. Similarly, while using the T_{min} in ATT metric, a daily minimum temperature greater than or equal to 20°C is considered as the threshold temperature, which largely represents the thermal discomfort during nighttime (Collins et al., 2000).

While using the RTT metric, T_{max} greater than or equal to 90th percentile (Fenner et al., 2019), 95th percentile (Pyrgou et al., 2020), or

97.5th percentile (Li et al., 2015; Sun et al., 2017) of the daily maximum temperature is considered as threshold temperature. Contrarily, while employing the T_{min} in RTT metric, the temperature greater than or equal to the 90th (Alexander et al., 2006) or 95th percentile (Scott et al., 2018) of the daily minimum temperature is reckoned as the threshold temperature. Some studies also used dual criteria to define the threshold temperature (Chew et al., 2021; Sun et al., 2017; Richard et al., 2021). For instance, in a study, the average number of days with T_{max} greater than 35°C, and T_{min} greater than 20°C were counted as heatwave days (Fischer and Schär, 2010). Similarly, a multiple threshold index has also been used in defining the threshold temperature. For instance, in a study, only those days were considered as heatwaves that were fulfilling the following criteria: (1) T_{max} greater than 97.5th percentile for at least three consecutive days, (2) average T_{max} greater than 97.5th percentile for the entire period, and (3) T_{max} greater than 81st percentile for every single day of the whole period (Meehl and Tebaldi, 2004).

While comparing the different temperature expressions, it was also concluded that T_{min}-based heatwaves might be suitable for agricultural studies, where the daily minimum temperature may seriously affect the plant's growth (Perkins and Alexander, 2013). Contrarily, T_{max}-based heatwave might be suitable for engineering studies, including UO, power, and transportation systems, whereas the coupled effect of T_{max} and T_{min} might be more advantageous in health-related studies. Similar to the threshold temperature, there is also a lack of consensus on the minimum duration of a hot spell to be considered a heatwave. Two consecutive days in the United States and $2-5$ consecutive days in Europe and Australia are considered as the minimum duration for heatwaves (Perkins and Alexander, 2013; Nairn and Fawcett, 2013). However, several studies have counted three consecutive days as the minimum duration for a heatwave episode (Meehl and Tebaldi, 2004; Perkins and Alexander, 2013; Fischer and Schär, 2010). Since heatwave is a regional-scale phenomenon, thus the geographical extension of the city is also important to consider. Enhanced anthropogenic heat may elevate the local temperature. Therefore it is important to consider those episodes that are occurring concurrently in multiple zones of the region to ensure the identification of accurate heatwaves. Excessive heat factor (EHF) is also used to define heatwaves (Nairn and Fawcett, 2013), which is the combined effect of excess heat (EHI_{sig}), and heat stress (EHI_{accl}) (Eq. 3.1). Heatwave intensity, duration, and spatial distribution are considered in this index. Further, both expressions—daily maximum and daily minimum temperatures—are used to define this index.

$$EHF = max \lfloor 1, EHU(accl.) \rfloor \times EHI(sig.) \tag{3.1}$$

EHI_{sig} in Eq. (3.2) is the excessive amount of heat from the amplified diurnal temperatures, which is not emitted at nighttime because of higher nocturnal temperatures. EHI_{sig} is defined as the temperature anomalies for 3 days and compared against the extreme temperature threshold (T_{95}—95th percentile temperature for the study period). T_i in the equation is the average daily temperature for the particular i-th day *and is calculated as the average of T_{min} and T_{max} over the 24-hours cycle.*

$$EHI(sig.) = \left[(T_i + T_{i-1} + T_{i-2})/3 \right] - T_{95} \tag{3.2}$$

Heat stress in Eq. (3.3) computes the temperature anomalies compared to the recent past. Usually, 3-days temperature anomalies are compared against the previous 30 days (Nairn and Fawcett, 2013). In both equations, besides the *i-th* day, the impact of two previous days ($i - 1$ and $i - 2$) is also considered. Further, the EHF should be positive for at least 3 consecutive days (i, $i + 1$, and $i + 2$) to be considered as a heatwave period.

$$EHI(accl.) = \left[(T_i + T_{i-1} + T_{i-2})/3 \right] - \left[(T_{i-3} + \cdots + T_{i-32})/30 \right] \tag{3.3}$$

Humidity index (HI)/Apparent temperature (AT) in Eq. (3.4) is also used to identify heatwave episodes. In this index, a few consecutive days with daily maximum HI greater than 65°C are considered heatwaves (Pyrgou et al., 2020; Rizvi et al., 2019). It is more useful in the cities where moist synoptic-scale weather conditions are more prevailing, as the index considers both parameters: temperature and humidity (Chew et al., 2021; Russo et al., 2017). The HI can be computed by utilizing the following equation.

$$AT = c_1 + c_2 T + c_3 RH + c_4 TRH + c_5 T^2 + c_6 RH^2 + c_7 T^2 RH$$
$$+ c_8 TRH^2 + c_9 T^2 RH^2 \tag{3.4}$$

$c_1 = -42.379; c_2 = 2.04901523; c_3 = 10.14333127; c_4 = -0.22475541;$
$c_5 = -6.83783 \times 10^{-3}; c_6 = -5.481717 \times 10^{-2}; c_7 = 1.22874 \times 10^{-3};$
$c_8 = 8.5282 \times 10^{-4}; c_9 = -0.199 \times 10^{-6},$ *and temperature is in* Fahrenheit

3.3 The combined effect of urban overheating and heatwaves on human health, economy, energy, and environment

UO impacts on mortality, morbidity, urban vulnerability, cooling energy, peak electricity demand, environmental quality, and the economy are very well-documented around the globe (Santamouris, 2015b). UO may synergistically interact with extreme heat events, which may further magnify its intensity. A projected increase in frequency, duration, and severity of heatwaves, as reported in many studies (Seneviratne et al., 2014; Horton et al., 2016), will be further disastrous for urban communities, and the sufferings and penalties in the form of medical bills, electricity bills, and disaster compensation will be massive. The combined impact of UO and heatwaves has gained increasing interest recently, and several studies have documented the impact of both phenomena in various sectors, either individually or holistically.

3.3.1 Mortality and morbidity

Built environments are extremely vulnerable to extreme heat events. Heatwaves have devastating impacts on human health, and it is reported as one of the major causes of climate-related mortality and morbidity around the globe (Basu, 2002; Patz et al., 2005). Summer heatwaves in Australia were reported as the most lethal natural hazard (Coates et al., 2014). In the United States, a 7.9% increase in daytime mortality and 2.2% increase in nighttime mortality was reported, with every 1°C rise in temperature during heatwaves (Zhao et al., 2018; Brooke Anderson and Bell, 2011). Over 70,000 deaths were reported in Europe during 2003 extreme heat events, where France, Italy, and Spain were the most affected countries with more than one-third of total deaths (Robine et al., 2008). Further, the aged people, women, children, people with chronic diseases, and those having low-socioeconomic status are reported to be at higher thermal risk during such extreme heat events (Tong et al., 2014). Heatwaves are also reported as one of the leading causes of climate-related hospitalization around the globe (Basu, 2002). In addition to hyperthermia (heat stroke, heat cramps, and exhaustion) and respiratory ailment, life-threatening ailments such as heat-related cardiovascular disease, including myocardial infarction, were found occurring with higher frequency during such extreme heat events (Peters and Schneider, 2020; Yin and Wang, 2017).

3.3.2 Energy

The combined impact of urbanization and heatwaves are also having adverse effects on cooling energy needs and the peak electricity demand. The average increase in the building cooling demand between 1970 and 2010 due to the combined effect of UO and the global climate change was 23%, while the average reduction in the heating demand was 19% (Santamouris, 2020). A 29%−86% increase in the cooling energy demand of typical residential, school, and office buildings was also reported in Sydney due to extreme temperatures (Garshasbi et al., 2020). Globally, a projected increase in the cooling energy demand of the residential and commercial buildings was estimated to be 275% and 750%, respectively, by 2050 (Santamouris, 2016). While studying the combined impact of UO and global climate change on the peak electricity demand in several global cities, an additional increase of 0.45% − 12.3% was reported with every 1°C rise in temperature (Santamouris, 2020). Similarly, the amplified temperature during extreme heat events may lower the efficiency of the power plants. A reduction in the efficiency of power plants due to the combined effect of UO and global climate change was reported to be 0.6%−2.0% with every 1°C rise in temperature (Linnerud et al., 2011). The lower efficiency of the power plants during extreme heat events will increase the electricity cost, which will also be sustained by the urbanites in addition to increased air conditioning usage charges. Further, the operation of powerplants for a prolonged time during extreme heat events will also enhance the atmospheric pollution, which will further increase the regional temperature and may cause many respiratory illnesses. A projected increase in heatwave's frequency and intensity, a projected increase in the cooling energy and peak electricity demand, and a reduction in power plant efficiency, will put an extra burden on the electric grids during heatwaves, and more blackout periods will be observed (Li and Bou-Zeid, 2013). The low-income population and people with preexisting medical conditions will be more vulnerable during these blackout periods.

3.3.3 Environment and the economy

Industrial and vehicular exhaust—NO_X and volatile organic compounds—chemically react with heat and sunlight to form the ozone, a pollutant that is harmful to human health and environmental quality. High temperatures during heatwaves promote ozone formation (Pyrgou et al., 2018). While exploring the heatwaves and UO impact on the environmental quality, it was found that ozone concentration may increase from 9.6% to 20% during heatwaves (Stathopoulou et al., 2008; Zhang et al., 1998). The number of days exceeding the threshold of ozone concentration increases by 10% due to the increase in the outdoor temperature by 1% (Diem et al., 2017). In western Sydney, a significant increase in ozone level was reported due to the combined effect of increased ozone formation during heatwaves and postheatwave pollutant transportation due to coastal wind activation (NSW Government, 2019). Higher ozone levels were also associated with higher hospital admission rates (760 hospital admissions due to respiratory illness in Sydney in 2007) and premature deaths (160 premature deaths in 2007 in Sydney) (Physick et al., 2014; Broome et al., 2015). Several studies have also investigated the financial impacts of heatwaves. The financial loss in terms of crop failure during summer 2003 heatwaves in Europe was reported to be approximately US$ 12.3 billion (Heck et al., 2004), while a financial loss of over 40 AUD billions during the 2019–2020 bushfires was estimated in Australia, with the decimation of 20 million acres, over 0.5 billion animal fatalities, and around 33 human deaths (Filkov et al., 2020).

3.4 UO interaction with heatwaves—quantification of energy budget equation

Urbanization modifies the land – atmosphere thermal and moisture exchange and influences the regional and continental-scale circulations (Vahmani and Ban-Weiss, 2016; Fischer et al., 2007). The UO and heatwaves may synergistically interact and influence the UO. Further, urbanization may aggravate regional warming and increase the frequency and severity of heatwaves (Zhao and Wu, 2017; Fischer et al., 2012). Generally, the energy budget equation is used to quantify the changes in UO magnitude. The urban – rural temperature contrast is evaluated in terms of sensible, latent, anthropogenic, advective, storage, and net all-wave radiative heat fluxes (Founda and Santamouris, 2017). The net all-wave radiative heat flux consists of incoming and outgoing shortwave and longwave radiations, while the available energy is estimated as the sum of sensible and latent heat fluxes.

3.4.1 Alteration in the radiative input during heatwaves

During heatwaves, regional high temperatures are induced by a large-scale stagnant high-pressure system that brings warm air from the troposphere

(Li and Bou-Zeid, 2013). Also, during the heatwaves, the clear sky conditions lead to exceptionally intense solar radiation in the region (Wang et al., 2017), which influence the sensible, latent, advective, heat storage, and anthropogenic heat fluxes. The intense solar radiation during spring and early summer in Europe before the extreme heat event in 2003 was also responsible for the early depletion of soil moisture (Fischer et al., 2007). These dry soil anomalies were also reported to strengthen the anticyclonic circulation, which was the major cause of extreme heatwave events in 2003.

3.4.2 Alteration in sensible and latent heat fluxes during heatwaves

The limited vegetated surfaces and availability of more impervious surfaces in the urban fabric alter the available energy partitioning, increase the sensible heat flux, and decrease the latent heat flux (Argüeso et al., 2014). Contrarily, in rural areas, the latent heat flux is higher due to the availability of more potentially plantable surfaces capable of retaining higher moisture content and the availability of higher tree canopy cover (Pyrgou et al., 2020). The intense solar radiation during heatwaves further increases the evaporation in both urban and rural areas, which further intensifies the urban − rural moisture contrast, and consequently the daytime thermal gradient (Zhao et al., 2018, 2014; Ngarambe et al., 2020). At nighttime, in the absence of solar radiation, convection and evaporation processes are comparatively weak. Hence, urban − rural moisture contrast is not an important synergistic interaction at nighttime.

The surface latent cooling potential in the rural regions is also influenced by the amount of precipitation in the region in that year, particularly before or during the summer period when most heatwave episodes are reported (Fischer et al., 2007). In the absence of sufficient upper soil moisture in the rural surfaces, the interactions between heatwaves and UO may get diminished. It happens as rural surfaces start partitioning more energy into sensible heat flux, similar to urban surfaces (Zhao et al., 2018; Fischer et al., 2007). An exacerbation in UO magnitude during heatwaves in Europe was reported due to the performance loss of urban greenery, which lost the evapotranspiration potential due to long-lasting extreme heat events (Ward et al., 2016).

Urban shading was also reported to lower the thermal gains (sensible heat flux) in the urban areas during the daytime and consequently reduce the daytime UO. Southern European cities were found to be better adapting during extreme heat events than cooler northern European cities, hypothetically due to urban/solar shading (Ward et al., 2016).

3.4.3 Alteration in advective heat fluxes during heatwaves

Wind flows from the high-pressure zone to the low-pressure zone. The temperature gradient between urban and rural surfaces promotes the secondary-air-circulation, which resultantly reduces the UO magnitude (Li and Bou-Zeid, 2013). As a rule of thumb, regional wind speed is inversely proportional to UO magnitude. The lower the regional wind speed, the lower is urban mixing—less advective cooling from surrounding rural areas—and the higher is UO magnitude (Morris and Simmonds, 2000). During heatwaves, further reduction in regional wind speed was reported, which was associated with high-pressure anticyclones (Li and Bou-Zeid, 2013; Ngarambe et al., 2020).

In a coastal city, localities adjacent to the coast are cooled by the steady sea breeze due to secondary-air-circulation. During heatwaves, secondary circulation becomes more influential due to a higher thermal gradient between urban surfaces and steady sea surfaces (Vahmani and Ban-Weiss, 2016; Founda and Santamouris, 2017; Ao et al., 2019). However, as the distance from the coast increases, the UO may exacerbate due to the lack of coastal winds penetration at the inland sites. Hence, the distance from the coast plays a pivotal role in magnifying the UO magnitude during heatwaves (Ramamurthy et al., 2017).

The dry air has the potential to absorb more moisture and increase the latent heat potential in the area, depending upon the soil moisture availability (Kool et al., 2018). However, advection from the heat source such as desert landmass not only amplifies the local temperatures drastically but can also affect the latent heat flux potential in the area by sweeping the ambient moisture (Khan et al., 2020, 2021b). In contrast to the high-speed regional winds, which reduce the UO, high-speed desert winds may increase UO by warming the area adjacent to the heat source (Yun et al., 2020). During heatwaves, advection from a heat source may become more intense, which may further intensify the UO (Khan et al., 2021b).

3.4.4 Alteration in anthropogenic heat fluxes during heatwaves

Typically, anthropogenic heat flux is higher in dense urban areas, consisting of more commercial buildings, and with higher automobile traffic.

Amplified anthropogenic heat flux was reported to exacerbate the daytime and nighttime UO during normal days (Taha, 1997). Additional usage of air-conditioning in the buildings and vehicles during extreme heat events, higher energy production from fossil fuels to meet the end-user energy needs during heatwaves may exacerbate UO magnitude further (Ao et al., 2019). Along with latent heat flux, amplified anthropogenic heat flux was reported as the prime contributor of exacerbated UO at daytime during heatwaves (Zhao et al., 2018).

3.4.5 Alteration in heat storage during heatwaves

Typically, urban fabric (e.g., concrete) absorbs and stores more heat during the daytime, compared to nonurban surfaces. This heat is emitted at nighttime in the form of sensible heat flux, and it is the major reason for pronounced nighttime UO in many cities (Ramamurthy et al., 2014). Due to the higher daytime radiative input during the heatwaves, more heat can be stored in the urban fabric, which may exacerbate the nighttime UO (Li et al., 2015; Imran et al., 2019). The thermal storage factor may extend the hot conditions in the urban areas, as the heat absorbed by the urban fabric may get released during the postheatwave period (Li and Bou-Zeid, 2013; Sun et al., 2017).

3.4.6 UO response to heatwaves in various cities

UO response to heatwaves is quite contradictory, and contrasting responses of surface energy budget to the additional incoming solar radiations were reported in various studies. For instance, no change in UO magnitude during heatwaves (Chew et al., 2021), a decline in UO magnitude during heatwaves (Kumar and Mishra, 2019), magnified UO at daytime (Pyrgou et al., 2020), exacerbated UO at nighttime (Imran et al., 2019), and amplified UO at both daytime and nighttime (Li and Bou-Zeid, 2013) were reported.

3.4.6.1 Exacerbated daytime UO during heatwaves

The amplified daytime UO response to heatwaves was reported in Athens (Founda and Santamouris, 2017), Shanghai (Ao et al., 2019), Nicosia (Pyrgou et al., 2020), New York (Ramamurthy and Bou-Zeid, 2017), and Karachi (Rizvi et al., 2019), which are coastal cities. Further, in those studies, advective heat flux and urban − rural moisture contrast were concluded as the prime contributor to the exacerbated daytime UO during heatwaves. Advective heat flux in the form of coastal winds was reported keeping the coastal areas cooler, whereas the inland sites were warmer due to the lack of penetration of coastal winds. In addition, advection from the coastal winds was also increasing the moisture content in the coastal sites, which was increasing the urban − rural moisture contrast, and affecting the UO magnitude. In Los Angeles, the higher daytime surface temperatures during heatwaves were reported at inland sites or in highly urbanized areas (Vahmani and Ban-Weiss, 2016). In Shanghai, China, exacerbated daytime UO was reported when analyses were performed in reference to the coastal station (Ao et al., 2019). Further, the peak UO magnitude difference between heatwaves and background conditions (ΔUO) in Shanghai was 1.3°C. In addition, the synoptic-scale weather conditions—lower regional wind speed and higher sea breeze—which were associated with heatwaves in Shanghai were concluded as the main synergistic interaction between UO and heatwaves.

Similarly, in Athens, magnified UO was reported during heatwaves at both daytime and nighttime; however, the daytime UO effect was more pronounced (peak average ΔUO = 3.5°C) (Founda and Santamouris, 2017). Further, advection from coastal winds was reported intensifying the UO magnitude in Athens during heatwaves. The max ΔUO in New York was reported during the daytime (ΔUO = 2.0°C), and the change in wind pattern from coastal southerlies to westerlies during heatwaves was the main contributor of amplified UO (Ramamurthy and Bou-Zeid, 2017). Another study in New York concluded similar results during heatwaves and also estimated the future UO magnitude to be 4°C higher than the current conditions (Tewari et al., 2019).

The urban − rural moisture contrast (ΔAH) was also reported as one of the prime contributors to the daytime UO intensification during heatwaves (Pyrgou et al., 2020; Zhao et al., 2018). For instance, in a temperate climate in the United States, amplified daytime UO response to heatwaves was reported in the current scenario, and ΔUO was 0.4°C with ambient temperature and 2.8°C with surface temperature (Zhao et al., 2018). Further, urban − rural moisture contrast and amplified anthropogenic heat flux were reported as major synergistic interactions between UO and heatwaves. Similarly, in Nicosia, Cyprus, an exacerbated UO during the daytime during heatwave was reported (ΔUO = 0.9°C−1.3°C) while performing the hourly analysis from 2007 to 2014. Further, evaporation of the dew formed on urban and rural surfaces was the main synergistic interaction during heatwaves in Nicosia.

The mechanism was justified with the explanation that the dew formed on the urban surfaces evaporated quickly, while the dew formed on the rural surfaces was absorbed by the nonurban surfaces, which consequently amplified the daytime UO (Pyrgou et al., 2020).

3.4.6.2 Exacerbated nighttime UO during heatwaves

The exacerbated nighttime UO during heatwaves was reported in Adelaide (Rogers et al., 2019), Szeged, Hungary (Unger et al., 2020), Seoul (Ngarambe et al., 2020), Athens (Katavoutas and Founda, 2019), Bucharest (Cheval et al., 2009), Berlin (Fenner et al., 2019), and in Melbourne (Imran et al., 2019), which were concluded as either non-coastal cities or inland sites. Further, higher heat storage flux, the amplified anthropogenic heat flux, and lower-regional wind speed were inferred as major synergistic interactions between UO and heatwaves.

In Berlin, the exacerbated response of UO was reported at nighttime (peak $\Delta UO = 1.0°C$). In Seoul, Korea, worsened UO during heatwaves was revealed in the afternoon to late-night (peak $\Delta UO = 4.5°C$), whereas the minimum ΔUO was reported in the early morning (Ngarambe et al., 2020). Further, intense synergies between UO and heatwaves in Seoul were reported under low-speed regional winds and in highly urbanized areas. In Melbourne and Adelaide exacerbated UO was reported at nighttime during heatwaves, despite being coastal cities (Rogers et al., 2019). The ΔUO was 1.4°C in Melbourne, whereas ΔUO was 1.2°C in Adelaide. In another study in Melbourne, while employing the weather research and forcasting (WRF) model, storage heat flux was reported as the major synergistic interaction, amplifying the temperatures at urban sites during heatwaves (Imran et al., 2019). In Szeged, Hungary, amplified UO was also reported at nighttime during heatwaves (Unger et al., 2020). In Washington and Baltimore using the WRF model, the nighttime UO effect was found more pronounced during heatwaves (peak $\Delta UO = 1.5°C-2.0°C$). Further, in contrast to other studies, urban − rural soil moisture deficit was primarily responsible for the difference (Ramamurthy and Bou-Zeid, 2017). The urban − rural moisture contrast increased two-fold during heatwaves in Baltimore and Washington, compared to non-heatwaves (Ramamurthy and Bou-Zeid, 2017).

3.4.6.3 Exacerbated UO at both daytime and nighttime

An exacerbated response of UO to heatwaves was also reported at both daytime and nighttime in several studies (Li and Bou-Zeid, 2013; Schatz and Kucharik, 2015; Li et al., 2015, 2016; He et al., 2020). For instance, in Beijing, China, a positive response of UO to heatwaves was reported at both daytime (peak average $\Delta UO = 0.4°C$) and nighttime (peak average $\Delta UO = 1.0°C$); however, the nighttime UO effect was more pronounced (Li et al., 2016). Enhanced urban wind speed at daytime and a reduced urban wind speed at nighttime were responsible for the intensified UO at daytime and nighttime, respectively. In another study performed in Beijing, similar results were presented, but instead of advective heat flux, the available energy contrast and the heat storage flux were largely responsible for exacerbated UO during heatwaves (Li et al., 2015). An additional study performed in Beijing using the WRF model (He et al., 2020) yielded the same results: exacerbated UO at daytime and nighttime with a more intense effect at nighttime (peak average $\Delta UO = 0.78°C$). Further, in the study, amplified daytime UO was due to urban − rural moisture contrast, whereas the pronounced nighttime UO was attributed to enhanced anthropogenic heat flux and enhanced warm advection. Exacerbated daytime and nighttime UO response to heatwave with a more prominent effect at nighttime was also reported in Baltimore (Li and Bou-Zeid, 2013). Urban-rural moisture contrast and low wind speed were the main synergistic interactions between UO and heatwaves in Baltimore. During extreme heat events in Madison, USA, UO during both daytime and nighttime was more intense than during average summer days, and the effect was more visible in highly populated areas (Schatz and Kucharik, 2015).

3.4.6.4 No change in urban overheating magnitude during heatwaves

In some studies, either no change or insignificant response of UO to heatwaves was reported. In Singapore, the relation of UO to heatwaves was investigated for one heatwave period, using the WRF simulation model for one urban − rural station. The UO magnitude during both heatwaves and nonheatwaves was not only the same, but also no significant changes were observed in soil moisture, heat storage, and wind speed (Chew et al., 2021). In Dijon, France, an exacerbated UO response was reported either before the heatwave period or within the first few days of heatwaves (Richard et al., 2021). Later, a declined response of UO was documented in the following heatwave days, and a statistically insignificant relationship was concluded between heatwaves and UO. Further, nocturnal rural temperatures were exorbitant compared to nocturnal urban temperatures during heatwaves, which was associated with the reduced soil moisture in Dijon, France. In Philadelphia, insignificant UO feedback to heatwave was associated with urban geometry (Ramamurthy and Bou-Zeid, 2017). In Phoenix, USA, a higher UO magnitude was reported during the postheatwave period (in both current and future climatic scenarios), and the synergistic interactions between UO and heatwaves were uncertain (Tewari et al., 2019).

3.4.6.5 A decline in urban overheating magnitude during heatwaves

In several studies, a declined UO during heatwaves was reported (Kumar and Mishra, 2019; Scott et al., 2018; Rogers et al., 2019; Brázdil and Budíková, 1999). In a study conducted in 54 US cities—with the prime focus on nighttime UO response to heatwaves—mostly amplified nighttime rural temperatures were recorded due to moist weather conditions (Scott et al., 2018). In another study, while using the land-surface temperatures in 89 major urban areas in India, a negative UO response was reported for 63% of the daytime and 74% of the nighttime (Kumar and Mishra, 2019). Further, reduced soil moisture in rural regions was concluded as the main contributor, which was associated with the harvesting before the summer. Further, most nonurban areas considered in the study were in the agricultural-dominated zones. In Perth, Australia, cooler urban areas during heatwaves can be attributed to advection from the coastal winds as mostly urban areas were located near the coast (Rogers et al., 2019).

3.4.7 Inconsistent response of urban overheating to heatwaves—important factors

Inconsistent responses of UO to heatwaves might be attributed to numerous factors. Different boundary conditions (e.g., coastal or noncoastal cities), the physical size of the city, PD, distance from the coast, and different weather conditions (e.g., wind patterns, cloud cover, radiative input, etc.) might be important parameters providing contrasting results. Further, inconsistent methods of UO estimation, inconsistent methods of heatwaves identification, ignoring the city size and sprawl, UO estimation for a few heatwaves episodes, and inconsistent methods of site selection (inland sites/coastal sites) may also be responsible for contrasting results.

3.4.7.1 Different boundary conditions

Different boundary conditions may lead to inconsistent results. For instance, if the analyses were performed for a coastal or a noncoastal city, this may yield a contrasting response of UO to heatwaves. Mostly, an exacerbated response of UO to heatwaves was reported during daytime for coastal cities (Founda and Santamouris, 2017; Pyrgou et al., 2020; Ao et al., 2019), while for noncoastal cities, UO was more pronounced during the night (Fenner et al., 2019; Basara et al., 2010). Similarly, the selection of urban and rural stations is also an important parameter that may also produce inconsistent results. In Shanghai, China, while investigating the interactions between UO and heatwaves, higher daytime UO was reported for coastal stations, whereas amplified nighttime UO was recorded when UO was computed in reference to the inland station (Ao et al., 2019). Further, UO-computed reference to the coastal station was quite high compared to UO-calculated reference in the inland site (Ao et al., 2019). Contrarily, in Melbourne and Adelaide, exacerbated nighttime UO response was noticed during heatwaves despite being coastal cities (Rogers et al., 2019). It was mainly due to the inland site selection to minimize the coastal wind's effect. Conclusively, advection from the coastal wind may significantly affect the response of UO to heatwaves in a coastal city (Li et al., 2016). Similarly, computing UO either as a thermal difference between urban and rural sites or as a temperature difference between various sites within the city may also be another variable. For instance, studies conducted in cities that have included nonurban LCLUs extensively (e.g., Berlin), and the studies conducted in the cities that have considered only city core (e.g., Barcelona) was one of the potential sources of inaccuracy as concluded in Ward et al. (2016).

3.4.7.2 Inconsistent urban overheating quantification methods

UO quantification by utilizing surface or ambient temperatures may also provide inconsistent results. For instance, the surface temperature may get reduced at nighttime due to higher convection (reduced surface UO), which may amplify the ambient temperatures (exacerbated ambient UO) (Zhao et al., 2018). Usually, a magnified daytime UO was reported when measured based on surface temperatures (Voogt and Oke, 2003), whereas the ambient temperatures mostly provided the pronounced nighttime UO due to the emission of stored heat in urban fabric (Oke, 1982).

Similarly, various expressions of temperature for UO estimation (hourly temperatures, daily maximum or minimum temperatures, daily average temperatures, daytime and nighttime average temperatures, monthly average temperatures) may also provide inconsistent results. In Singapore, 3-day averages for all meteorological parameters were computed to compare the UO during heatwaves and nonheatwaves (Chew et al., 2021). Using averages of temperature, relative humidity, and wind pattern for the whole heatwave/nonheatwave period may have provided inconsistent results in Singapore. Meteorological parameters such as wind patterns alter very quickly, particularly during extreme heatwave events (Li et al., 2016), and daily, monthly, or annual averages may not elucidate the diurnal variations. Consequently, it may provide inconsistent results, particularly in the cities where the interactions between UO and heatwaves largely depend upon synoptic-scale weather conditions.

3.4.7.3 Inconsistent heatwaves calculation methods

The unavailability of a universal definition of heatwaves may also be another variable. For heatwaves identification, some studies employed daily maximum temperatures (Founda and Santamouris, 2017), whereas others used daily minimum temperatures (Kumar and Mishra, 2019; Scott et al., 2018). Mostly, contrasting results were reported while utilizing the daily minimum temperature for heatwaves identification, as concluded in Fenner et al. (2019). While defining a heatwave period, the spatial extension of the city is an important parameter to consider, as explained earlier. Numerous studies have used one meteorological station, either urban or rural, to identify heatwave episodes (Pyrgou et al., 2020; Ao et al., 2019; Li et al., 2015), which may also be a reason for inconsistent results. Limiting the study to one urban and rural site and not considering the spatial extension of the city could also be a limitation of the study in Nicosia, where, in contrast to other coastal studies, the impact of advective heat flux was insignificant (Pyrgou et al., 2020).

Similarly, some studies used absolute temperatures to delineate the threshold temperature criterion (Founda and Santamouris, 2017; Zhao et al., 2018), whereas others utilized relative temperatures (Li et al., 2015). Further, the duration of heatwaves could be another variable as various studies have used two consecutive days, three consecutive days (Founda and Santamouris, 2017; Zhao et al., 2018), or four consecutive days (Pyrgou et al., 2020) to define the heatwave duration. In Dijon, France, selected threshold temperatures (daily max = 32°C, and daily min = 16°C) were less than regional health alerts (daily max = 34°C, and daily minimum 19°C) for two consecutive days (Richard et al., 2021). The less-intense heatwave period might have generated inconsistent results as no statistically significant relationship was developed between temperature anomalies and UO in Dijon, France.

Similarly, considering interactions between UO and heatwaves for one episode or a limited period might be another variant. For instance, a sudden change in the climatic conditions before the selected heatwave episode may yield inaccurate results. Temperature anomalies during extreme heat events in summer 2003 in Europe were attributed to 25% reduced soil moisture during spring, which was due to low precipitation (Fischer et al., 2007). Several studies have used one heatwave episode to investigate the synergies between UO and heatwaves, which is also reckoned as the limitation of those studies (Li and Bou-Zeid, 2013; Basara et al., 2010; Schatz and Kucharik, 2015; Rizvi et al., 2019).

3.5 Synoptic climatology

Synoptic climatology is the study of the local and regional scale climate that fundamentally examines the properties and behavior of the atmosphere over a particular region. In the synoptic scale, also known as the large scale or cyclonic scale, the weather patterns are observed in a range of 1000 km-above (approximately 4000 km) at a horizontal length scale. Some parts of the Earth receive more solar radiation, whereas others receive less due to the earth's rotation around the sun. The uneven distribution of solar radiation generates a global circulation pattern. The air masses—exceptionally large bodies of air—are driven by these global circulation patterns (MetLink Royal Meteorological Society, 2020). An air mass covers the entire troposphere and has uniform thermal and humidity properties in any horizontal direction and at a given altitude, and its characteristics are determined by the location where it is formed. However, the characteristics of an air mass may change as it moves from one place to the other. When two air masses of opposite characteristics collide, fronts are formed (Koutsoyiannis and Langousis, 2011). When a lightweight warm air mass strikes a static heavyweight cold air mass, warm fronts are formed. In the warm fronts, warm air mass moves upward after colliding with cold air mass, and horizontal clouds are formed. Contrarily, when a cold air mass strikes a lightweight static warm air mass, the vertical movement of warm air mass is further steeper, and vertical clouds are formed. The phenomenon is called a cold front (Koutsoyiannis and Langousis, 2011).

In the hypothetical case of the absence of Earth's rotation—if the Earth were stationary—the air circulation would be from the high-pressure (HP) poles region to the low-pressure (LP) equator. However, instead of a straight pattern, the circulating air deflects to the right in the northern hemisphere and to the left in the southern hemisphere due to the Earth's rotation (Koutsoyiannis and Langousis, 2011). This phenomenon is referred to as the Coriolis effect. A vertical upward or downward movement of air is defined in terms of cyclonic and anticyclonic circulations. A cyclonic circulation is an LP system, where winds from the outer HP side rotate to the inner LP side due to the Coriolis effect, that is, convergence at surface and divergence at the upper level (Koutsoyiannis and Langousis, 2011). Due to these Coriolis forces, the rotation of the cyclones is clockwise in the southern hemisphere and anticlockwise in the northern hemisphere. In the cyclonic movement, the warm ground air rises upward due to the convergence, cools down when reaching the higher point, and releases moisture, which results in cloud formation. The anticyclonic conditions are the reverse phenomenon of cyclonic conditions. In anticyclonic conditions, from the central HP zone at the surface, the winds diverge outside to the LP zone (Colucci, 2003). As a replacement, the converging winds in the upper atmosphere descend to the surface, get warmer while coming down, absorb moisture, and determine clear sky conditions. Anticyclonic circulations are clockwise in the northern hemisphere and anticlockwise in the southern hemisphere due to the Coriolis effect.

3.5.1 Classification

In urban-climatological or bioclimatological studies, the weather typing classification (WTC) and circulation-pattern-based classification (CPC) both have been utilized (Sheridan and Lee, 2018; Yarnal, 1993; Lee, 2017). The CPCs are based on pressure patterns and have been more studied and utilized in Europe (Huth, 2010; Philipp et al., 2010). Instead, the WTCs are based on atmospheric conditions and have been more developed and employed in the United States (Kalkstein et al., 1996; Sheridan, 2002; Lee, 2015a). Further, in urban climatology, approaches such as circulation-to-environment (Hardin et al., 2018; Zhang et al., 2014) and environment-to-circulation (Morris and Simmonds, 2000) have both been used to investigate the association between surface temperature and synoptic-scale weather conditions. In the circulation-to-environment approach, the UO dataset is organized according to synoptic-scale weather conditions, whereas in the environment-to-circulation approach, synoptic-scale circulations are organized according to UO groups.

In CPC, synoptic-scale conditions are classified according to directional (wind flow) and nondirectional (cyclonic and anticyclonic circulation) types. In Szeged, Hungary, 13 synoptic-scale circulations were defined based on cyclonic and anticyclonic movements (Unger, 1996), and both environment-to-circulation and circulation-to-environment approaches were used. Similarly, in Poznan, Poland, the Niedzwiedz's calendar containing 21 cyclonic and anticyclonic circulation-type was utilized (Półrolniczak et al., 2017). In Athens, Greece, atmospheric circulations were stratified into eight a priori categories at 850 hPa by employing five parameters: temperature, equivalent potential temperature, geopotential height, wind speed, and wind direction (Mihalakakou et al., 2002). In Birmingham, UK, 11 Lamb weather types (WTs) were used to define the synoptic-scale conditions, in which eight were directional type (wind flows), two were nondirectional (anticyclonic and cyclonic), and one was unclassified (Zhang et al., 2014). In Melbourne, Australia, while employing the environment-to-circulation approach, against each UO group, mean sea level pressure charts were compared with the mean monthly conditions to find the daily synoptic-scale circulations (Morris and Simmonds, 2000).

In WTC, mostly synoptic conditions are stratified into daily air masses. In WTC, spatial synoptic classification (SSC) (Sheridan, 2002), and Gridded weather typing classification (GWTC) (Lee, 2015a, 2020) are the most conspicuous. In both classifications, temperature and humidity are partitioned to formulate daily air masses by using meteorological variables, including ambient temperature, dew-point temperature, mean sea-level pressure, wind speed, wind direction, and cloud cover. From the traditional air masses system—continental polar (cP), maritime polar (mP), continental tropical (cT), and maritime tropical (mT), the SSC was initially classified into six (Kalkstein et al., 1996), and then into seven WTs (Sheridan, 2002): dry polar (DP), dry moderate (DM), dry tropical (DT), moist polar (MP), moist moderate (MM), moist tropical (MT), and transitional (TR). The first letter in the nomenclature represents the humid characteristics of air masses, while the second letter represents the thermal characteristics. The continental or dry air masses are originated from the land, while the ocean/sea is the source of maritime/moist air masses. Similarly, tropical air masses are originated from the tropical side of the region and are hot, while polar air masses are emerged from the polar side and are cold in nature. Further details on SSC can be found at (http://sheridan.geog.kent.edu/ssc.html), and the definitions of SSC air masses are also provided in Table S3.1. The GWTC is a newly developed WTC that is stratified into 11 WTs: four WTs with extreme characteristics—humid cold (HC), humid warm (HW), dry cool (DC), and dry warm (DW), four moderate/average WTs either in terms of temperature or humidity—humid (H), dry (D), cool (C), and warm (W), two transitional—cold frontal passage (CFP), and warm frontal passage (WFP), and seasonal (S) WTs. The GWTC also considers the geographical and seasonal relevance, which assists in reducing the seasonal variability of air masses, and makes the classification easily transportable to the other locations. Thus the same air mass may occur in different seasons and at various locations with relatively different characteristics (Lee, 2020). Further details on GWTC can be found at (https://www.personal.kent.edu/~cclee/gwtc2global.html), and the definitions of GWTC air masses are also provided in Table S3.1. In addition to SSC, and GWTC, in Buenos Aires, six air masses—very cold (C++), southern (S), warm (W+), southwestern (SW), very warm (W++), southeastern (SE)—were defined using the K-means clustering method. The air masses were also linked to the circulation patterns to comprehend the association between air masses and the circulation field (Bejarán and Camilloni, 2003).

3.5.2 Synoptic-scale weather conditions and urban overheating

Globally, an increase of 0.85°C in average temperature has been reported over the past century (IPCC, 2014). The outdoor thermal comfort is not only affected by the temperature but other meteorological variants, including wind pattern, ambient moisture, atmospheric pressure, and cloud cover, holistically interact and adversely affect human thermal comfort. A significant change in sea-level pressure (Gillett and Stott, 2009) and generally a reduction in wind speed is also reported in recent decades (Roderick et al., 2009). Further, higher evaporation rates were also associated with the higher temperature, which has consequently increased the land-based precipitation in mid-latitude along with the higher humidity (Santer et al., 2007; Willett et al., 2007). In North America, warm air masses are reported occurring more frequently every year (e.g., humid-warm: +6.1%, dry-warm: +2.9%), particularly in recent years, whereas a drop in the frequency of cool air masses is also reported (e.g., dry-cold: -4.5%, Cold: -5.7%) (Lee and Sheridan, 2018).

Urbanization may interact with synoptic-scale weather conditions and affect the UO magnitude. Atmospheric properties above the urban areas are altered due to modified land-atmosphere thermal and moisture exchange properties and radiative and aerodynamic behavior of the surfaces. Urban expansion along with large-scale global warming can adversely affect public health, energy usage, and ecological quality (Santamouris, 2020). The combined effect of both these phenomena also has detrimental effects on the economy by putting an extra burden, for instance, on the health sector and the power grid.

The synergies between local-scale UO and large-scale weather conditions are a continuous seesaw, and both conditions may affect each other at different times and locations. Synoptic-scale weather conditions may alter UO magnitude by modifying the wind pattern, cloud cover, humidity, and outgoing longwave radiation. Contrarily, rapid urbanization and the surface roughness in the urban fabric were also reported to reduce the wind speed (Zhang et al., 2010, 2011; Oke, 1987). In a study, 30%–40% lower mean annual wind speed was reported in cities compared to their rural counterparts (Lee, 1984). The influence of regional climate (synoptic-scale weather conditions) on European cities, resulting in exacerbated UO during heatwaves, was also concluded in Ward et al. (2016). In several studies, the urban-rural moisture contrast was reported as the major contributor exacerbating the daytime UO magnitude (Oke, 1982), and it is predominantly governed by synoptic-scale weather conditions (Dixon and Mote, 2003) and the surface cover (Henry et al., 1985). A positive response between soil moisture and synoptic-scale weather conditions was also reported in France during the summer heatwave of 2003, where drought conditions were associated with tropospheric circulations (Fischer et al., 2007). Clear sky conditions, low wind speed, and low atmospheric pressure were also associated with exacerbated UO at nighttime (Oke, 1982; Schatz and Kucharik, 2014).

Wind pattern and cloud cover are considered the most important meteorological parameters, causing differential heating and cooling rates between urban and rural areas due to the change in ventilation and insolation conditions (Półrolniczak et al., 2017; Bejarán and Camilloni, 2003). Low pressures induced by UO in the highly urbanized areas foster the airflow from the high-pressure rural areas that result in reduced UO magnitude (Grady Dixon and Mote, 2003). However, lower regional wind speed results in higher UO magnitude, as explained earlier (Morris and Simmonds, 2000). Similarly, lower cloud cover increases the radiative losses at nighttime, and quick radiative cooling in rural areas creates an urban − rural thermal gradient (Oke and Runnalls, 2000). The wind speed and cloud cover were concluded to be inversely proportional to the fourth root of UO in Melbourne, though the relational magnitude may vary from city to city (Morris et al., 2001). Radiative cooling at nighttime in rural areas is enhanced under clear (cloudless), calm (low wind speed), and dry conditions, compared to humid conditions (Chow and Roth, 2006). Mostly, these clear and calm conditions were also associated with anticyclonic circulation patterns (Szegedi and Kircsi, 2003). Under clear sky conditions at nighttime, not only radiative cooling is enhanced at rural sites, but the longwave emission in the urban fabric from daytime thermal storages also amplifies the urban ambient temperatures, which amplifies the nocturnal UO (Imran et al., 2019; Tan et al., 2010). In addition to thermal storage, urban shading/geometry also plays a pivotal role in keeping the cities warmer at nighttime by decelerating the radiative losses due to decreased sky view factor (Shahmohamadi et al., 2011). Instead, lower urban temperatures early in the day were also reported around the buildings due to urban shading, which may result in urban cooling (UC) at daytime (Zhang et al., 2014; Alonso et al., 2003). Urban shading was reported as one of the important reasons behind amplified nighttime UO and daytime UC in New York (Hardin et al., 2018). In contrast to dry conditions, moist conditions at daytime reduce the latent heat flux potential, particularly in the rural areas, and more available energy is partitioned into sensible heat flux, which may also lead to UC at daytime (Hardin et al., 2018). Further, an association between moist/cloudy conditions and cyclonic circulation patterns was also reported Unger (1996). In addition to meteorological factors, the topography of the area also affects the interactions between UO and synoptic-scale weather conditions. Leeward flow in the valley was reported to elevate the temperatures drastically through adiabatic warming (Adams et al., 2020). As reported in UO and heatwaves synergies section, advection from heat sources was also reported as the major cause of urban − rural thermal contrast in desert cities (Yun et al., 2020). Advection from coastal winds was also seen as the major contributor behind exacerbated UO magnitude due to lack of coastal winds penetration at inland sites (Vahmani and Ban-Weiss, 2016; Founda and Santamouris, 2017).

Synoptic-scale classifications, including CPC and the WTC, have been employed to identify the interactions between local and large-scale phenomena. While using the CPC, in Szeged, Hungary, anticyclonic circulation and a cloudless sky and calm wind speed were reported exacerbating UO, whereas lower UO was reported under cyclonic conditions (Unger, 1996). Similarly, in Debrecen, Hungary, mobile measurements captured a magnified UO (mean/max UO: 2.3°C) under anticyclonic conditions, whereas UO diminished under cyclonic conditions (Szegedi and Kircsi, 2003). In Poznan, Poland, exacerbated UO (average UO: 1.2°C) was reported mostly at nighttime under anticyclonic circulations. In contrast, UC was mostly reported during the daytime in the colder part of the year under cyclonic circulations (Półrolniczak et al., 2017). In Athens, Greece, high-pressure ridges (anticyclonic categories) were mostly favoring the UO formation, whereas cold air advection from northerly component winds terminated the UO (Mihalakakou et al., 2002). In Birmingham, while associating the surface and ambient UO with synoptic-scale circulation, maximum ambient UO (max/mean UO: 7.0/2.5°C) was reported at nighttime under anticyclonic conditions (Zhang et al., 2014). Similarly, higher surface UO (UO: 4.16°C) in Birmingham was reported in the city center at nighttime under cloudless anticyclonic conditions. In Melbourne, Australia, anticyclonic conditions were responsible for the warmest 17% UO events (mean group UO: 3.56°C) (Morris and Simmonds, 2000). Further, these anticyclonic circulations in Melbourne were associated with weak warm and dry N to NE airflow, which resulted in clear and calm conditions. The weakest UO in Melbourne was observed under northwesterly airflow instead of cyclonic conditions.

While utilizing the WTC (SSC) in northeastern cities of the United States (Baltimore, Philadelphia, Boston, and New York), the most intense UO was reported at nighttime (peak average UO: 3.5°C in New York) under DT (hot and dry) WT. The absolute maximum temperatures at both urban and rural sites were recorded under moist weather conditions (Hardin et al., 2018). Similar results in the same northeastern US cities were reported in another study, where all dry air masses (DM, DP, DT) were responsible for higher UO magnitude (2°C−5°C) compared to all moist air masses (Sheridan et al., 2000). Further, the UO effect was more pronounced during summer and exacerbated more in the second half of the century due to higher air-conditioning usage. In Atlanta, while utilizing the SSC, DT (hot and dry) WT at nighttime and MT (hot and moist) WT at daytime were reported as influential synoptic conditions exacerbating the UO magnitude (Dixon and Mote, 2003). Further, the UO magnitude in Atlanta under DT conditions at nighttime was comparatively higher (peak average UO: 3.84°C) compared to MT WT at daytime. In the same study, moist air masses were also linked to UHI-associated precipitation. In Buenos Aires, Argentina, exacerbated UO at nighttime during winter (max/mean UO: 2.8°C) was attributed to cold air masses, which were associated with a cold-core anticyclone, low wind speed, and low cloud cover (Bejarán and Camilloni, 2003). In Phoenix, USA, using the monthly mean minimum temperature for June from 1990 to 2014, 2°C−4°C spatial UO effect was reported under DT (dry and hot) WT along with clear and calm conditions (Brazel et al., 2007).

In addition to urban climatology, bioclimatological response to synoptic-scale weather conditions is also very well-documented. While investigating the association between synoptic-scale weather conditions and mortality rate during summer, DT, and MT WTs in New York (Sheridan and Lin, 2014) and Rome, MT conditions in Shanghai were found the most oppressive WTs (Sheridan, 2002). In Canada, MT and DT WTs were also associated with a higher level of air pollution and consequently with higher mortality rates (Vanos et al., 2013). In the 19 US cities, DP conditions during winter and DT conditions during summer were associated with increased cardiovascular mortality (Lee, 2015b). Further, it is also reported that moist air masses affect human thermal comfort more in the humid climate (Schatz and Kucharik, 2015; Zhao et al., 2014), while DT conditions are in the arid region. Therefore MT air masses in coastal cities, whereas DT air masses in dry regions were associated with higher mortality rates.

3.6 A case study from Sydney, Australia

Synergies between UO, heatwaves, and synoptic-scale weather conditions were also investigated in Sydney, Australia. Sydney is geographically the largest (12,367.7 km^2 land area) and the most populous (over 5.3 million people) city in Oceania (Australian Bureau of Statistics), with its climate classified as cfa (humid-subtropical) under the Koppen − Geiger classification (Australian Bureau of Statistics ABS Climate and the Sydney, 2000). The Sydney Central Business District (CBD) is located in the proximity of the Tasman Sea in the east, while the metropolitan area extends by ∼50 km to the west and is exposed to one of the largest arid biomes (Yun et al., 2020; Australian Bureau of Statistics, 2011; Byrne et al., 2008). Further, the city is largely growing westward as to the north and south it is surrounded by national parks. Two more city centers are being developed in Western Sydney, which is projected to host over 50% of the total city's population by 2036 (Ivan Demography and destiny in Western Sydney, 2020).

While investigating the interactions between local and large-scale phenomena, the city was divided into three zones: Eastern Sydney, Inner Sydney, and Western Sydney. As the geographical extent of the city cannot be illustrated by one urban/rural station, six stations in three zones were considered, as shown in Fig. 3.1. This stratification was made based on distance from the coast. Observatory Hill (OBS Hill), located in Sydney CBD and in the proximity of the coast, was considered in eastern Sydney. Further, OBS Hill (urban station) was also taken as the reference station as recommended in several other studies (Santamouris et al., 2017; Khan et al., 2020; Yun et al., 2020). Inner Sydney stations comprise Olympic Park and Canterbury, approximately 8—12 km away from the nearest coast, while Western Sydney includes the stations of Penrith Lakes, Campbelltown, and Liverpool. The minimum distance of any Western Sydney site from the nearest coast is around 25 km, whereas the Penrith Lakes is 50 km away from the coast. The tree canopy cover is the least in Sydney CBD and increasing from Eastern to Western Sydney: 15% in the Sydney CBD, 15%—17% in Inner Sydney, and 25%—35% in Western Sydney (Jacbos et al., 2014). Similarly,

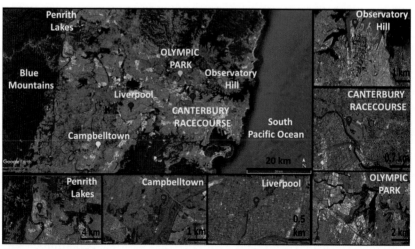

Figure 3.1 Location of weather stations and site characteristics. *Image source: Google map.*

potentially plantable surfaces are also lower in the Sydney CBD and higher in Inner and Western Sydney: 13% in Sydney CBD, 22%–32% in Inner Sydney, and 40%–50% in Western Sydney (Jacbos et al., 2014). Finally, the PD is the highest in Sydney CBD (over 6000 people/km^2) and lowering from Inner (2500–4000 people/km^2) to Western Sydney (500–650 pop/km^2). Further details about site location and characteristics can be found in Khan et al. (2021b), and a summary is also provided in Table 3.1.

Semihourly ground data obtained from the Australian Bureau of Meteorology (BOM) from 1999 to 2017 have been used to investigate the interactions between local and regional-scale phenomena (Australian Government Bureau of Meteorology Climate Data, 2020). The meteorological parameters employed in the analysis were ambient temperatures at 2 m height, wind speed, wind direction, and absolute humidity (AH). Validation procedures proposed in Estévez et al. (2011) were applied to remove the outliers and null values. Linear interpolation and triangulation methods were utilized to infill the gaps. Hourly averages have been then computed from semihourly data. Inland sites (suburbs) were thermally more affected, compared to Sydney CBD. Therefore UO was computed as the temperature difference between suburbs and Sydney CBD ($\Delta T = T_{suburbs} - T_{urb}$). To investigate the interaction between UO and heatwaves, qualitatively hourly analyses were performed separately for all heatwave events. The hourly analyses were performed to comprehend the diurnal variations in meteorological conditions, particularly in wind patterns. Sunrise and sunset times were used to define the daytime and nighttime durations.

Heatwaves were identified by employing the RTT metric (95th percentile of daily max temperature). RTT metric was used due to local's climatic acclimatization considerations. Minimum three consecutive days were considered to identify a heatwave and background condition's duration. Only heatwave episodes occurring concurrently at all sites were counted, to consider the spatial extension criterion of heatwave definition. Heatwave episodes studied during the investigation can be found in Khan et al. (2021b). While examining the interactions between UO and synoptic-scale weather conditions, GWTC was employed. Further, higher 5% of daily maximum ΔT (UO) and lower 5% of daily minimum ΔT (UC) were compared with daily synoptic-scale weather conditions. Daily max ΔT and daily minimum ΔT were employed instead of the daily average ΔT to investigate both extremes (UO and UC). Daily max

Table 3.1 Site data and land characteristics of the sites.

Station location (Australian Government Bureau of Meteorology Climate Data, 2020)	Zone	Station surrounding	Surface cover (%) (Jacbos et al., 2014)	Water source presence	Area classification	Distance from CBD (km)	Distance from the coast (km)	Population density (pers./km^2) (Australian Bureau of Statistics)
(A) Penrith Lakes, BOM-067113								
Penrith Lakes regional park (150°40'42" E, 33°43'10" S, 24.7 m amsl)	Western Sydney	- Greenery - Bare soil - Lakes	- 25% tree canopy cover - 15.3% nonplantable - 54% potentially plantable	- Aquapark and aqua golf facility nearby - The Nepean River to the west of the site	Primarily residential and commercial buildings	55 km northwest of CBD	49 km southeast of the station	484
(B) Campbelltown, BOM—068257								
The Australian Botanic garden (150° 46'25" E, 34° 3'41" S, 112 m amsl)	Western Sydney	- Greenery - Trees - Bare soil	- 34.2% tree canopy cover - 12% nonplantable - 38.5% potentially plantable	Lake Nadungamba near the station, Annan lake to the northwest of the station (1.43 km)	Residential/ semirural area	46 km southwest of CBD	28 km southeast of the station	509.4

(Continued)

Table 3.1 (Continued)

Station location (Australian Government Bureau of Meteorology Climate Data, 2020)	Zone	Station surrounding	Surface cover (%) (Jacbos et al., 2014)	Water source presence	Area classification	Distance from CBD (km)	Distance from the coast (km)	Population density (pers./km²) (Australian Bureau of Statistics)
(C) Liverpool (NSW government)								
Raine park (150° 54′ 21″ E, 33° 55′ 58″ S, 22 m amsl)	Western Sydney	- Greenery - Roads	- 23.3% tree canopy cover - 16.3% nonplantable - 51.1% potentially plantable	None	Mixed residential and commercial area	28.9 km southwest of CBD	23 km southeast of the station	668.82
(D) Olympic Park, BOM—066212 (city of Auburn)								
Olympic park archery center (151° 4′18″ E, 33° 50′2″ S, 4 m amsl)	Inner Sydney	- Greenery - Creeks - Parramatta river	- 15.4% tree canopy cover - 50.4% nonplantable - 31.1% potentially plantable	- Paramatta river - Many creeks in the surrounding - Duck river	Mixed use (residential, commercial, industrial, recreational areas, and parkland uses)	12.7 km, northwest of CBD	13.5 km, southeast of the station	2527.88

(E) Canterbury, BOM—066194

Canterbury racecourse park (151° 6'48" E, 33° 54'21" S, 3 m amsl)	Inner Sydney	- Greenery - Ponds - Cooks River	- 17.5% tree canopy cover - 54.5% nonplantable - 22.8% potentially plantable	- Cooks River - Pond	Primarily residential and light industrial	17 km, southwest of CBD	7.5 km, southeast of the station	4303

(F) OBS Hill, BOM—066062

Observatory Hill (151°12'18" E, 33°51'39" S, 39 m amsl)	Eastern Sydney	- Roads - Greenery - High rise buildings	- 15.2% tree canopy cover - 69.1% nonplantable - 13.2% potentially plantable	- Site near coast	Commercial buildings, parks, etc.	N/A	0.6 km, northeast of the station	6160

daytime ΔT and daily min nighttime ΔT were also compared with daily synoptic-scale weather conditions to understand the diurnal variations in comparison with synoptic-scale weather conditions.

3.6.1 Interaction between urban overheating and heatwaves in Sydney

While investigating the synergies between UO and heatwaves in Sydney, higher threshold temperatures for heatwave identification were computed for Western Sydney than for the coastal suburbs (Khan et al., 2021b). Further, frequent and severe heatwave episodes, particularly in recent years (2015 onward), were computed also in Sydney's coastal areas. Indeed the daytime UO magnitude during heatwaves, compared to background conditions (ΔUO), was comparatively higher in Western Sydney (peak average ΔUO: 8°C) than Inner Sydney (peak average ΔUO: 4°C−4.5°C). Further, the UO magnitude during heatwaves was reported increasing every year after 2015, particularly in Western Sydney. Contrarily, alleviated nocturnal temperatures were observed in the suburbs during heatwaves, and the UO magnitude was either negative or approaching zero at nighttime. While examining the AH during heatwaves, Western Sydney was reported as the most affected zone. AH was plummeting in the daytime and regrowing in the afternoon. The drop in AH at noon (compared to morning) was 20%−55% in Western Sydney, 15%−35% in Inner Sydney, and 10%−25% in coastal Sydney as heatwave days were progressing. Conversely, nocturnal AH at all sites was uninterrupted, and it displayed according to site characteristics. Also, the suburban − urban moisture contrast was inversely proportional to UO during both heatwave and background conditions.

The investigation of the wind pattern revealed that during the daytime in heatwave conditions, Sydney was exposed to the influence of two opposite advective mechanisms. Western Sydney was predominantly influenced by desert winds, driven from the western side of the city, increasing temperatures in Western Sydney, and occasionally affecting the inner suburbs of Sydney. Conversely, the Sydney CBD was mostly influenced by coastal winds. The desert winds were increasing the temperatures in Western Sydney due to warm-air advection, which was also sweeping the ambient moisture when blowing with higher wind speed. Consequently, site AH was diminishing at noon in the suburbs, particularly in Western Sydney. Conversely, a steady sea breeze maintained lower temperatures in the Sydney CBD than inland. Western Sydney, during heatwaves, was under the influence of desert winds for 70%—80% of the time, whereas Inner Sydney is for 54% of the time. During the background condition, only 20%—35% in Western Sydney, and 10%—15% in Inner Sydney, desert winds were influential. Conclusively, two opposite heating and cooling advective mechanisms during the daytime during heatwaves were responsible for an exacerbated UO in Sydney.

On the other hand, advection was not so influential during the night due to low wind speed. Instead, Western Sydney and the inner suburbs cooled down quickly overnight due to radiative cooling, which was attributed to their high proportion of nonurban surfaces (higher than in the more densely built coastal suburbs and CBD). At the same time, the nocturnal temperatures at Sydney CBD were either comparatively higher due to higher urbanized surfaces or almost the same as of suburbs due to slight advection from coastal winds. In conclusion, the distance from the coast was identified as the primary contributor to the exacerbated daytime UO magnitude in Sydney during heatwaves. The desert winds became more influential with increasing distance from the coast. Further, more urbanized surfaces in Western Sydney would make the region more vulnerable, increasing the frequency of heatwaves due to modified land-— atmosphere moisture exchanges. Also, they would further reduce the coastal wind's penetration in the suburbs, increasing the dominance of the desert winds.

3.6.2 Interaction between urban overheating and synoptic-scale weather conditions in Sydney

While examining the interactions between UO and synoptic-scale weather conditions in Sydney, it was observed that over 50% of the time (1999—2017) daytime UO was positive in all three zones of Sydney, while for 90% of nighttime UO was negative (Khan et al., 2021a). At daytime, HW (humid and warm), W (warm), and H (humid) WTs were more influential. HW and H conditions were slightly more dominant in Western Sydney, whereas the W conditions proved more aggressive in Inner Sydney. At nighttime, negative UO (UC) was primarily observed under DW (dry and warm), W, and D (dry) WTs. Tropical maritime Tasman air masses (warm, moist, and unstable) were responsible for HW conditions in the region, whereas DW and W conditions were attributed to tropical continental air masses (hot, dry, and unstable) formed over central Australia. Temperate maritime weather patterns were reported responsible for H conditions in the region. HW and H conditions at daytime were more dominant in Western Sydney as they reduced the evapotranspiration potential in the region, partitioned more energy into sensible heat flux, and increased the ambient temperatures. On the other hand, coastal winds during HW and H conditions kept the temperature steady in Sydney CBD and exacerbated the UO. W conditions in inner Sydney were comparatively more dominant (than HW/H) due to comparatively few nonurban surfaces. Therefore advection from tropical continental winds made inner Sydney more vulnerable.

DW conditions were more influential at nighttime and were responsible for UC. DW, D conditions are associated with clearing sky and calm meteorological conditions. Under clear and calm meteorological conditions at night, the radiative cooling process became more enhanced in the suburbs, particularly in Western Sydney. Conversely, comparatively amplified nocturnal temperatures in Sydney CBD were attributed to higher nighttime longwave emission from daytime heat storage, which resulted in UC. HW and W WTs were also responsible for exacerbated UO during heatwaves, which was associated with the dualistic synoptic system in Sydney.

3.7 Discussion and conclusion

The local, regional, and large-scale weather conditions are affected by urbanization due to the change in sensible, latent, radiative surface fluxes and the overall dynamic response of the land − atmosphere system. UO, heatwaves, and synoptic-scale weather conditions may synergistically interact and severely affects the communities. A contradictory response of UO to heatwaves was reported in the literature, where daytime UO exacerbation (Khan et al., 2021b), nighttime UO exacerbation (Fenner et al., 2019), UO exacerbation at both daytime and nighttime (He et al., 2020), the inconclusive response of UO to heatwaves (Richard et al., 2021), and a declined UO response during heatwaves (Scott et al., 2018) were all reported. The exacerbated daytime UO response to heatwaves was mostly reported for coastal cities (Founda and Santamouris, 2017; Pyrgou et al., 2020; Ao et al., 2019; Ramamurthy and Bou-Zeid, 2017; Rizvi et al., 2019). Further, advective heat flux, urban − rural moisture contrast, and distance from the coast were reported as the major synergistic interactions between UO and heatwaves in those studies.

Advection from the coastal winds was the major source of thermal gradient between coastal and inland sites during heatwaves, which was also affecting the moisture contrast between both sites as reported in Athens

(Founda and Santamouris, 2017), Shanghai (Ao et al., 2019), and Los Angeles (Vahmani and Ban-Weiss, 2016). Further, higher UO magnitude during heatwaves was also reported for coastal cities, compared to noncoastal cities or inland sites, which was also attributed to advection from the coastal winds. Advection from steady sea breeze may keep coastal areas cooler during heatwaves, compared to inland sites, and magnify the UO magnitude enormously (Vahmani and Ban-Weiss, 2016; Ao et al., 2019). For instance, in Shanghai, China, a higher UO magnitude was documented during heatwaves when UO was quantified reference to the coastal site instead of the inland site (Ao et al., 2019). In contrast to advection from coastal winds, advection from the heat source (e.g., desert) during heatwaves may elevate the temperatures of adjacent sites swiftly and cause thermal imbalance in the region. Similar to coastal winds, the impact of advection from a heat source on UO magnitude during heatwaves was extremely strong. For instance, the activation of the dualistic synoptic system simultaneously on the opposite side of Sydney exacerbated the UO magnitude drastically during heatwaves (Khan et al., 2020, 2021b). The dualistic synoptic system was the warm-air advection from the continental side of the city and the cold-air advection from the coast. The UO magnitude was approximately 8°C higher in Sydney during heatwaves than in nonheatwaves periods (Khan et al., 2020). In coastal cities, the distance from the coast also plays an important role during heatwaves, and site temperature increases as the distance from the coast increases. Amplified daytime temperatures were recorded for inland sites in Sydney (Khan et al., 2021b) and Los Angeles (Vahmani and Ban-Weiss, 2016).

Urban − rural moisture contrast is not only regulated by synoptic-scale weather conditions as stated above but by surface characteristics as well. The urban − rural moisture contrast caused by the intense solar radiation during heatwaves was also reported as an important contributor to exacerbated daytime UO. The moisture contrast at daytime is attributed to differential available energy partitioning at both urban and rural surfaces as rural surfaces exhibited higher latent heat flux potential during heatwaves than urban counterparts reported in Nicosia (Pyrgou et al., 2020). However, in contrast to other coastal sites, the impact of advective heat flux was insignificant in Nicosia. The inconsistent results in Nicosia might be attributed to ignoring the spatial extension of the city, which is an important parameter in identifying the heatwaves as concluded in (Khan et al., 2021b). Similarly, in Melbourne and Adelaide (Rogers et al., 2019), the nighttime UO response was more prominent than in other coastal cities. The contrasting results can be attributed to the inconsistent method of UO quantification and inland site selection, which curtailed the impact of the coast and advective heat flux, as reported in Fenner et al. (2019).

Exacerbated nighttime UO magnitude during heatwaves was mostly reported in noncoastal cities or at inland sites (Fenner et al., 2019; Rogers et al., 2019; Ngarambe et al., 2020; Imran et al., 2019; Unger et al., 2020; Katavoutas and Founda, 2019; Cheval et al., 2009). Further, the differential heat storage capacity of urban/rural surfaces, amplified anthropogenic heat flux, and reduced advective heat flux were presented in the literature as the major synergistic interactions between UO and heatwaves at nighttime. Higher longwave emission at nighttime in the urban fabric due to higher daytime heat storage during heatwaves increases the nocturnal urban ambient temperature. In addition, trapped longwave radiations in urban fabric attributed to urban shading further scorch the nocturnal urban temperatures. Contrarily, radiative cooling in rural areas reduces the nocturnal temperatures and magnifies the nocturnal UO during heatwaves. Higher emission of stored heat at nighttime and the higher anthropogenic heat fluxes were largely considered the major interaction between UO and heatwaves in the several US cities that exhibited amplified nighttime UO during heatwaves (Zhao et al., 2018). Similarly, lower wind speed at nighttime during heatwaves results in lower urban mixing, that is, reduces secondary air circulation and exacerbates the UO. The reduced regional wind speed was an important synergistic interaction in Shanghai (Ao et al., 2019) and Seoul (Ngarambe et al., 2020), which intensified the combined effect of UO and heatwaves at nighttime during heatwaves. Further, an association between heat storage ratio and lower regional wind speed is also established during heatwaves. For instance, in the United Kingdom (London and Swindon), reduced regional wind speed was associated with enhanced heat storage ratio, whereas in Beijing, high-speed winds restricted the heat storage ratio during extreme heat events, compared to normal days (Sun et al., 2017).

An inconclusive response of UO (Tewari et al., 2019; Chew et al., 2021; Richard et al., 2021) and declined response of UO (Kumar and Mishra, 2019; Scott et al., 2018; Brázdil and Budíková, 1999) during heatwaves is also discussed in several studies. The contrasting results might be attributed to inconsistent methods of UO quantification and inconsistent methods of heatwaves identification. In Singapore, insignificant changes in UO magnitude and inconsiderable variation in the energy budget response, particularly in latent heat flux, storage heat flux, and advective heat flux during heatwaves and nonheatwaves, might be attributed to numerous factors: (1) inconsistent method of UO estimation where 3-days averages for all meteorological parameters during heatwaves and non-heatwaves were computed, (2) limitation of study to one heatwave/non-heatwave episode, and (3) ignoring the spatial extension of the city while defining the heatwaves (Chew et al., 2021). Similarly, in Dijon, France, considering less intense temperature thresholds for heatwave identification, compared to regional health alerts, might be a reason for inconclusive results (Richard et al., 2021). A declined response of UO to heatwaves in 54 cities of the United States (Scott et al., 2018) and 89 urban regions of India (Kumar and Mishra, 2019) might be attributed to the inconsistent method of heatwave identification, since in contrast to other studies, to define the heatwaves' threshold temperatures daily minimum temperatures were used instead of daily maximum temperatures (Fenner et al., 2019).

Interactions between the UO magnitude and large-scale weather conditions were investigated by utilizing both CPC and WTC. Exacerbated UO magnitude was reported under anticyclonic conditions while utilizing the CPC (Morris and Simmonds, 2000; Zhang et al., 2014; Unger, 1996; Półrolniczak et al., 2017; Mihalakakou et al., 2002; Szegedi and Kircsi, 2003), whereas warm and dry conditions were responsible for magnified UO magnitude while employing the WTC as reported in Hardin et al. (2018); Grady Dixon and Mote (2003); Brazel et al. (2007). Contrarily, cyclonic conditions in CPC (Unger, 1996; Półrolniczak et al., 2017) and moist conditions in WTC (Hardin et al., 2018) were associated with lower UO or UC. However, in Athens (Mihalakakou et al., 2002) and Melbourne (Morris and Simmonds, 2000), instead of cyclonic conditions, UC was associated with northerly cold air advection and northwesterly airflows, respectively. Further in the literature, anticyclonic conditions in CPC (Szegedi and Kircsi, 2003) and dry weather conditions in WTC (Hardin et al., 2018; Brazel et al., 2007) are linked to clear (cloudless sky) and calm (low wind speed) meteorological conditions, which brings the undisturbed radiations and magnify the UO magnitude (Unger, 1996; Unger et al., 2001). Contrarily, cyclonic conditions in CPC and moist conditions in WTC were mostly associated with cloudy conditions and lower UO magnitude (Unger, 1996). Under clear, calm, and dry conditions at nighttime, quick radiative cooling at rural sites and enhanced longwave emission in urban fabric results in exacerbated UO as reported in northeastern US states (Hardin et al., 2018) and in Szeged, Hungary (Unger, 1996). Contrarily, under cloudy conditions at nighttime, slow radiative cooling at rural sites, and declined longwave emission in the urban fabric, diminishes the UO. Similarly, under moist conditions during the day, latent heat flux potential at rural sites is compromised, magnifies the rural temperatures, and results in UC as reported in New York (Hardin et al., 2018). Exacerbated rural temperatures during the day under moist conditions were also reported in Sydney (Khan et al., 2021a).

Nomenclature

UO	Urban overheating
UC	Urban cooling
ΔUO	UO magnitude difference between heatwaves and background conditions
AH	Absolute humidity
ΔAH	Urban − rural moisture contrast
LCLU	Land-cover, and Land-use classification
ATT	Absolute temperature threshold
RTT	Relative temperature threshold
T_{max}	Daily maximum temperature
T_{min}	Daily minimum temperature
AT	Apparent temperature
HI	Humidity index
EHF	Excessive heat factor
EHI (sig)	Excess Heat
EHI (accl)	Heat Stress
HP	High pressure
LP	Low pressure
MSLP	Mean sea level pressure
CPC	Circulation-pattern-based classification
WTC	Weather typing classification
SSC	Spatial synoptic classification
GWTC	Gridded weather typing classification
WT	Weather type
cP	Continental Polar
mP	Maritime Polar
cT	Continental tropical
mT	Maritime tropical
DP	Dry polar
DM	Dry moderate
DT	Dry tropical
MP	Moist polar
MM	Moist moderate
MT	Moist tropical
TR	Transitional
HC	Humid cold
HW	Humid warm
DC	Dry cool

DW	Dry warm
H	Humid
D	Dry
C	Cool
W	Warm
CFP	Cold frontal passage
WFP	Warm frontal passage
S	Seasonal
CBD	Central business district

Supplementary material

Table S3.1 Definition of air masses in spatial synoptic classification (SSC) available at Sheridan (2021).

Sr. no.	Air mass	Definition
1	Dry Polar (DP)	is synonymous with the traditional cP air mass classification. This air mass is generally advected from polar regions around a cold-core anticyclone and is usually associated with the lowest temperatures observed in a region for a particular time of year, as well as clear, dry conditions
2	Dry Moderate (DM)	air is mild and dry. It has no traditional analog but is often found with the zonal flow in the middle latitudes, especially in the lee of mountain ranges. It also arises when a traditional air mass such as cP or mT has been advected far from its source region and has thus modified considerably
3	Dry Tropical (DT)	weather type is similar to the cT air mass; it represents the hottest and driest conditions found at any location. There are two primary sources of DT: either it is advected from the desert regions, such as the Sonoran or the Sahara Desert, or it is produced by rapidly descending air, whether via orography (such as the chinook) or strong subsidence
4	Moist Polar (MP)	air is a large subset of the mP air mass; weather conditions are typically cloudy, humid, and cool. MP air appears either by inland transport from a cool ocean or as a result of frontal overrunning well to the south of the region. It can also arise in situ as a modified cP air mass, especially downwind of the Great Lakes

(Continued)

(Continued)

Sr. no.	Air mass	Definition
5	Moist Moderate (MM)	is considerably warmer and more humid than MP. The MM air mass typically appears in a zone south of MP air, still in an area of overrunning but with the responsible front much nearer. It can also arise within an mT air mass on days when high cloud cover suppresses the temperature
6	Moist Tropical (MT)	analogous to the traditional mT air mass, is warm and very humid. It is typically found in warm sectors of mid-latitude cyclones or in a return flow on the western side of an anticyclone; as one approaches the tropics, this weather type dominates
7	Transition (TR)	days are defined as days in which one weather type yields to another, based on large shifts in pressure, dew point, and wind over the course of the day

Table S3.2 Definition of air masses in Gridded weather typing classification (GWTC) available at Lee (2021).

Sr. no.	Air mass	Symbol	Definition
1	Humid Cool	HC	Cooler and more humid than normal for the location and time of year
2	Humid	H	More humid than normal for the location and time of year
3	Humid Warm	HW	Warmer and more humid than normal for the location and time of year
4	Cool	C	Cooler than normal for the location and time of year
5	Seasonal	S	Near-normal conditions for the location and time of year
6	Warm	W	Warmer than normal for the location and time of year
7	Dry Cool	DC	Cooler and drier than normal for the location and time of year
8	Dry	D	Drier than normal for the location and time of year

(Continued)

(Continued)

Sr. no.	Air mass	Symbol	Definition
9	Dry Warm	DW	Warmer and drier than normal for the location and time of year
10	Cold Front Passage	CFP	Transitional weather day often with a drop in temperatures and dew points and rising sea-level pressure
11	Warm Front Passage	WFP	Transitional weather day often with increasing temperatures and dew points and lowering sea-level pressure

References

Adams, R.E., Lee, C.C., Smith, E.T., Sheridan, S.C., 2020. The relationship between atmospheric circulation patterns and extreme temperature events in North America. International Journal of Climatology.

Akbari, H., Bell, R., Brazel, T., Cole, D., Estes, M., Heisler, G., et al., 2008. Urban heat island basics. In Reducing Urban Heat Islands: Compendium of Strategies | Heat Island Effect US EPA.

Alexander, L.V., Zhang, X., Peterson, T.C., Caesar, J., Gleason, B., Klein Tank, A.M.G., et al., 2006. Global observed changes in daily climate extremes of temperature and precipitation. Journal of. Geophysical Research: Atmospheres 111, D05109.

Alonso, M.S., Labajo, J.L., Fidalgo, M.R., 2003. Characteristics of the urban heat island in the city of Salamanca, Spain. Atmosfera 16.

Ao, X., Wang, L., Zhi, X., Gu, W., Yang, H., Li, D., 2019. Observed synergies between urban heat islands and heat waves and their controlling factors in Shanghai, China. Journal of Applied Meteorology and Climatology 58, 1955−1972.

Argüeso, D., Evans, J.P., Fita, L., Bormann, K.J., 2014. Temperature response to future urbanization and climate change. Climate Dynamics 42, 2183−2199.

Australian Bureau of Statistics 3218.0- Regional Population Growth, Australia, 2017−2018, Available online: https://www.abs.gov.au/AUSSTATS/abs@.nsf/Lookup/3218.0Main + Features12017-18?OpenDocument (accessed 15.01.20).

Australian Bureau of Statistics (ABS) Climate and the Sydney, 2000. Olympic Games Available online: https://www.abs.gov.au/AUSSTATS/abs@.nsf/Previousproducts/1301.0.Feature Article32000 (accessed 10.02.20).

Australian Bureau of Statistics Australian Statistical Geography Standard: Volume 4—Significant Urban Areas, Urban Centres and Localities, Section of State. 2011, 4, 29−31.

Australian Government Bureau of Meteorology Climate Data Online Available online: http://www.bom.gov.au/climate/data/index.shtml?bookmark = 201 (accessed 15.02.20).

Basara, J.B., Basara, H.G., Illston, B.G., Crawford, K.C., 2010. The impact of the urban heat island during an intense heat wave in oklahoma city. Advances in Meteorology 2010, 1−10.

Basu, R., 2002. Relation between elevated ambient temperature and mortality: a review of the epidemiologic evidence. Epidemiologic Reviews 24, 190−202.

Bejarán, R.A., Camilloni, I.A., 2003. Objective method for classifying air masses: an application to the analysis of Buenos Aires' (Argentina) urban heat island intensity. Theoretical and Applied Climatology 74, 93—103.

Bohnenstengel, S.I., Evans, S., Clark, P.A., Belcher, S.E., 2011. Simulations of the London urban heat island. Quarterly Journal of the Royal Meteorological Society 137, 1625—1640.

Brazel, A., Gober, P., Lee, S., Grossman-Clarke, S., Zehnder, J., Hedquist, B., et al., 2007. Determinants of changes in the regional urban heat island in metropolitan Phoenix (Arizona, USA) between 1990 and 2004. Climate Research 33, 171—182.

Brooke Anderson, G., Bell, M.L., 2011. Heat waves in the United States: mortality risk during heat waves and effect modification by heat wave characteristics in 43 U.S. communities. Environmental Health Perspectives 119, 210—218.

Broome, R.A., Fann, N., Cristina, T.J.N., Fulcher, C., Duc, H., Morgan, G.G., 2015. The health benefits of reducing air pollution in Sydney, Australia. Environmental Research 143, 19—25.

Brázdil, R., Budíková, M., 1999. An urban bias in air temperature fluctuations at the Klementinum, Prague, The Czech Republic, In: Proceedings of the Atmospheric Environment; Elsevier Science Ltd., Vol. 33, pp. 4211—4217.

Byrne, M., Yeates, D.K., Joseph, L., Kearney, M., Bowler, J., Williams, M.A.J., et al., 2008. Birth of a biome: insights into the assembly and maintenance of the Australian arid zone biota. Molecular Ecology 17, 4398—4417.

Chen, K., Bi, J., Chen, J., Chen, X., Huang, L., Zhou, L., 2015. Influence of heat wave definitions to the added effect of heat waves on daily mortality in Nanjing, China. The Science of the Total Environment 506—507, 18—25.

Cheval, S., Dumitrescu, A., Bell, A., 2009. The urban heat island of Bucharest during the extreme high temperatures of July 2007. Theoretical and Applied Climatology 97, 391—401.

Chew, L.W., Liu, X., Li, X.-X., Norford, L.K., 2021. Interaction between heat wave and urban heat island: a case study in a tropical coastal city, Singapore. Atmospheric Research 247, 105134.

Chow, W.T.L., Roth, M., 2006. Temporal dynamics of the urban heat island of Singapore. International Journal of Climatology .

Coates, L., Haynes, K., O'Brien, J., McAneney, J., de Oliveira, F.D., 2014. Exploring 167 years of vulnerability: an examination of extreme heat events in Australia 1844—2010. Environmental Science & Policy 42, 33—44.

Collins, D.A., Della-Marta, P.M., Plummer, N., Trewin, B.C., 2000. Trends in annual frequencies of extreme temperature events in Australia. Australian Meteorological Magazine 49, 277—292.

Colucci, S.J., 2003. Anticyclones. Encyclopedia of Atmospheric Sciences. Elsevier, pp. 142—146.

Cui, L., Shi, J., 2012. Urbanization and its environmental effects in Shanghai, China. Urban Climate 2, 1—15.

Diem, J.E., Stauber, C.E., Rothenberg, R., 2017. Heat in the southeastern United States: characteristics, trends, and potential health impact. PLoS One.

Dixon, P.G., Mote, L.T., 2003. Patterns and causes of Atlanta's urban heat island—initiated precipitation. Journal of Applied Meteorology.

Elsayed, I.S.M., 2012. Mitigation of the urban heat island of the city of Kuala Lumpur, Malaysia. Middle-East Journal of Scientific Research 11, 1602—1613.

Estévez, J., Gavilán, P., Giráldez, J.V., 2011. Guidelines on validation procedures for meteorological data from automatic weather stations. Journal of Hydrology 402, 144—154.

Fenner, D., Holtmann, A., Meier, F., Langer, I., Scherer, D., 2019. Contrasting changes of urban heat island intensity during hot weather episodes. Environmental Research Letters 14, 124013.

Filkov, A.I., Ngo, T., Matthews, S., Telfer, S., Penman, T.D., 2020. Impact of Australia's catastrophic 2019/20 bushfire season on communities and environment. Retrospective analysis and current trends. Journal of Safety Science and Resilience 1, 44–56.

Fischer, E.M., Oleson, K.W., Lawrence, D.M., 2012. Contrasting urban and rural heat stress responses to climate change. Geophysical Research Letters 39, 8.

Fischer, E.M., Schär, C., 2010. Consistent geographical patterns of changes in high-impact European heatwaves. Nature Geoscience 3.

Fischer, E.M., Seneviratne, S.I., Vidale, P.L., Lüthi, D., Schär, C., 2007. Soil moisture-atmosphere interactions during the 2003 European summer heat wave. Journal of Climate 20, 5081–5099.

Founda, D., Santamouris, M., 2017. Synergies between urban heat island and heat waves in Athens (Greece), during an extremely hot summer (2012). Scientific Reports 7, 10973.

Garshasbi, S., Haddad, S., Paolini, R., Santamouris, M., Papangelis, G., Dandou, A., et al., 2020. Urban mitigation and building adaptation to minimize the future cooling energy needs. Solar Energy 204, 708–719.

Gillett, N.P., Stott, P.A., 2009. Attribution of anthropogenic influence on seasonal sea level pressure. Geophysical Research Letters 36, L23709.

Grady Dixon, P., Mote, T.L., 2003. Patterns and causes of Atlanta's urban heat island-initiated precipitation. Journal of Applied Meteorology 42, 1273–1284.

Hardin, A.W., Liu, Y., Cao, G., Vanos, J.K., 2018. Urban heat island intensity and spatial variability by synoptic weather type in the northeast U.S. Urban Climate 24, 747–762.

Heck, P., Zanetti, A., Enz, R., Green, J., Suter, S., 2004. Natural catastrophes and man-made disasters in 2003. Sigma Rep. 1/2004, Swiss Re Tech. Rep., Zurich, Switzerland, 10 pp.

Henry, J.A., Dicks, S.E., Marotz, G.A., 1985. Urban and rural humidity distributions: relationships to surface materials and land use. Journal of Climatology. 5, 53–62.

He, X., Wang, J., Feng, J., Yan, Z., Miao, S., Zhang, Y., et al., 2020. Observational and modeling study of interactions between urban heat island and heatwave in Beijing. Journal of Cleaner Production. 247, 119169.

Horton, R.M., Mankin, J.S., Lesk, C., Coffel, E., Raymond, C., 2016. A review of recent advances in research on extreme heat events. Current Climate Change Reports 2, 242–259.

Huth, R., 2010. Synoptic-climatological applicability of circulation classifications from the COST733 collection: first results. Physics and Chemistry of the Earth, Parts A/B/C 35, 388–394.

Hu, Z., Yu, B., Chen, Z., Li, T., Liu, M., 2012. Numerical investigation on the urban heat island in an entire city with an urban porous media model. Atmospheric Environment (Oxford, England: 1994).

Imran, H.M., Kala, J., Ng, A.W.M., Muthukumaran, S., 2019. Impacts of future urban expansion on urban heat island effects during heatwave events in the city of Melbourne in southeast Australia. Quarterly Journal of the Royal Meteorological Society 145, 2586–2602.

IPCC Climate Change 2014: Synthesis Report. Contribution of Working Groups I, II and III to the Fifth Assessment Report of the Intergovernmental Panel on Climate Change, 2014. ISBN 9789291691432.

Ivan Demography and destiny in Western Sydney. Available online: https://blog.id.com.au/2018/population/population-trends/demography-and-destiny-in-western-sydney/ (accessed 30. 10.20).

Jacbos, B., Mikhailovich, N., Delaney, C., 2014. Benchmarking Australia's urban tree canopy: an i-Tree assessment, Final Report 2014, prepared for Horticulture Australia Limited by the Institute for Sustainable Futures, University of Technology Sydney.

Kalkstein, L.S., Nichols, M.C., David Barthel, C., Scott Greene, J., 1996. A new spatial synoptic classification: application to air-mass analysis. International Journal of Climatology 16, 983−1004.

Katavoutas, G., Founda, D., 2019. Response of urban heat stress to heat waves in Athens (1960−2017). Atmosphere (Basel) 10, 483.

Khan, H., Asif, M., 2017. Impact of green roof and orientation on the energy performance of buildings: a case study from Saudi Arabia. Sustainability 9, 640.

Khan, H., Asif, M., Mohammed, M., 2017. Case study of a nearly zero energy building in Italian climatic conditions. Infrastructures 2, 19.

Khan, H.S., Paolini, R., Santamouris, M., Caccetta, P., 2020. Exploring the synergies between orban overheating and heatwaves (HWs) in Western Sydney. Energies 2020 13, 470.

Khan, H.S., Santamouris, M., Kassomenos, P., Paolini, R., Caccetta, P., Petrou, I., 2021a. Spatiotemporal variation in urban overheating magnitude and its association with synoptic air-masses in a coastal city. Scientific Reports 11, 6762.

Khan, H.S., Santamouris, M., Paolini, R., Caccetta, P., Kassomenos, P., 2021b. Analyzing the local and climatic conditions affecting the urban overheating magnitude during the heatwaves (HWs) in a coastal city: a case study of the greater Sydney region. The Science of the Total Environment 755, 142515.

Kool, D., Ben-Gal, A., Agam, N., 2018. Within-field advection enhances evaporation and transpiration in a vineyard in an arid environment. Agricultural and Forest Meteorology 255, 104−113.

Koutsoyiannis, D., Langousis, A., 2011. Precipitation, Treatise on Water Science, Vol. 2. Elsevier, pp. 27−77, ISBN 9780444531933.

Kumar, R., Mishra, V., 2019. Decline in surface urban heat island intensity in India during heatwaves. Environmental. Research Communications 1, 031001.

Lee, D.O., 1984. Urban climates. Progress in Physical Geography: Earth and Environment 8, 1−31.

Lee, C.C., 2015a. The development of a gridded weather typing classification scheme. International Journal of Climatology.

Lee, C.C., 2015b. A systematic evaluation of the lagged effects of spatiotemporally relative surface weather types on wintertime cardiovascular-related mortality across 19 US cities. International Journal of Biometeorology.

Lee, C.C., 2017. Reanalysing the impacts of atmospheric teleconnections on cold-season weather using multivariate surface weather types and self-organizing maps. International Journal of Climatology 37, 3714−3730.

Lee, C.C., 2020. The gridded weather typing classification version 2: a global-scale expansion. International Journal of Climatology.

Lee, C.C., 2021. Gridded weather typing classification (GWTC). Available online: https://www.personal.kent.edu/~cclee/gwtc2global.html (accessed 09.04.21).

Lee, C.C., Sheridan, S.C., 2018. Trends in weather type frequencies across North America. npj Climate and Atmospheric Science.

Li, D., Bou-Zeid, E., 2013. Synergistic interactions between urban heat islands and heat waves: the impact in cities is larger than the sum of its parts. Journal of Applied Meteorology and Climatology. 52, 2051−2064.

Li, Y., Schubert, S., Kropp, J.P., Rybski, D., 2020. On the influence of density and morphology on the urban heat island intensity. Nature Communications.

Li, D., Sun, T., Liu, M., Wang, L., Gao, Z., 2016. Changes in wind speed under heat waves enhance urban heat islands in the Beijing metropolitan area. Journal of Applied Meteorology and Climatology. 55, 2369—2375.

Li, D., Sun, T., Liu, M., Yang, L., Wang, L., Gao, Z., 2015. Contrasting responses of urban and rural surface energy budgets to heat waves explain synergies between urban heat islands and heat waves. Environmental Research Letters 10, 054009.

Linnerud, K., Mideksa, T.K., Eskeland, G.S., 2011. The impact of climate change on nuclear power supply. The Energy Journal 32.

Liu, X., Tian, G., Feng, J., Ma, B., Wang, J., Kong, L., 2018. Modeling the warming impact of urban land expansion on hot weather using the weather research and forecasting model: a case study of Beijing, China. Advances in Atmospheric Sciences 35, 723—736.

Meehl, G.A., Tebaldi, C., 2004. More intense, more frequent, and longer lasting heat waves in the 21st century. Science (New York, N.Y.) 305, 994—997.

MetLink Royal Meteorological Society UK Climate—Metlink Weather & Climate Teaching Resources. Available online: https://www.metlink.org/secondary/key-stage-4/air/ (accessed 30.12.20).

Mihalakakou, G., Flocas, H.A., Santamouris, M., Helmis, C.G., 2002. Application of neural networks to the simulation of the heat island over Athens, Greece, using synoptic types as a predictor. Journal of Applied Meteorology 41, 519—527.

Morris, C.J.G., Simmonds, I., 2000. Associations between varying magnitudes of the urban heat island and the synoptic climatology in Melbourne, Australia. International Journal of Climatology 20, 1931—1954.

Morris, C.J.G.G., Simmonds, I., Plummer, N., 2001. Quantification of the influences of wind and cloud on the nocturnal urban heat island of a large city. Journal of Applied Meteorology 40, 169—182.

Nairn, J., Fawcett, R., 2013. Defining heatwaves, heatwave defined as a heat impact event servicing all community and business sectors in Australia. CAWCR Technical Report. http://cawcr.gov.au/publications/technicalreports/CTR_060.pdf.

Ngarambe, J., Nganyiyimana, J., Kim, I., Santamouris, M., Yun, G.Y., 2020. Synergies between urban heat island and heat waves in Seoul: the role of wind speed and land use characteristics. PLoS ONE 15, e0243571.

Nichol, J.E., Fung, W.Y., Lam, K., Wong, M.S., 2009. Urban heat island diagnosis using ASTER satellite images and 'in situ' air temperature. Atmospheric Research 94, 276—284.

NSW Government, 2019. Summer ozone episode from 11 to 12 February 2017. https://www.environment.nsw.gov.au/research-and-publications/publications-search/summer-ozone-episode-from-11-to-12-february-2017.

Oke, T.R., 1982. The energetic basis of the urban heat island. Quarterly Journal of the Royal Meteorological Society 108, 1—24.

Oke, T.R., 1987. Boundary Layer Climates, second ed., 435. *Routledge.*

Oke, T.R., Mills, G., Christen, A., Voogt, J.A., 2017. Urban Climates. Cambridge University Press, Cambridge, ISBN 9781139016476.

Oke, T.R., Runnalls, K.E., 2000. Dynamics and controls of the near-surface heat island of Vancouver, British Columbia. Physical Geography .

Patz, J.A., Campbell-Lendrum, D., Holloway, T., Foley, J.A., 2005. Impact of regional climate change on human health. Nature 438, 310—317.

Perkins, S.E., 2015. A review on the scientific understanding of heatwaves—their measurement, driving mechanisms, and changes at the global scale. Atmospheric Research 164—165, 242—267.

Perkins, S.E., Alexander, L.V., 2013. On the measurement of heat waves. Journal of Climate 26, 4500—4517.

Peters, A., Schneider, A., 2020. Cardiovascular risks of climate change. Nature Reviews Cardiology.

Philipp, A., Bartholy, J., Beck, C., Erpicum, M., Esteban, P., Fettweis, X., et al., 2010. Cost733cat—a database of weather and circulation type classifications. Physics and Chemistry of the Earth.

Physick, W., Cope, M., Lee, S., 2014. The impact of climate change on ozone-related mortality in Sydney. International Journal of Environmental Research and Public Health.

Półrolniczak, M., Kolendowicz, L., Majkowska, A., Czernecki, B., 2017. The influence of atmospheric circulation on the intensity of urban heat island and urban cold island in Poznań, Poland. Theoretical and Applied Climatology 127, 611−625.

Pyrgou, A., Hadjinicolaou, P., Santamouris, M., 2018. Enhanced near-surface ozone under heatwave conditions in a Mediterranean island. Scientific Reports 8.

Pyrgou, A., Hadjinicolaou, P., Santamouris, M., 2020. Urban-rural moisture contrast: regulator of the urban heat island and heatwaves' synergy over a Mediterranean city. Environmental Research 182, 109102.

Ramamurthy, P., Bou-Zeid, E., 2017. Heatwaves and urban heat islands: a comparative analysis of multiple cities. Journal of Geophysical Research 122, 168−178.

Ramamurthy, P., Bou-Zeid, E., Smith, J.A., Wang, Z., Baeck, M.L., Saliendra, N.Z., et al., 2014. Influence of subfacet heterogeneity and material properties on the urban surface energy budget. Journal of Applied Meteorology and Climatology. 53, 2114−2129.

Ramamurthy, P., Li, D., Bou-Zeid, E., 2017. High-resolution simulation of heatwave events in New York City. Theoretical and Applied Climatology.

Richard, Y., Pohl, B., Rega, M., Pergaud, J., Thevenin, T., Emery, J., et al., 2021. Is urban heat island intensity higher during hot spells and heat waves (Dijon, France, 2014−2019)? Urban Climate 35, 100747.

Rizvi, S.H., Alam, K., Iqbal, M.J., 2019. Spatio-temporal variations in urban heat island and its interaction with heat wave. Journal of Atmospheric and Solar-Terrestrial Physics 185, 50−57.

Rizwan, A.M., Dennis, L.Y.C., Liu, C., 2008. A review on the generation, determination and mitigation of urban heat island. Journal of Environmental Sciences 20, 120−128.

Robine, J.M., Cheung, S.L.K., Le Roy, S., Van Oyen, H., Griffiths, C., Michel, J.P., et al., 2008. Death toll exceeded 70,000 in Europe during the summer of 2003. Comptes Rendus—Biologies 331, 171−178.

Roderick, M.L., Hobbins, M.T., Farquhar, G.D., 2009. Pan evaporation trends and the terrestrial water balance. I. Principles and observations. Geography Compass.

Rogers, C.D.W., Gallant, A.J.E., Tapper, N.J., 2019. Is the urban heat island exacerbated during heatwaves in southern Australian cities? Theoretical and Applied Climatology 137, 441−457.

Russo, S., Sillmann, J., Sterl, A., 2017. Humid heat waves at different warming levels. Scientific Reports 7, 1−7.

Sailor, D.J., 2013. Energy buildings and urban environment, Climate Vulnerability, Vol. 3. Elsevier, pp. 167−182.

Santamouris, M., 2015a. Analyzing the heat island magnitude and characteristics in one hundred Asian and Australian cities and regions. The Science of the Total Environment 512−513, 582−598.

Santamouris, M., 2015b. Regulating the damaged thermostat of the cities—status, impacts and mitigation challenges. Energy and Buildings 91, 43−56.

Santamouris, M., 2016. Cooling the buildings—past, present and future. Energy and Building 128, 617−638.

Santamouris, M., 2020. Recent progress on urban overheating and heat island research. Integrated assessment of the energy, environmental, vulnerability and health impact. Synergies with the global climate change. Energy and Buildings 207, 109482.

Santamouris, M., Haddad, S., Fiorito, F., Osmond, P., Ding, L., Prasad, D., et al., 2017. Urban heat island and overheating characteristics in Sydney, Australia. An analysis of multiyear measurements. Sustainability 9, 712.

Santer, B.D., Mears, C., Wentz, F.J., Taylor, K.E., Gleckler, P.J., Wigley, T.M.L., et al., 2007. Identification of human-induced changes in atmospheric moisture content. Proceedings of the. National Academy of Sciences of the United States of America 104, 15248−15253.

Schatz, J., Kucharik, C.J., 2014. Seasonality of the urban heat island effect in Madison, Wisconsin. Journal of Applied Meteorology and Climatology.

Schatz, J., Kucharik, C.J., 2015. Urban climate effects on extreme temperatures in Madison, Wisconsin, USA. Environmental Research Letters 10, 094024.

Scott, A.A., Waugh, D.W., Zaitchik, B.F., 2018. Reduced urban heat island intensity under warmer conditions. Environmental Research Letters 13, 064003.

Seneviratne, S.I., Donat, M.G., Mueller, B., Alexander, L.V., 2014. No pause in the increase of hot temperature extremes. Nature Climate Change 4, 161−163.

Shahmohamadi, P., Che-Ani, A.I., Maulud, K.N.A., Tawil, N.M., Abdullah, N.A.G., 2011. The impact of anthropogenic heat on formation of urban heat island and energy consumption balance. Urban Studies Research.

Sheridan, S.C., 2002. The redevelopment of a weather-type classification scheme for North America. International Journal of Climatology.

Sheridan, S., 2021. Spatial synoptic classification (SSC). Available online: http://sheridan. geog.kent.edu/ssc.html (accessed 09.04.21).

Sheridan, S.C., Kalkstein, L.S., Scott, J.M., 2000. An evaluation of the variability of air mass character between urban and rural areas. Biometeorol. Urban Climatol. Turn Millenn. 487−490.

Sheridan, S.C., Lee, C.C., 2018. Temporal trends in absolute and relative extreme temperature events across North America. Journal of. Geophysical Research: Atmospheres.

Sheridan, S.C., Lin, S., 2014. Assessing variability in the impacts of heat on health outcomes in New York City over time, season, and heat-wave duration. Ecohealth 11, 512−525.

Stathopoulou, E., Mihalakakou, G., Santamouris, M., Bagiorgas, H.S., 2008. On the impact of temperature on tropospheric ozone concentration levels in urban environments. In: Proceedings of the Journal of Earth System Science.

Stone, B., Norman, J.M., 2006. Land use planning and surface heat island formation: a parcel-based radiation flux approach. Atmospheric Environment (Oxford, England: 1994).

Sun, T., Kotthaus, S., Li, D., Ward, H.C., Gao, Z., Ni, G.-H., et al., 2017. Attribution and mitigation of heat wave-induced urban heat storage change. Environmental Research Letters 12, 114007.

Szegedi, S., Kircsi, A., 2003. The effects of the synoptic conditions on development of the urban heat island in Debrecen, Hungary. Acta Climatologica et Chorologica 36−37, 111−120.

Taha, H., 1997. Urban climates and heat islands: albedo, evapotranspiration, and anthropogenic heat. Energy and Buildings 25, 99−103.

Tan, J., Zheng, Y., Tang, X., Guo, C., Li, L., Song, G., et al., 2010. The urban heat island and its impact on heat waves and human health in Shanghai. International Journal of Biometeorology 54, 75−84.

Tewari, M., Yang, J., Kusaka, H., Salamanca, F., Watson, C., Treinish, L., 2019. Interaction of urban heat islands and heat waves under current and future climate conditions and their mitigation using green and cool roofs in New York City and Phoenix, Arizona. Environmental Research Letters 14, 034002.

Tong, S., FitzGerald, G., Wang, X.-Y., Aitken, P., Tippett, V., Chen, D., et al., 2015. Exploration of the health risk-based definition for heatwave: a multi-city study. Environmental Research 142, 696–702.

Tong, S., Wang, X.Y., Yu, W., Chen, D., Wang, X.Y., 2014. The impact of heatwaves on mortality in Australia: a multicity study. BMJ Open 4, e003579.

Unger, J., 1996. Heat island intensity with different meteorological conditions in a medium-sized town: Szeged, Hungary. Theoretical and Applied Climatology 54, 147–151.

Unger, J., Skarbit, N., Kovács, A., Gál, T., 2020. Comparison of regional and urban outdoor thermal stress conditions in heatwave and normal summer periods: a case study. Urban Climate 32, 100619.

Unger, J., Sümeghy, Z., Gulyás, Á., Bottyán, Z., Mucsi, L., 2001. Land-use and meteorological aspects of the urban heat island. Meteorological Applications.

Vahmani, P., Ban-Weiss, G.A., 2016. Impact of remotely sensed albedo and vegetation fraction on simulation of urban climate in WRF-urban canopy model: a case study of the urban heat island in Los Angeles. Journal of. Geophysical Research: Atmospheres 121, 1511–1531.

Vanos, J.K., Cakmak, S., Bristow, C., Brion, V., Tremblay, N., Martin, S.L., et al., 2013. Synoptic weather typing applied to air pollution mortality among the elderly in 10 Canadian cities. Environmental Research 126, 66–75.

Voogt, J., Oke, T., 2003. Thermal remote sensing of urban climates. Remote Sensing of Environment 86, 370–384.

Wang, J., Yan, Z., Quan, X.-W., Feng, J., 2017. Urban warming in the 2013 summer heat wave in eastern China. Climate Dynamics 48, 3015–3033.

Ward, K., Lauf, S., Kleinschmit, B., Endlicher, W., 2016. Heat waves and urban heat islands in Europe: a review of relevant drivers. The Science of the Total Environment 569–570, 527–539.

Willett, K.M., Gillett, N.P., Jones, P.D., Thorne, P.W., 2007. Attribution of observed surface humidity changes to human influence. Nature 449, 710–712.

World Urbanization Prospects: The 2018 Revision, UN, 2019.

Wu, J., 2008. Making the case for landscape ecology: an effective approach to urban sustainability. Landscape Journal 27, 41–50.

Yarnal, B., 1993. Synoptic climatology in environmental analysis. A primer. Environment International 19, 529.

Yin, Q., Wang, J., 2017. The association between consecutive days' heat wave and cardiovascular disease mortality in Beijing, China. BMC Public Health 17, 223.

Yun, G.Y., Ngarambe, J., Duhirwe, P.N., Ulpiani, G., Paolini, R., Haddad, S., et al., 2020. Predicting the magnitude and the characteristics of the urban heat island in coastal cities in the proximity of desert landforms. The case of Sydney. The Science of the Total Environment 709, 136068.

Zhang, F., Cai, X., Thornes, J.E., 2014. Birmingham's air and surface urban heat islands associated with lamb weather types and cloudless anticyclonic conditions. Progress in Physical Geography 38, 431–447.

Zhang, N., Gao, Z., Wang, X., Chen, Y., 2010. Modeling the impact of urbanization on the local and regional climate in Yangtze River Delta, China. Theoretical and Applied Climatology.

Zhang, J., Rao, S.T., Daggupaty, S.M., 1998. Meteorological processes and ozone exceedances in the northeastern United States during the 12–16 July 1995 episode. Journal of Applied Meteorology.

Zhang, N., Zhu, L., Zhu, Y., 2011. Urban heat island and boundary layer structures under hot weather synoptic conditions: a case study of Suzhou City, China. Advances in Atmospheric Sciences 28, 855–865.

Zhao, L., Lee, X., Smith, R.B., Oleson, K., 2014. Strong contributions of local background climate to urban heat islands. Nature 511, 216–219.

Zhao, L., Oppenheimer, M., Zhu, Q., Baldwin, J.W., Ebi, K.L., Bou-Zeid, E., et al., 2018. Interactions between urban heat islands and heat waves. Environmental Research Letters 13, 9326.

Zhao, D., Wu, J., 2017. Contribution of urban surface expansion to regional warming in Beijing, China. Journal of Applied Meteorology and Climatology. 56, 1551–1559.

CHAPTER 4

Multiscale modeling techniques to document urban climate change

Negin Nazarian[1,2,3], Mathew Lipson[2] and Leslie K. Norford[4]
[1]School of Built Environment, Faculty of Arts, Design and Architecture, University of New South Wales (UNSW), Sydney, NSW, Australia
[2]City Futures Research Centre, University of New South Wales (UNSW), Sydney, NSW, Australia
[3]ARC Centre of Excellence for Climate Extreme, Sydney, NSW, Australia
[4]Department of Architecture, Massachusetts Institute of Technology, MA, United States

4.1 Introduction: why model urban and intra-urban climate change?

The World Meteorological Organization (WMO) reports that between 1970 and 2019, 79% of disasters worldwide involved weather, water, and climate-related hazards, noting an increase of 9% between 2010 and 2019 compared to the previous decade (World Meteorological Organization, 2020). This motivated four priority areas identified by the WMO World Weather Research Program, two of which closely focused on urban-related challenges: high-impact weather and urbanization (Masson et al., 2020b).

As the increase in urbanization and disruptions to weather patterns due to climate change continues, it is paramount that urban climate change is documented, understood, and further predicted such that we minimize adverse impacts on human life. In this context, technological advances in climate models play a critical role in analyzing and addressing the impacts of urban climate change. Urban climate models—focused on various scales and perspectives—provide important means to analyze complex physical processes forming urban climates and further quantify the ways climate change and urbanization have resulted, and will result, in local-scale modification of climate in the built environment. In this context, numerical, statistical, and physical scale models have made significant contributions to understanding and addressing climate challenges identified in cities. These contributions include the following.

Urban Climate Change and Heat Islands
DOI: https://doi.org/10.1016/B978-0-12-818977-1.00004-1
123

a. Better understanding of urban climate impact on humans and environments: Before addressing the need for urban climate modeling, it is important to acknowledge why the urban boundary layer (the lower atmosphere directly affected by the presence of cities and the primary focus of urban climate models) is deemed important to study and evaluate. While only 3% of the world's land surface is covered with urban areas, cities are responsible for local and global climate changes, impacting urban environments in which the majority of the planet's population lives. This emphasis on *people* shapes our modeling approach and leads to five major focus areas in modeling the impact of urban and intra–urban climate variability:

1. *Energy consumption*, dynamically interacting with local climate changes and through the release of anthropogenic waste heat (Santamouris, 2014).
2. *Urban wind and ventilation*, modified by urban form, fabric, and design (Allegrini et al., 2015).
3. *Thermal exposure*, exacerbated by Urban Heat Island (UHI) phenomena (Oke, 1982).
4. *Air quality*, dominated by urban pollutant sources such as transportation and urban energy use (Karagulian et al., 2015).
5. *High-impact weather*, with local and synoptic-scale weather events (heatwaves, cyclones, extreme wind, and rain) impacting different urban regions.

   Urban climate models assist us to isolate, dissect, and analyze complex interactions between the land, atmosphere, and built environment leading to these implications and address critical questions regarding human health, performance, and well-being.

b. Better planning through scenario-building and climate-projection analysis: In addition to attaining a better understanding of physical processes and the state of the current and past climate, modeling assists us in quantifying the interaction between larger-scale climate change drivers and patterns with city-scale weather and climate implications. Urban climate models allow us to test and draw conclusions on future projected climate systems and the intersection that they are expected to have with the urban population. This consequently paves the way for assessing mitigation and adaptation strategies that aim to respond to urban and intra-urban climate impacts on human life. With the help of urban climate modeling, the physical hypothesis of various mitigation and adaptation strategies can be tested in the complex urban climate system that constitutes thermodynamic, aerodynamic, radiative, and moisture dynamics.

c. Providing spatial and temporal distribution of urban climate parameters: The urban microclimate is shaped by background climate, urban form and fabric, and human activities, which have all been rapidly changing over the last century. The modification of land cover, the presence of built materials, the shape of urban structures, and anthropogenic heat and emissions influence airflow, the thermal environment, and moisture availability in cities (Grimmond et al., 1986; Nazarian et al., 2018c; Roth and Lim, 2017; Sanaieian et al., 2014). Urban climate changes have large spatial and temporal distribution across different city neighborhoods, that is, intraurban climate variability (dos Santos et al., 2017; Harlan et al., 2006). The most robust method to document such intraurban variabilities in microclimate is to deploy a high-resolution sensor network that provides spatial and temporal microclimate data. However, such networks—even with current advances in Internet-of-Things technologies and low-cost sensing (Pantelic et al., 2021)—have proven to be difficult to implement, leaving a significant gap in our knowledge of three-dimensional atmospheric processes across various scales. Urban climate modeling is therefore helpful in filling this data gap, providing critical tools to understand the three-dimensional and ever-evolving distribution of urban climate processes across various scales.

Focusing on these critical characteristics, this chapter aims to provide an overview of the state-of-the-art in modeling urban and intraurban climate variability at various scales in cities and documenting the interaction with larger-scale climate change impacts. To achieve this objective, we first discuss various modeling techniques that assist us in evaluating urban climate processes at various scales and levels of complexities. Forcing on numerical models, we then discuss how key physical processes (such as vegetation and anthropogenic heats and water wastes) are infused in state-of-the-art modeling techniques across different scales. Lastly, we provide a more in-depth review of modeling approaches applied to simulate various implications of urban and intraurban climate, ranging from thermal exposure to high-impact weather.

4.2 Modeling techniques to document urban and intraurban climate variability and change

Modeling urban systems can be challenging due to the complexity of the urban landscape and fabric, and the wide range of scales over which important processes act. Here, we review various physical, statistical, and computational modeling techniques, applied across street-to-regional scales and with different levels of complexity.

4.2.1 Scale models

Scaled physical models can be used to simulate urban environment effects in laboratory-controlled or outdoor natural weather conditions. Geometric, dynamic, or thermal similarity between the scale model analog and real-world environments are required depending on the process being studied. For example, geometric similarity is required for studies of radiative exchange, dynamic similarity is required for studies of atmospheric flow, and thermal similarity is required in studies of surface − atmosphere energy exchange.

Within a laboratory, *wind tunnels* can be used to study airflow in urban landscapes, for example, to show how individual building shape affects pedestrian-level comfort (Blocken et al., 2016), how street canyons and intersections affect near-surface flow and dispersion (Ahmad et al., 2005; Barlow and Belcher, 2002), or how flow and temperature interact in stable or unstable conditions (Allegrini et al., 2013; Uehara et al., 2000). *Water flumes and channels* have been used for similar purposes, modeling flow inside and around urban street canyons (Chew et al., 2017; Li et al., 2008). Static *water tanks* can simulate near-calm conditions, with the advantage that they can be more easily stratified by differential heating, or by altering the density of fluids with additives like salt (Yin et al., 2017). Water tanks have been used to study larger-scale buoyancy-induced urban circulations over single or adjoining urban areas (Fan et al., 2018; Lu et al., 1997).

Outdoor models rely on natural weather for their forcing so are typically exposed to greater variability than laboratory experiments. Smaller outdoor models can be used to study radiation exchange and temperature evolution (Chen et al., 2020; Mills, 1997). However, in order to study airflow outdoors model scales must be large to satisfy dynamic similarity, for example, the ⅕ scale COSMO experiment (Kanda et al., 2006), or even purely natural urban analogs (Wang et al., 2018a). Research using physical scale models provides a critical connection to real-world processes, with results often forming the theoretical and empirical underpinning for statistical or numerical techniques described below.

4.2.2 Statistical methods

Statistical empirical methods use existing observed or modeled data to develop regressions or other statistical relations and apply them in other situations. (Garuma, 2018; Landsberg, 1981). The scale of processes represented by a statistical model can vary widely; from single urban surfaces through to entire urban environments. Implementation of statistical models in mesoscale weather forecasting systems or downscaling for local weather analysis provide means to generate rapid predictions and evaluations without excessive computation load, which is often needed for emergency weather response as well as multiscale analyses of urban climate. Statistical methods are also the key technique that allows regional, mesoscale, and microscale models to adapt the information provided by climate projections to spatial and time scales appropriate for studying urban climate implications. Statistical downscaling methods are used for obtaining climate forcing data at smaller scales that assist in simulating urban rainfall (Maraun and Widmann, 2018; Willems, 2012) as well as UHI and building energy consumption (Schoetter et al., 2020a) in the face of climate change.

Statistical methods can be inaccurate if relations do not appropriately capture the complexity or variety of processes being modeled, for example, simple regressions often poorly represent extreme conditions or complex systems. In addition, as statistical methods rely on previously observed or modeled data and are not based on "first-principle" physical laws, the approach is not necessarily valid where conditions differ from training data, for example, with alternative urban morphology or with an altered global or regional climate. Nonetheless, at microscales (e.g., wall-atmosphere convective heat transfer), statistical methods are widely used in numerical modeling of urban climate described below.

4.2.3 Numerical methods

The most common approach of urban climate modeling, and most thoroughly explored here, is numerical simulation. Numerical models follow a series of physical and mathematical formulations evolving through time to simulate important drivers of urban climate including local structure, fabric, and metabolism of the built-up areas that result in the modification of atmospheric turbulence, surface energy, and water balances. Key considerations in numerical modeling are two aspects of scale: domain size and resolution. A large model domain allows more dynamic interactions and feedbacks across scales, while higher resolution allows more heterogeneity and local interactions.

All numerical models simplify landscapes through abstractions, each making different design decisions based on the scales of the processes being considered along with computational and informational constraints. Simpler models that use higher levels of abstraction are faster to run, but may exclude processes or dynamics important for the subject at hand. More complex models require greater computational resources and require more information to appropriately configure them, but can still perform worse than simpler models, particularly if input parameters and boundary conditions are not correctly prescribed (Best and Grimmond, 2015). It is therefore critical in the design of any experiment that an appropriate model type and scale are chosen, the model can be appropriately configured with available information, and it can be evaluated as being fit for the intended purpose (Baklanov et al., 2009; Krayenhoff et al., 2021).

A broad distinction between urban model types are the *energy balance models* (EBM) applied to surfaces and urban canopy processes, which do not use the Navier − Stokes equations to describe fluid flow, and *computational fluid dynamic* (CFD) models, which do. Here we first discuss simpler models designed for use in large domain weather and climate simulations up to global scales. Urban models used in larger domains are necessarily simpler, with lower resolution, than models designed for understanding microscale impacts. We progressively work toward more complex models with higher resolution and computational expense, and which are unsuitable for large domain simulations. Fig. 4.1 summarizes the schematic representation of various approaches to modeling urban environments, from the simplest one-dimensional slab models used as lower boundary conditions in larger simulations, through to building resolving models with fluid dynamics in three dimensions. A sample of existing models in each category is given in Table 4.1.

4.2.3.1 Surface and urban canopy energy balance models
One-dimensional approaches (slab or bulk models)
One-dimensional urban models represent various urban areas as horizontal, slab-like materials interacting with the atmospheric boundary layer.

urban representation

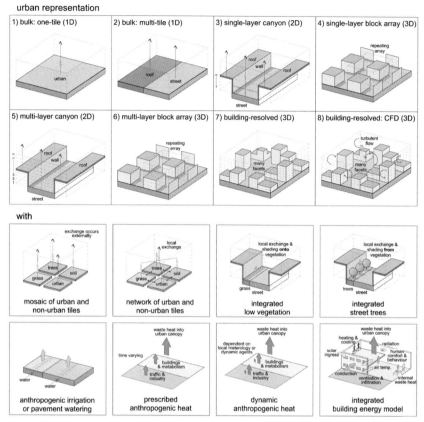

Figure 4.1 A range of approaches to numerical modeling for urban environments, from the simplest one-dimensional slab models used as lower boundary conditions in larger simulations, through to building resolving models with fluid dynamics in three dimensions. Each approach has advantages and limitations. Classes 1–6 would typically represent a single grid in a larger domain simulation, while classes 7 and 8 often represent the entire simulation domain. Models can also integrate important evapotranspiration and anthropogenic fluxes of heat and water in various ways (lower boxes).

These *bulk models* are attractive where the model domain is necessarily large, or where computational efficiency is a high priority. For these reasons, bulk models have been used in regional or global climate change modeling, and in numerical weather prediction where computational efficiency is critical (Best, 2006). The simplest represents all urban surfaces together in one tile as shown in Fig. 4.1 (1) (Best, 2005; Fortuniak, 2003; Myrup, 1969; Wouters et al., 2015). For bulk surface − atmosphere energy exchange, one-tile schemes can perform well compared with

Table 4.1 Examples of numerical urban climate models categorized by scale and level of complexity (as shown in Fig. 4.1).

	Urban representation	Example models
Bulk		
1	One–tile	SUEB (Fortuniak, 2003); BEST1T (Best, 2005); TERRA-URB (Wouters et al., 2015), UrbClim (Ridder et al., 2015)
2	Multitile	MORUSES (Porson et al., 2010); SUEWS (Järvi et al., 2011)
Building averaged		
3	Single-layer canyon	TEB (Masson, 2000); SLUCM (Single Layer Urban Canopy Model; Kusaka et al., 2001); CLMU (Oleson et al., 2008); VUCM (Lee and Park, 2008); UCLEM (Lipson et al., 2018); CAT (Erell and Williamson, 2006); UWG (Bueno et al., 2013)
4	Single-layer block array	SUMM (Kanda et al., 2005); SPUC (Aoyagi and Seino, 2011)
5	Multilayer canyon	BEP (Martilli et al., 2002); BEP-BEM (Salamanca et al., 2010); BEP-TREE (Krayenhoff et al., 2020); UT&C (Meili et al., 2020), VCWG (Moradi et al., 2019); TEB multilayer (Schoetter et al., 2020b), MLUCM (Nazarian et al., 2020a)
6	Multilayer block array	CM (Kondo et al., 2005)
Building resolved		
7	3D energy balance models	TUF-3D (Krayenhoff and Voogt, 2007); TUF-IOBES (Yaghoobian and Kleissl, 2012); VTUF-3D (Nice et al., 2018)
8	Computational fluid dynamics (CFD)	ENVI-Met (Huttner and Bruse, 2009); PALM4U (Maronga et al., 2020); ANSYS Fluent; OpenFOAM (Jasak, 2009)

more complex models (Best and Grimmond, 2015). However, experimental data suggest that representing an urban area as a single material poorly captures the diurnal evolution of the local urban energy balance because thermal and radiative processes over roof surfaces are fundamentally different from higher-dimensional areas such as street canyons (Porson et al., 2009). For this reason, bulk models have been extended to include multiple tiles within a grid cell (each tile representing a surface type in one dimension, such as roof, street canyon, or vegetation, Fig. 4.1 (2)) and blending their output fluxes externally (in the first atmospheric level) or internally through a simple resistance network (Järvi et al., 2011; Porson et al., 2010).

Whether bulk models include one or multiple tiles, they only consider the vertical exchanges of heat, radiation, and momentum between the surface and atmosphere, and consist of a surface heat budget equation for simulating surface temperature (Kusaka et al., 2012). The impact of varying urban geometry on aerodynamic roughness, thermal inertia, moisture availability, and albedo is reduced to single parameter values for each tile. Because these bulk parameters represent an aggregate of many structures, and compress vertical urban geometry to a flat plane, it becomes difficult to assign appropriate values to them as those values cannot be directly measured. Bulk parameters are therefore often "tuned" for each model, a process of adjusting parameters to reduce differences between model outputs and available observations. This can be effective, but presents a major limitation of one-dimensional models—how to configure parameters in situations outside of the environment for which they were tuned? Approaches have been developed to address this through additional empirical and theoretical formulations that translate higher-dimensional information into bulk parameters (Porson et al., 2010; e.g., Wouters et al., 2016), or define bulk parameters from satellite observations and other data sources (Lindberg et al., 2018).

Even where one-dimensional models are able to reasonably simulate bulk surface − atmosphere exchange of energy, moisture, and momentum, they cannot explicitly represent conditions where people live (at street level or within buildings). Their use is therefore often limited to acting as lower boundary conditions in large-scale weather and climate simulations where microscale, pedestrian-level conditions are not required.

Building-averaged approaches (canyon and block array models)

In an attempt to represent impacts of building morphology and street-level processes more explicitly while avoiding the high computational cost needed to resolve individual buildings, a middle class of urban models have been developed, which represents "building-averaged" neighborhood-scale conditions in two and three dimensions. These urban canopy models (UCMs) are now used extensively in mesoscale urban weather modeling systems such as weather research and forecasting model (WRF) (Chen et al., 2011) and are even implemented in a handful of global climate modeling systems (Katzfey et al., 2020; Oleson et al., 2008).

The simplest of this class is based on a two-dimensional street canyon configuration and includes a single well-mixed atmospheric layer between buildings (Kusaka et al., 2001; Masson, 2000; Oleson et al., 2008). *Single-layer canyon models* [Fig. 4.1 (3−4)] greatly simplify the equations used to describe shading from buildings, radiation trapping, and in−canyon airflow, and need only three facets (roof − wall − street) to represent the urban system. More complex single-layer canyon models can include additional interactions such as sunlit and shaded walls (Thatcher and Hurley, 2012), evapotranspiration from low vegetation (Lemonsu et al., 2012), shading, and radiative trapping from street trees (Lee and Park, 2008), or the exchange of air and energy into building volumes (Lipson et al., 2018).

Single-layer canyon model studies report critical advantages over slab approaches including (1) better representation of urban boundary layers in regional-scale meteorological models (Trusilova et al., 2013), (2) ability to assess the interaction between scales and processes [such as the interaction between canopy temperature and building energy (Bueno et al., 2011a) or impact of UHI on regional atmospheric pollution (Sarrat et al., 2006)], and (3) quantification of intraurban variability in such microclimate parameters as air temperature (Erell and Williamson, 2007)—without excessive computational resources.

Expanding the two-dimensional representation of urban physics, another category of UCMs is focused on the vertical exchange of heat, energy, and turbulence within and above the urban canopy [Fig. 4.1 (5−6)]. In contrast with single-layer models that assume well-mixed air within the canopy, multilayer urban canopy models [MLUCMs (Martilli et al., 2002)] recognize the vertical variability of urban fluxes and have several layers within the canopy with different source and sink terms considered in horizontal and vertical surfaces. These considerations permit a more process-based treatment of urban canopy physics by solving prognostic momentum, turbulent kinetic energy (TKE), and energy equations, and therefore allow prediction of vertical profiles of urban meteorological variables in urban streets. Several examples of MLUCMs, Building Effect Parameterization (BEP) (Martilli et al., 2002), and Canopy Model (CM) (Kikegawa et al., 2003), or the multilayer Town Energy Budget (TEB) (Schoetter et al., 2020b), have been incorporated in several regional meteorological models and were shown to accurately predict urban canopy temperature (Chen et al., 2011). Compared to Single Layer Urban Canopy Model (SLUCMs) and slab models in mesoscale analysis, multilayer urban canopy representations were shown to better predict the vertical profiles of air temperature, wind speed, and TKE, particularly as it accounts for the turbulence and multireflection in the urban canopy (Jandaghian and Berardi, 2020; Nazarian et al., 2020a; Teixeira et al., 2019). They are therefore used successfully to assess high-resolution intraurban variability in urban climate by incorporating detailed maps of local climate zones (Stewart and Oke, 2012) in various cities (Brousse et al., 2016; Ribeiro et al., 2021). Nonetheless, it should be noted that computational resources required for multilayer models are higher than for SLUCMs and slab models, with Jandaghian and Berardi (2020) estimating an average 20% increase when multilayer parameterization is used.

An advantage of canyon representations over bulk schemes is that model parameters such as model building morphology and materials are more directly related to real-world analogs, making it easier to explore changes to urban systems. However, canyon models still rely heavily on empirical parameterizations and abstraction, effectively representing infinitely long streets and buildings (within the limit of each grid cell). A building-averaged approach can be extended to three dimensions through a repeating *block array model* (Kanda et al., 2005). This increases the representation to include four walls, a roof, and a road (i.e., six facets). Although a seemingly promising approach balancing computational expense with realistic geometry, single-layer block array models have been implemented in only a few mesoscale models (e.g., Aoyagi and Seino, 2011). Multilayer block array models (e.g., Kondo et al., 2005) have also been coupled with larger-scale models and used in studies of urban weather processes (Nakajima et al., 2021).

Building-resolved approaches

A final level of complexity for models that exclude the Navier − Stokes equations is *building resolved multilayer* EBMs [Fig. 4.1 (7)]. One of the early examples of this approach is the microscale three-dimensional urban EBM TUF-3D (Krayenhoff and Voogt, 2007) developed for assessing the distribution of temperature and energy balance on urban facades. The key advantage of building-resolved approaches is the ability to assess the realistic distribution of surface heating as well as determining the interbuilding radiation exchange, which is critical for thermal comfort and energy balance analyses in the urban canopy. For instance, building-resolving numerical approaches can assess the indoor and outdoor energy exchange driven by surface heating (Yaghoobian and Kleissl, 2012), quantify the impact of placement, height, and density of urban vegetation in modifying microscale urban climate (Nice et al., 2018), evaluate mitigation strategies such as cool walls (Nazarian et al., 2019b), and assess pedestrian comfort in the urban canopy (Lachapelle et al., 2020; Nazarian et al., 2018c). However, due to their high computation cost as well as complex and three-dimensional interbuilding processes considered, building-resolving models are often limited to idealized configurations or study areas no larger than a neighborhood. Therefore before opting for the most detailed representation of surface energy balance in these models, it is critical to consider if the subject at hand warrants investing the high computational cost and assumptions made in urban representation.

4.2.3.2 Computational fluid dynamic models
Three-dimensional modeling of urban airflow and thermal environment

Three-dimensional simulations of airflow and thermal exchange occur primarily at two scales: micro and regional scale. At the regional scale, it has not been possible to explicitly resolve building- and street-scale processes and, at the same time, span a domain large enough to assess regional climatic impacts in cities and communities. Accordingly, the representation of urban canopy processes is done through energy balance parameterization methods previously described, with only larger-scale atmospheric fluid processes resolved in three dimensions. At the microscale, however, decades of research on CFD [Fig. 4.1 (8)] have enabled high-resolution (sub-meter), building-resolving simulations of urban climate processes, particularly focusing on the detailed distribution of urban airflow, turbulence, and thermal environment (Toparlar et al., 2017). CFD models are developed to resolve the transfer of heat and mass and assess their interaction with individual obstacles such as buildings, vegetation, and other roughness elements. Accordingly, they enable us to simulate the three-dimensional time-dependent horizontal and vertical wind and temperature fields (Nazarian et al., 2018a), pedestrian wind comfort and safety (Blocken et al., 2012), and distribution and dispersal of pollutants in the street canopy (Tominaga and Stathopoulos, 2013). CFD models of urban flow can further be categorized in three major methods based on the approximate form of the governing Navier − Stokes equations solved and the turbulence (subgrid-scale) model used: (1) Reynolds-averaged

Navier – Stokes (RANS) models (modeling time-averaged turbulence representation, therefore requiring lower computational cost), (2) Large Eddy Simulations (LES) models (parameterizing the smallest length scales, which are the most computationally expensive to resolve, and resolving larger-scale eddies), and (3) Direct Numerical Simulations (DNS) models (solving the Navier–Stokes equations without any turbulence parameterizations at any scale). While DNS models are often used for enhancing our theoretical understanding of turbulent flow in urban canopies, their computational cost is prohibitive (Coceal et al., 2007), so LES and RANS models are frequently used for assessing critical implications of urban climate on human life (Giometto et al., 2017; Nazarian et al., 2018b). The performance of RANS and LES models, and their choice in urban climate assessments, vary depending on the application area but is also based on study parameters such as the building configuration under study (Blocken, 2018).

Overall, although microscale CFD simulations offer the most detailed numerical solutions of urban climate processes, it is important to note that their performance and accuracy highly depend on initial intraurban and boundary conditions, which are often unknown with sufficient density or are computationally expensive to implement for long-term analyses. One of the recent methods to address the latter concern is to apply a clustering method to yearly weather files with the aim of obtaining representative boundary conditions for urban microclimatic models, or typical-day weather situations (Acero et al., 2020). Clustered weather day types or CFD calculations limited to a small set of wind directions both attempt to support rapid design and analysis by reducing the number of CFD calculations. Another approach is to simply perform the calculations faster, via trade-offs of accuracy for speed, accessing a computer's graphical processing unit (GPU), or employing machine learning techniques. Chronis et al. (2011) note that the use of the reduced-computation-order fast fluid dynamics method (Jin et al., 2013) to solve the Navier – Stokes equations in a parallel-computing environment can reduce computation times by factors of 1000. Deep neural networks have out-performed other models of the stress tensor that represent turbulence (Duraisamy et al., 2019; Nathan Kutz, 2017) and a convolutional neural network estimated 2D and 3D nonuniform steady laminar flow two orders of magnitude faster than a GPU-based solver and four orders of magnitude faster than a CPU-based solver, both using the Lattice Boltzmann method (Guo et al., 2016).

4.2.3.3 Inclusion of vegetation in numerical models

Soil, vegetation, and tree canopies play critical roles in determining local urban climate conditions (Bowler et al., 2010) and surface − atmosphere energy exchanges (Grimmond et al., 2011) through water storage and evapotranspiration. Many of the first-generation UCMs did not include an explicit representation of soil, vegetation, or tree canopies (Kusaka et al., 2001; Martilli et al., 2002; Masson, 2000), and so excluded most evapotranspiration processes. A traditional approach to overcome these dry urban landscapes has been to tile an impervious urban scheme with one or a series of nonurban (vegetated) schemes, blending their outputs externally and re-applying new atmospheric conditions at the next time-step. While common, and broadly representing subgrid heterogeneity of pervious and impervious surfaces, this approach does not capture important local-scale interactions between vegetation and urban elements, like overshadowing onto, or shadowing from, vegetation canopies, trapping of longwave radiation within street canyons, alteration of wind flow and turbulent mixing in and above the urban canopy and vertically distributed, in-canopy exchange of water, heat, and pollutants.

In response, the last decade has seen concerted community efforts focused on better integrating local-scale evapotranspiration processes into existing and new UCMs, including:

1. Multitile bulk schemes with integrated grass, tree, and soil tiles that capture local water exchange and evapotranspiration processes through an exchange network, for example, surface urban energy and water balance scheme (SUEWS) (Järvi et al., 2011).
2. Canyon schemes with integrated low vegetation that include overshadowing from buildings and can better represent the true shape of canyons with street vegetation, for example, TEB (Lemonsu et al., 2012), and urban climate and energy model (UCLEM) (Thatcher and Hurley, 2012).
3. Single-layer models with a tree canopy within the canyon that include shading of trees onto walls and roads, and longwave trapping by tree canopies, for example, vegetated urban canopy model (VUCM) (Lee and Park, 2008), a single-layer urban canopy model (ASLUM) (Wang et al., 2021), or TEB (Redon et al., 2020).
4. Multilayer models able to represent the vertical distribution of tree foliage and environmental processes at various levels of the urban canopy, with particular interest at pedestrian level, for example, BEP-Tree (Krayenhoff et al., 2020) or UT&C (Meili et al., 2020).

5. Three-dimensional, multilayer building resolving EBMs, for example, VTUF-3D (Nice et al., 2018), that are mostly focused on vegetative physiological processes and shading effects of trees that affect the energy balance of urban canopy.

6. Three-dimensional CFD models that resolve airflow and radiation through street trees and further analyze the radiation impact of tree foliage through ray tracing, for example, ENVI-met (Huttner and Bruse, 2009), PALM4U (Maronga et al., 2020), but of course are more computationally expensive to run and cannot generally be used in larger-scale meteorological simulations.

Table 4.2 summarizes the above approaches, indicating the level of complexity incorporated in several commonly used models, as well as the range of computational resources required.

4.2.3.4 Inclusion of anthropogenic waste heat and water fluxes in numerical models

Anthropogenic heat is the waste produced when energy is consumed. It is generated from all electrical appliances, from building heating and cooling, combustion for water heating, cooking, vehicle engines, and metabolic heat from our bodies. Water is also emitted into urban atmospheres by human activities, including combustion, metabolism, and irrigation. Anthropogenic heat fluxes vary greatly in their magnitude, with neighborhood-scale mean values ranging from $5-500 \text{ Wm}^2$, with daily, weekly, and seasonal cycles. Anthropogenic emissions increase temperature and humidity and can generate buoyancy-induced flows and reduce atmospheric stability. Anthropogenic emissions, therefore, influence thermal conditions at the pedestrian level, but also convection-induced breezes, local air quality, and precipitation events, all important for the comprehensive assessment of urban climate change across scales.

There are many methods that can be used to include anthropogenic emissions of heat and water in urban climate models. For heat, the simplest method is to inject an energy flux into the urban canopy air layer based on estimates of total energy consumption. Water fluxes into the atmosphere and soil can also be estimated, although with more difficulty (Sailor, 2011). These *prescribed* anthropogenic fluxes have been implemented in urban climate models as time-invariant, or with diurnal, weekly, monthly, or seasonal cycles. Total energy consumption can be estimated based on an inventory of local data for electricity, gas, and vehicle fuel use.

Table 4.2 A selection of models and their capability to represent vegetation (refer to text for model references). The range of complexity means a typical simulation year can take from seconds for simpler models, up to months for the three-dimensional CFD models (indicative only, dependent on timestep, domain, inclusions, and computational resources).

Example models:	BEST1T	SUEWS	UCLEM	TEB	ASLUM	UT&C	BEP	VTUF-3D	ENVImet	PALM4U
Urban representation (Fig. 4.1)	1	2	3	3	3	5	5	7	8	8
Externally tiled vegetation	Y	Y	Y	Y	Y	Y	Y			
Internally tiled vegetation		Y	Y	Y	Y	Y	Y	Y	Y	Y
Low canyon vegetation			Y	Y	Y	Y	Y	Y	Y	Y
Green roof vegetation					Y	Y			Y	Y
Street tree canopy				Y	Y	Y	Y	Y	Y	Y
Indicative run-time (1 year)	seconds	minutes	minutes	minutes	minutes	hours	hours	hours	months	months

Alternative estimates of anthropogenic heat are available with 1 km and hourly resolution globally (Dong et al., 2017). These regional or global products take a "top-down" approach, where country-level statistics for annual energy usage are separated spatially by population density and night light data, and temporally by empirically based assumptions of diurnal and seasonal variation. Fluxes may be separated into the building, transport, metabolic, and industrial fluxes, each with their own prescribed variation.

Although a prescribed approach is relatively common in urban modeling, a disadvantage is that the dynamic interplay between local climate and anthropogenic inputs is not captured. Energy consumed in heating and cooling buildings, or water used in irrigation, depends on local climatological conditions, but their input also affects local climatological conditions, for example, by increasing air temperatures and humidity, and introducing local buoyancy-driven flows. An alternative to a prescribed top-down approach is to include "bottom-up" submodules that dynamically model anthropogenic fluxes depending on local conditions. *Dynamic* approaches include:

1. Applying linear regressions of energy consumption on local air temperature and applying the related anthropogenic heat flux into the urban canopy, for example, SUEWS (Järvi et al., 2011).
2. Calculating energy required to maintain a prescribed temperature for the inside wall and roof surfaces and releasing a corresponding heat flux into the urban canopy, for example, original TEB (Masson, 2000), the community land model urban (CLMU-4) (Oleson et al., 2008), and ATEB (Thatcher and Hurley, 2012).
3. Integrating relatively simple building energy models (BEM) within urban climate models that are able to be coupled with larger-scale meteorological models, for example, CM-BEM (Kikegawa et al., 2003), BEP-BEM (Salamanca et al., 2010), TEB-BEM (Bueno et al., 2011b), UCLEM (Lipson et al., 2018), CLMU-5 (Oleson and Feddema, 2020), and DCEP-BEM (Jin et al., 2021).

Fully featured BEMs with detailed representation of building fabric, geometry, heating, and cooling systems (e.g., EnergyPlus) can be used to resolve individual buildings in a neighborhood, or a series of archetypal building types, and to scale up results to city-wide predictions of energy use (Reinhart and Cerezo Davila, 2016). These detailed BEMs can account for shading from neighboring buildings or vegetation and the use of archetypes reduces the computational burden of the neighborhood- to city-scale building energy modeling, but the use of these BEMs is typically domain-limited and does not capture dynamic interactions of energy-use/climate at the city or larger scales. The simpler BEMs mentioned above, integrated within building-averaged UCMs and coupled with larger-scale meteorological models, can capture interactions across scales. Many studies have used these integrated BEMs to evaluate interactions of energy use and urban climate in current and future climate conditions (Kikegawa et al., 2014; Lipson et al., 2019; Ortiz et al., 2018; Salamanca et al., 2014; Takane et al., 2019; Wang et al., 2018b).

The simplicity of integrated BEMs/UCMs and a building-averaged approach does present problems, particularly in capturing a range of human behaviors related to energy use, and the appropriate representation of both conditioned and unconditioned spaces that exist in neighborhoods and across cities. This can lead to an overestimation of total energy use, especially when all internal spaces are assumed to be conditioned (Takane et al., 2017). Recent progress in this regard includes parameterizing a variety of building types and human behaviors (Schoetter et al., 2017), a statistical representation of human behaviors with a subgrid tiling of conditioned and nonconditioned spaces (Lipson et al., 2019, 2018), and the integration of dynamic agent-based modeling for building occupancy, heating—cooling demands, appliance and vehicle use (Capel-Timms et al., 2020). In these cases, the overestimation of energy use often found in typical building-averaged approaches was reduced, and a better representation of energy demand variability was possible.

Few global climate models currently include any explicit representation of urban areas, let alone dynamic anthropogenic fluxes or integrated building energy modeling (Daniel et al., 2019). There are now growing calls for better integration of urban resolving models within global climate models (Sharma et al., 2021). The wide range of necessary scales presents significant technical, computational, and data availability challenges, although some recent advancements have been made. The CLMU urban model was recently extended with a BEM and run globally with the Community Earth System Model from 1850 to 2005, finding total simulated anthropogenic heat emissions related to heating and air conditioning close to global estimates (Oleson and Feddema, 2020). The Australian town energy budget (ATEB) urban model, using a simpler fixed internal building temperature parameterization, was integrated with the Conformal Cubic Atmospheric Model (CCAM) and run globally with climate projections from 1980 to 2019, finding the inclusion of an urban parameterization had climate impacts within and around cities, even in areas far from any urbanization

(Katzfey et al., 2020). UCLEM, which extends ATEB with a BEM, was used in an ensemble of single column climate projections from 2000 to 2099, quantifying the increase in peak electricity demands in Melbourne from a warming climate, but finding human behaviors and ownership patterns had greater impact (Lipson et al., 2019). These integrated modeling systems show great promise in assessing climate interactions across broad scales of space and time. However, a major challenge for global urban simulations remains the availability of the input data needed to describe cities in models (Masson et al., 2020a).

4.2.4 Summary and review of modeling techniques

Modeling techniques for assessing urban climate processes and quantifying the impact of current and projected climates on the urban boundary layer have evolved dramatically over the last three decades. To date there is no single model that can address the wide range of objectives and implications on human life, and therefore the choice of appropriate level of detail is based on the application area the model is used. Depending on the scales and parameters of interest (such as UHI, pedestrian wind speed, or surface energy exchange), different modeling techniques are deemed suitable following "Fitness-for-Purpose" guidelines (Baklanov et al., 2009). For instance, single-layer UCMs may prove sufficient or computationally viable to assess the impact of various urban land cover in regional and global climate analyses, but fall short in providing detailed air pollution or heat exposure analysis at the pedestrian level. Accordingly, it is critical that each modeling method is considered independently based on the problem at hand.

Here, in addition to summarizing different methods and scales in representing different urban climate processes, we provide a comprehensive list of review articles (Table 4.3) that detail the development and performance of different modeling approaches, as well as identifying the primary urban impact that they tend to focus on.

4.3 Modeling urban climate's impact on human life

In addition to developing techniques and methodologies to model urban climate change across different scales and with different levels of complexity, modeling approaches have been used to address critical questions in the face of climate change. This section aims to review some of the key modeling approaches applied to simulate various implications of urban

Table 4.3 Summary of review articles detailing various techniques for urban climate modeling at different scales.

Reference	Title	Technique	Scale	Impact focus	Description
Plate (1999)	Methods of investigating urban wind fields—physical models	Physical	Reduced-scale/ microscale	Urban ventilation.	Review of wind tunnel experiment on geometrically similar models of the urban area
Ahmad et al. (2005)	Wind tunnel simulation studies on dispersion at urban street canyons and intersections—a review	Physical	Street scale	Air quality, urban ventilation	Review of wind tunnel experiments in light of street ventilation, turbulence, traffic emission. and air quality
Kanda (2006)	Progress in the scale modeling of urban climate: Review	Physical	Reduced-scale/ microscale	Air quality, urban ventilation	Review of scale modeling studies, both in laboratory and outdoor conditions. Discussion on similarity requirements, as well as 40 scale model studies on flow, dispersion, and energy exchange
Willems et al. (2012)	Climate change impact assessment on urban rainfall extremes and urban drainage: Methods and shortcomings	Statistical	Global to urban catchment scale	High-impact weather	Review of methods for statistical downscaling of urban climate model output to urban catchments and assessing climate change impacts on rainfall in urban areas
Brown (2000)	Urban parameterizations for mesoscale meteorological models	Numerical/ statistical	Micro to regional scale	Thermal exposure, urban ventilation	Review of early developments of parameterizations to improve urban surface characteristics in mesoscale models
Hamdi et al. (2020)	The state-of-the-art of urban climate change modeling and observations	Numerical/ statistical	Regional to global scale	Thermal exposure	Review of urban climate models and interaction with global- and regional-scale change and detailing urban observations that assist in validating these models in different climates

Reference	Title	Numerical/ statistical	Scale	Application	Description
Masson et al. (2020b)	Urban climates and climate change	Numerical/ statistical	Micro to regional scale	Weather and climate	Comprehensive review of urban climate change across scales, including descriptions of numerical and statistical modeling methods, observation, and important urban climate processes
Garuma (2018)	Review of urban surface parameterizations for numerical climate models	Numerical	Microscale	Thermal exposure, urban ventilation	Review of urban surface parameterization development for mesoscale models and identifying datasets and processes needed for urban weather and climate prediction
Martilli (2007)	Current research and future challenges in urban mesoscale modeling	Numerical	Regional scale	Thermal exposure, urban ventilation.	Review of techniques used to parameterize urban-induced dynamical and thermal effects in mesoscale models
Masson et al. (2020a)	City-descriptive input data for urban climate models: Model requirements, data sources, and challenges	Numerical	Neighborhood scale	Thermal exposure, urban ventilation.	Review of methods for collecting and inputting descriptive data required to run mesoscale and microscale urban climate models
Grimmond et al. (2009)	Urban surface energy balance models: Model characteristics and methodology for a comparison study	Numerical	Neighborhood scale	Weather and climate	Description and comparison of urban surface energy balance models used primarily in large-scale weather and climate simulations
Ching (2013)	A perspective on urban canopy layer modeling for weather, climate, and air quality applications	Numerical	Regional scale	Air quality, urban ventilation	Brief review and perspective on means and science parameterizations for urban focused modeling in major modeling systems
Toparlar et al. (2017)	A review on the CFD analysis of urban microclimate	Numerical CFD	Microscale	Thermal exposure, urban ventilation	Review of wind tunnel experiment on geometrically similar models of the urban area

and intraurban climate on human life, while detailing emerging directions for future research in the field.

4.3.1 Urban climate and climate change interaction

When modeling the interaction between global climate change and local urban climate processes, there are often two main research questions that are of focus. The first aims to assess the inclusion of urban land covers and their future expansions in the regional and global climate analyses, while the second research area focuses on quantifying the impact of projected climate change on various urban climate processes and implications on human life.

To determine the impact of urban representation on global climate change analysis (through the use of UCMs described above), it is critical to note a few factors:

1. *Resolution of climate models.* Particularly in global climate analyses, the grid resolutions of traditional models ($\sim$100 km) are significantly larger than most urban areas, and urban areas only cover 3% of the world's land surface. This led to a traditional assumption in the climate community that the fraction of urban areas is too small to have an impact on larger-scale circulations that dominate global climate analyses, and so urban parameterizations are often not included. More recent global models are able to reach higher grid resolutions of 10−50 km, in which the representation of urban areas becomes more critical and misrepresentation of the magnitude or the extent of urban land cover can result in model bias (Hertwig et al., 2021; Katzfey et al., 2020). Accordingly, with more advances in computing power that are leading to higher resolution general circulation models (GCMs), it is critical that more UCMs are incorporated in GCMs such that previous assumptions for neglecting cities are rigorously tested and more representative forcing data for regional and microscale urban climate models are achieved.

2. *Area of influence.* While it is widely acknowledged that the change in land cover in urban areas results in *local* changes in urban temperature and surface fluxes, there is limited analysis regarding nonlocal, regional, or larger-scale circulation impacts of urban areas. A few global-scale climate simulations with urban canopy parameterizations (Georgescu et al., 2013; Jacobson and Ten Hoeve, 2012; Katzfey et al., 2020) have suggested that urban areas can have non-local impacts on regions

distant from urban areas (and in the same order of magnitude as the impact of non-urban land cover inputs on regional effects). There is now growing interest in quantifying the climate impact of cities at regional and global scales, as demonstrated through the recent endorsement of a 5-year Flagship Pilot Program on Urban Environments and Regional Climate Change (URB-RCC) through the CORDEX community.

3. *Contribution of cities to global climate change.* Cities interact with global climate change not only through greenhouse gases (GHG) emissions, but also modified land cover with continued urban expansion and anthropogenic heat wastes. Particularly in highly dense cities, anthropogenic heat is reported to be as important as solar forcing for urban regional climate and local canopy temperature (Salamanca et al., 2014; Schoetter et al., 2020b). However, at the global-scale, anthropogenic heat contribution is significantly smaller, an estimated $0.039 \, W/m^2$ globally averaged compared with current GHG forcing of $\sim 3 \, W/m^2$ (Dong et al., 2017). The increase in urban development, on the other hand, appears to be a significant contributor to local warming, representing impacts with similar order-of-magnitude as GHG contributions that lead to local urban-induced warmings as high as 4°C for the maximum expansion scenario (Georgescu et al., 2014, 2013; Zhao et al., 2021). The dynamic interaction between GHG-induced and urban-induced warming is nonlinear (Krayenhoff et al., 2018). This indicates that, particularly in higher-resolution simulations with the focus on local urban climate change, it is critical that urban expansion scenarios are considered together with GHG forcings such that we achieve a more comprehensive and accurate outlook on climate projections.

More investigations are done in the second research area, that is, assessing the impact of climate change on urban climate and subsequently human life, focusing on critical aspects of projected urban temperature (Chapman et al., 2017), heat exposure impact (Broadbent et al., 2020a), cooling demand (Takane et al., 2020), intensity and duration of extreme heat events, and other high impact weather events such as rainfall intensity (Li et al., 2020). Such analyses further lead to a more holistic assessment of mitigation and adaptation strategies for minimizing the negative impacts of current and future climate change (Broadbent et al., 2020b; Haddad et al., 2020). However, we note that a significant majority of studies considering impacts of future climate change on human health (e.g., heat stress) are based on outputs of climate models that do not include urban canopy parameterizations. In addition, the simplicity in underlying assumptions leads to inaccurate representation of outcomes in a changing climate, as noted by Vanos et al. (2020) in the case of projecting heat-health impacts.

4.3.2 Urban ventilation

Urban ventilation is a measure of city breathability, "i.e., city potential of removing and diluting pollutants, heat, moisture and other scalars" (Buccolieri et al., 2010), and depends on two key factors: (1) physical characteristics of the city (encompassing terrain and urban form) and (2) time-varying meteorological forcings (including mesoscale circulations as well as surface/air heating). These factors create a dynamic interaction between the mechanical- and buoyancy-driven ventilation—the latter arising due to temperature (and therefore air density) differences in the urban canopy. The combined effect then dominates the exchange and dispersion of pollutants, heat, and moisture from urban canyons, which has been the subject of modeling for three decades.

As cities grapple with air quality challenges and are expected to experience elevated temperature due to climate change and urban–induced local warming, it is critical that modeling approaches are used as means to better understand and predict ventilation capacity and breathability of compact cities (Buccolieri et al., 2015; Ng et al., 2011) in current and projected climates. Furthermore, as urban design characteristics (such as urban morphology, building shape, built-up materials, and vegetation cover) have a critical role in dominating urban airflow, modeling ventilation is further instrumental to curate the knowledge-base for climate-resilient urban design, informing us of the impact of various design parameters on urban microclimate. Such assessments can be applicable in both existing cities and new developments. In the case of the former, modeling urban ventilation can quantify the effectiveness of architectural interventions, such as canopy-level wind enhancement using wind catchers (Chew et al., 2017) or roof edge roughness (Aliabadi et al., 2017). For new developments, on the other hand, there are opportunities for developing city-wide or project-specific plans for enhanced ventilation. Modeling at the neighborhood and street scale has enabled us to propose and evaluate "wind corridors" (Gu et al., 2020; Hsieh and Huang, 2016) or building porosities (Chew and Norford, 2018; Yuan and Ng, 2012) that allow for more efficient transfer of pollutants and heat from high-density urban environments.

Overall, there are three main focus areas when modeling urban ventilation: (1) quantifying the mean and turbulent transfer of heat and mass within and above the urban canopy, which directly contribute to air quality and heat exposure at the canopy level (Giometto et al., 2017; Nazarian et al., 2018a,b); (2) determining wind flow around and through obstacles (buildings, trees, and city infrastructure) in time-varying environmental conditions (Zhang et al., 2020); and (3) assessing pedestrian-level wind comfort and safety (Blocken et al., 2016, 2012). These outcomes can then be evaluated across the city, neighborhood, street, and building scales. At larger scales, synoptic or regional circulations (induced by proximity to water or elevated terrains) can interact in surprising ways with existing or proposed urban fabric, affecting the ability of cities to ventilate themselves (Sharma et al., 2016). When assessing ventilation at the neighborhood scale, roughness parameters can be identified that characterize the collective impacts of urban morphology and obstacles on wind patterns and profiles through urban canopies. For describing the roughness of the urban surfaces, two parameters are commonly calculated in urban neighborhoods: plan area density (the percentage of the site occupied by building plan area) and frontal area density (the ratio of the frontal area and the total site area) (Grimmond and Oke, 1999). These parameters can be used, through statistical modeling or parameterizations, to present a map of neighborhood ventilation capacity through identifying ventilation paths or urban porosities (Gál and Unger, 2009; Wong et al., 2010). Such parameterizations of ventilation capacity are often informed through building-resolving CFD simulations [Fig. 4.1 (8)] that investigate the impact of urban morphology as well as street and building shape (such as street height-to-width ratio, roof shape, as well as street trees) on the three-dimensional airflow patterns in our cities. Numerous simulations are done, commonly in idealized urban-like arrays and for neutral conditions, to demonstrate that different building layouts, shapes, and density dominate urban airflow patterns (Blocken, 2015, 2014; Li et al., 2006). While these are valuable in shaping our understanding of urban ventilation at the street scale, there are certain limitations that should be the subject of future modeling efforts. First, a systematic review of the field is needed, particularly to compare the impacts of each morphological and design parameter on both street- and roof-level ventilation indicators. Second, time-varying environmental conditions, such as thermal forcing (often dominated by solar heating of urban facades (Allegrini et al., 2013; Nazarian and Kleissl, 2016)), highly dynamic wind conditions (speed and direction (Claus et al., 2012; Wise et al., 2018)) as well as traffic flow and moving vehicles are significant contributors to airflow distributions, while being significantly less studied in the field. With the advancement in computational resources, it is critical that the compounding impacts of time-varying environmental conditions are studied together with morphological parameters. Third, although CFD simulations are currently capable of simulating realistic urban configurations and street trees (Geletič et al., 2021), more analyses are yet needed to evaluate how they compare with simulations with idealized configurations, particularly as the latter are the basis for parameterizations of urban flow. Lastly, the interaction between outdoor and indoor ventilation is done through reduced-scale modeling or CFD analysis. Although the main focus of assessing ventilation at the building scale is to determine the natural airflow through buildings, it is critical to note that the accuracy of modeling largely relies on how the "Influence Region," that is, the neighborhood region surrounding the building, is characterized (Tong et al., 2016) and therefore, assessing the ventilation capacity of stand-alone buildings is not an effective method of informing urban design. Accordingly, more multiscale modeling of urban ventilation is needed such that natural ventilation through buildings can be accurately assessed.

4.3.3 Thermal environment and exposure in the built environment

In the face of climate change and increased urbanization, there is a growing demand for urban climate model outputs to be incorporated into current and future design and planning decisions such that we improve human exposure to heat and cold. In this context, we need to consider modeling objectives as assessing both thermal environment and thermal exposure, the latter defined as the contact between people and their immediate thermal environment that results in physiological strain, change in thermal sensation and comfort, or change in various aspects of life quality and lifestyle (Nazarian and Lee, 2021).

Regarding the thermal environment, modeling studies at city and neighborhood scales have greatly assisted in documenting the impact of the built environment on local temperature, as indicated by (1) the difference of urban canopy air and surface temperature from rural surroundings

(UHI) [modeling approaches and findings detailed in Mirzaei (2015)], and (2) intraurban temperature variability in different Local Climate Zones (Brousse et al., 2016; Mussetti et al., 2020). At a larger scale, modeling has documented local warming exacerbated in the face of climate change due to urban expansion (Georgescu et al., 2013), and further complemented observational analyses that showed synergies between UHI and extreme heat events based on different background climates and warming scenarios (Zhao et al., 2018), which can be exacerbated during daytime in future climates.

Modeling methods for thermal exposure (including thermal comfort and heat stress analysis) are further developed across multiple scales with different levels of complexity. This includes calculation of thermal comfort indices (TCIs) for a single person subject to specified environmental conditions through web-based or open-source algorithms (Tartarini et al., 2020; Tartarini and Schiavon, 2020) that can be incorporated in building-resolving numerical models [Fig. 4.1 (7)] such as the Ladybug Tools (Mackey et al., 2017) for annual calculations of TCIs at a specified location. Such calculations can include the impact of designed shading devices or other elements that obstruct direct solar radiation and limit the view of the sky.

However, it should be noted that unlike air temperature variation in the street canyon that is rather small at the pedestrian level, thermal exposure is highly variable in space due to microscale variation of the radiant environment and wind in the urban canopy. To simulate the two- and three-dimensional distribution of the radiative environment at the pedestrian level (through calculation of Mean Radiant Temperature, MRT), several building-resolving energy balance or CFD models [Fig. 4.1 (7−8)] are deployed, such as RayMan (Matzarakis et al., 2010), SOLWEIG (Lindberg et al., 2008), and ENVI-met (Bruse and Fleer, 1998). All noted models are able to incorporate complex urban geometries and a representation of urban trees for MRT calculations, while varying greatly on (1) the ability to output two-dimensional distribution of MRT at pedestrian height, (2) calculations of urban surface temperature and view factors, and (3) assumptions made for underlying shortwave and longwave radiation exchange with urban surfaces and the atmosphere. Recent comparisons of these models with local measurements across different urban forms, sky view factors, and seasons (Crank et al., 2020; Gál and Kántor, 2020) indicate that models can produce simulation errors larger than a suggested accuracy of $+$ 5°C for heat stress studies, which can be attributed to the calculations of surface temperature and reflected shortwave radiation from urban facades. Therefore in-situ validation is critical when using simulations in micrometeorological or morphological extremes that can significantly affect human health and well-being. On the other hand, accounting for wind speed variability in thermal exposure calculations requires detailed consideration of local morphology, which is done with CFD (Envi-Met and PALM-4U). Alternatively, modular (multimodel) approaches can be used that deploy more than one modeling technique for comprehensiveness and accuracy (Nazarian et al., 2018c) or improve the computationally expensive processes that are not attractive in a fast-paced design process that emphasizes rapid, evidence-based design iterations (Kastner and Dogan, 2019).

Assessing spatially and temporally differentiated TCIs requires care because of the wide range of conditions that people experience moving through urban environments. Metrics for outdoor thermal comfort and thermal exposure, such as outdoor thermal comfort autonomy, can succinctly describe overall conditions and assist in providing micro- to city-scale analysis, which can intelligently inform urban design (Nazarian et al., 2019a).

Furthermore, it is critical to note that calculating 3D distribution of TCIs does not account for how people use outdoor spaces (Melnikov et al., 2017). An agent-based modeling approach to simulate thermal perception and its impact on adaptive behavior addresses this, where different forms of adaptation (speed, thermal attraction, and repulsion) and vision-based route adjustment are considered (Melnikov et al., 2017). Agent-based response to thermal exposure can also be incorporated within a system-dynamics framework, Gagge's two-node thermal model (Doherty and Arens, 1988; Foda and Sirén, 2011; Gagge et al., 1986) to include the thermal dynamics of the human body (Melnikov et al., 2018). This presents a computationally efficient model that is suitable for simulating thermal comfort along designated routes.

In addition to modeling efforts made at the human to neighborhood scale, there is a significant interest in modeling thermal exposure at the city and regional sales. This is often achieved by using a mesoscale meteorological model while focusing on the 2 m air temperature and humidity. An example of such analysis can be found when various mitigation strategies, such as water bodies (Theeuwes et al., 2013), urban greening (Papangelis et al., 2012), and green and cool roofs (Sun et al., 2016) are used to assess the localized or city-level cooling, UHI, or thermal comfort.

However, these studies are often limited in modeling thermal exposure in three main aspects:

1. Incomplete representation of thermal environment parameters (i.e., taking air temperature/humidity as the proxy for thermal environment without considering MRT and wind speed or calculating TCIs).
2. Limited representation of urban canopy (i.e., considering single-layer CMs that do not consider the vertical or horizontal variability of thermal environment).
3. Limited information regarding thermal exposure time and intensity.

In order to address the first limitation, Hwang et al. (2019) deployed the mesoscale model WRF with multilayer UCM BEP + BEM (Salamanca and Martilli, 2010) and coupled the outputs with the biometeorological Man-Environment Heat Exchange model to calculate a thermal comfort index. They concluded that the human heat balance models are needed in mesoscale modeling to provide more reliable information about thermal comfort for vulnerable groups and particularly in complex urban environments. To address the second common limitation, several studies have used outputs of mesoscale models to inform microscale analysis of thermal comfort (Berardi et al., 2020). Although comprehensive, these cross-scale coupling efforts are computationally expensive and challenging to achieve. As a more efficient alternative Nazarian et al. (2020b) extended the UCM to estimate the spatial variability of MRT in the street canopy and, using previous microscale simulations of urban flow, parameterized the wind speed variability with urban density. Together with the mesoscale assessment of urban temperature, these parameterizations provide a comprehensive set of environmental parameters for thermal comfort and therefore, the variability in TCIs can be estimated at the neighborhood scale.

4.4 Conclusions

Numerical, statistical, and physical scale models have made significant contributions to understanding climate processes in the urban boundary layer, and have been developed extensively to address environmental challenges (such as heat, air quality, and energy) across different scales. Here, we summarized state-of-the-art methods in representing urban climate processes across different scales and detailed the representation of vegetation and anthropogenic heat and water wastes in urban areas as critical components dynamically affecting urban physics. In addition, it is critical that modeling methods are considered based on the scales and problem at hand and following the "Fitness-for-Purpose" guidelines (Baklanov et al., 2009). Accordingly, we provided a detailed overview of numerical models deployed for addressing critical research questions in the field, such as assessing the interaction between global and local climates as well as quantifying ventilation and thermal exposure in different urban configurations and background climates.

Overall, we find that rigorous evaluation of model physics is critical for assuring the reliability and utility of urban climate models, but is hindered by the limited availability of observational data at appropriate temporal and spatial scales (Krayenhoff et al., 2021). Future research is critical to extend numerical models for a more comprehensive representation of urban processes, and more effectively deploy multiscale models in addressing emerging research questions on urban climate change. An equally important direction is developing standardized databases for evaluating model physics and application for representing urban canopy processes relevant to the problem at hand.

References

Acero, J.A., Koh, E.J.K., Pignatta, G., Norford, L.K., 2020. Clustering weather types for urban outdoor thermal comfort evaluation in a tropical area. Theoretical and Applied Climatology 139, 659–675.
Ahmad, K., Khare, M., Chaudhry, K.K., 2005. Wind tunnel simulation studies on dispersion at urban street canyons and intersections—a review. Journal of Wind Engineering and Industrial Aerodynamics 93, 697–717. Available from: https://doi.org/10.1016/j.jweia.2005.04.002.
Aliabadi, A.A., Krayenhoff, E.S., Nazarian, N., 2017. Effects of roof-edge roughness on air temperature and pollutant concentration in urban canyons. Bound – Layer Meteorology 164, 249–279.
Allegrini, J., Dorer, V., Carmeliet, J., 2013. Wind tunnel measurements of buoyant flows in street canyons. Builduing and Environment 59, 315–326. Available from: https://doi.org/10.1016/j.buildenv.2012.08.029.
Allegrini, J., Dorer, V., Carmeliet, J., 2015. Influence of morphologies on the microclimate in urban neighbourhoods. Journal of Wind Engineering and Industrial Aerodynamics 144, 108–117. Available from: https://doi.org/10.1016/j.jweia.2015.03.024.
Aoyagi, T., Seino, N., 2011. A square prism urban canopy scheme for the NHM and its evaluation on summer conditions in the Tokyo Metropolitan Area, Japan. Journal of Applied Meteorology and Climatology 50, 1476–1496. Available from: https://doi.org/10.1175/2011JAMC2489.1.
Baklanov, A., Ching, J., Grimmond, C.S.B., Martilli, A., 2009. Model urbanization strategy: summaries, recommendations and requirements. In: Baklanov, A., Sue, G., Alexander, M., Athanassiadou, M. (Eds.), Meteorological and Air Quality Models for Urban Areas. Springer Berlin Heidelberg, Berlin, Heidelberg, pp. 151–162. Available from: https://doi.org/10.1007/978-3-642-00298-4_15.

Barlow, J.F., Belcher, S.E., 2002. A wind tunnel model for quantifying fluxes in the urban boundary layer. Boundary – Layer Meteorology 104, 131–150.

Berardi, U., Jandaghian, Z., Graham, J., 2020. Effects of greenery enhancements for the resilience to heat waves: a comparison of analysis performed through mesoscale (WRF) and microscale (Envi-met) modeling. The Science of the Total Environment 747, 141300. Available from: https://doi.org/10.1016/j.scitotenv.2020.141300.

Best, M.J., 2005. Representing urban areas within operational numerical weather prediction models. Boundary – Layer Meteorology 114, 91–109. Available from: https://doi.org/10.1007/s10546-004-4834-5.

Best, M.J., 2006. Progress towards better weather forecasts for city dwellers: from short range to climate change. Theoretical and Applied Climatology 84, 47–55. Available from: https://doi.org/10.1007/s00704-005-0143-2.

Best, M.J., Grimmond, C.S.B., 2015. Key conclusions of the first international urban land surface model comparison project. Bulletin of the American Meteorological Society 96, 805–819. Available from: https://doi.org/10.1175/BAMS-D-14-00122.1.

Blocken, B., 2014. 50 years of computational wind engineering: past, present and future. Journal of Wind Engineering and Industrial Aerodynamics 129, 69–102. Available from: https://doi.org/10.1016/j.jweia.2014.03.008.

Blocken, B., 2015. Computational fluid dynamics for urban physics: importance, scales, possibilities, limitations and ten tips and tricks towards accurate and reliable simulations. Building and Environment 91, 219–245. Available from: https://doi.org/10.1016/j.buildenv.2015.02.015.

Blocken, B., 2018. LES over RANS in building simulation for outdoor and indoor applications: a foregone conclusion? Building Simulation 11, 821–870. Available from: https://doi.org/10.1007/s12273-018-0459-3.

Blocken, B., Janssen, W.D., van Hooff, T., 2012. CFD simulation for pedestrian wind comfort and wind safety in urban areas: general decision framework and case study for the Eindhoven University campus. Environmental Modelling & Software 30, 15–34. Available from: https://doi.org/10.1016/j.envsoft.2011.11.009.

Blocken, B., Stathopoulos, T., van Beeck, J.P.A.J., 2016. Pedestrian-level wind conditions around buildings: review of wind-tunnel and CFD techniques and their accuracy for wind comfort assessment. Building and Environment 100, 50–81. Available from: https://doi.org/10.1016/j.buildenv.2016.02.004.

Bowler, D.E., Buyung-Ali, L., Knight, T.M., Pullin, A.S., 2010. Urban greening to cool towns and cities: a systematic review of the empirical evidence. Landscape and Urban Planning 97, 147–155. Available from: https://doi.org/10.1016/j.landurbplan.2010.05.006.

Broadbent, A.M., Krayenhoff, E.S., Georgescu, M., 2020a. The motley drivers of heat and cold exposure in 21st century US cities. Proceedings of the National Academy of Sciences of the United States of America 117, 21108–21117. Available from: https://doi.org/10.1073/pnas.2005492117.

Broadbent, A.M., Scott Krayenhoff, E., Georgescu, M., 2020b. Efficacy of cool roofs at reducing pedestrian-level air temperature during projected 21st century heatwaves in Atlanta, Detroit, and Phoenix (USA). Environmental Research Letters. 15, 084007. Available from: https://doi.org/10.1088/1748-9326/ab6a23.

Brousse, O., Martilli, A., Foley, M., Mills, G., Bechtel, B., 2016. WUDAPT, an efficient land use producing data tool for mesoscale models? Integration of urban LCZ in WRF over Madrid. Urban Climate 17, 116–134. Available from: https://doi.org/10.1016/j.uclim.2016.04.001.

Brown, M.J., 2000. Urban parameterizations for mesoscale meteorological models. Mesoscale Atmospheric Dispersion 9, 193–255.

Bruse, M., Fleer, H., 1998. Simulating surface–plant–air interactions inside urban environments with a three dimensional numerical model. Environmental Modelling &

Software 13, 373—384. Available from: https://doi.org/10.1016/S1364-8152(98) 00042-5.

Buccolieri, R., Sandberg, M., Di Sabatino, S., 2010. City breathability and its link to pollutant concentration distribution within urban-like geometries. Atmospheric Environment (Oxford, England: 1994) 44, 1894—1903. Available from: https://doi.org/10.1016/j. atmosenv.2010.02.022.

Buccolieri, R., Salizzoni, P., Soulhac, L., Garbero, V., Di Sabatino, S., 2015. The breathability of compact cities. Urban Climate 13, 73—93. Available from: https://doi.org/ 10.1016/j.uclim.2015.06.002.

Bueno, B., Norford, L., Pigeon, G., Britter, R., 2011a. Combining a detailed building energy model with a physically-based urban canopy model. Boundary — Layer Meteorology 140, 471—489. Available from: https://doi.org/10.1007/s10546-011-9620-6.

Bueno, B., Pigeon, G., Norford, L.K., Zibouche, K., 2011b. Development and evaluation of a building energy model integrated in the TEB scheme. Geoscientific Model Development Discussions 4, 2973—3011. Available from: https://doi.org/10.5194/ gmdd-4-2973-2011.

Bueno, B., Norford, L., Hidalgo, J., Pigeon, G., 2013. The urban weather generator. Journal of Building Performance Simulation. Available from: https://doi.org/10.1080/ 19401493.2012.718797.

Capel-Timms, I., Smith, S.T., Sun, T., Grimmond, S., 2020. Dynamic anthropogenic activities impacting heat emissions (DASH v1.0): development and evaluation. Geoscientific Model Development 13, 4891—4924. Available from: https://doi.org/ 10.5194/gmd-13-4891-2020.

Chapman, S., Watson, J.E.M., Salazar, A., Thatcher, M., McAlpine, C.A., 2017. The impact of urbanization and climate change on urban temperatures: a systematic review. Landscape Ecology 32, 1921—1935. Available from: https://doi.org/10.1007/s10980-017-0561-4.

Chen, F., Kusaka, H., Bornstein, R., Ching, J., Grimmond, C.S.B., Grossman-Clarke, S., et al., 2011. The integrated WRF/urban modelling system: development, evaluation, and applications to urban environmental problems. International Journal of Climatology 31, 273—288. Available from: https://doi.org/10.1002/joc.2158.

Chen, G., Wang, D., Wang, Q., Li, Y., Wang, X., Hang, J., et al., 2020. Scaled outdoor experimental studies of urban thermal environment in street canyon models with various aspect ratios and thermal storage. The Science of the Total Environment 726, 138147. Available from: https://doi.org/10.1016/j.scitotenv.2020.138147.

Chew, L.W., Nazarian, N., Norford, L., 2017. Pedestrian-level urban wind flow enhancement with wind catchers. Atmosphere 8, 159. Available from: https://doi.org/ 10.3390/atmos8090159.

Chew, L.W., Norford, L.K., 2018. Pedestrian-level wind speed enhancement in urban street canyons with void decks. Building and Environment 146, 64—76.

Ching, J.K.S., 2013. A perspective on urban canopy layer modeling for weather, climate and air quality applications. Urban Climate 3, 13—39. Available from: https://doi.org/ 10.1016/j.uclim.2013.02.001.

Chronis, A., Turner, A., Tsigkari, M., 2011. Generative fluid dynamics: integration of fast fluid dynamics and genetic algorithms for wind loading optimization of a free form surface, In: Proceedings of the Symposium on Simulation for Architecture and Urban Design, SimAUD '11. Society for Computer Simulation International, San Diego, CA, USA, 29—36.

Claus, J., Coceal, O., Thomas, T.G., Branford, S., Belcher, S.E., Castro, I.P., 2012. Wind-direction effects on urban-type flows. Boundary — Layer Meteorology 142, 265—287. Available from: https://doi.org/10.1007/s10546-011-9667-4.

Coceal, O., Dobre, A., Thomas, T.G., Belcher, S.E., 2007. Structure of turbulent flow over regular arrays of cubical roughness. Journal of Fluid Mechanics 589, 375−409. Available from: https://doi.org/10.1017/S002211200700794X.

Crank, P.J., Middel, A., Wagner, M., Hoots, D., Smith, M., Brazel, A., 2020. Validation of seasonal mean radiant temperature simulations in hot arid urban climates. The Science of the Total Environment 749, 141392. Available from: https://doi.org/ 10.1016/j.scitotenv.2020.141392.

Daniel, M., Lemonsu, A., Déqué, M., Somot, S., Alias, A., Masson, V., 2019. Benefits of explicit urban parameterization in regional climate modeling to study climate and city interactions. Climate Dynamics 52, 2745−2764. Available from: https://doi.org/ 10.1007/s00382-018-4289-x.

Doherty, T., Arens, E.A., 1988. Evaluation of the physiological bases of thermal comfort models. ASHRAE Trans. 94 Part 1.

Dong, Y., Varquez, A.C.G., Kanda, M., 2017. Global anthropogenic heat flux database with high spatial resolution. Atmospheric Environment (Oxford, England: 1994) 150, 276−294. Available from: https://doi.org/10.1016/j.atmosenv.2016. 11.040.

dos Santos, A.R., de Oliveira, F.S., da Silva, A.G., Gleriani, J.M., Gonçalves, W., Moreira, G.L., et al., 2017. Spatial and temporal distribution of urban heat islands. The Science of the Total Environment 605−606, 946−956. Available from: https:// doi.org/10.1016/j.scitotenv.2017.05.275.

Duraisamy, K., Iaccarino, G., Xiao, H., 2019. Turbulence modeling in the aAge of data. Annual Review of Fluid Mechanics 51, 357−377. Available from: https://doi.org/ 10.1146/annurev-fluid-010518-040547.

Erell, E., Williamson, T., 2006. Simulating air temperature in an urban street canyon in all weather conditions using measured data at a reference meteorological station. International Journal of Climatology 26, 1671−1694. Available from: https://doi.org/ 10.1002/joc.1328.

Erell, E., Williamson, T., 2007. Intra-urban differences in canopy layer air temperature at a mid-latitude city. International Journal of Climatology: A Journal of the Royal Meteorological Society 27, 1243−1255.

Fan, Y., Li, Y., Yin, S., 2018. Interaction of multiple urban heat island circulations under idealised settings. Building and Environment 134, 10−20. Available from: https://doi. org/10.1016/j.buildenv.2018.02.028.

Foda, E., Sirén, K., 2011. A new approach using the Pierce two-node model for different body parts. International Journal of Biometeorology 55, 519−532. Available from: https://doi.org/10.1007/s00484-010-0375-4.

Fortuniak, K., 2003. A slab surface energy balance model (SUEB) and its application to the study on the role of roughness length in forming an urban heat island. Acta Universitatis Wratislaviensis, Studia Geograficzne 2542, 368−377.

Gagge, A.P., Fobelets, A.P., Berglund, L.G., 1986. A standard predictive Index of human reponse to thermal enviroment. Transactions/American Society of Heating, Refrigerating and Air-Conditioning Engineers 92, 709−731.

Gál, T., Unger, J., 2009. Detection of ventilation paths using high-resolution roughness parameter mapping in a large urban area. Building and Environment 44, 198−206. Available from: https://doi.org/10.1016/j.buildenv.2008.02.008.

Gál, C.V., Kántor, N., 2020. Modeling mean radiant temperature in outdoor spaces. A comparative numerical simulation and validation study. Urban Climate 32, 100571. Available from: https://doi.org/10.1016/j.uclim.2019.100571.

Garuma, G.F., 2018. Review of urban surface parameterizations for numerical climate models. Urban Climate 24, 830−851. Available from: https://doi.org/10.1016/j. uclim.2017.10.006.

Geletič, J., Lehnert, M., Krč, P., Resler, J., Krayenhoff, E.S., 2021. High-resolution modelling of thermal exposure during a hot spell: a case study using PALM-4U in Prague, Czech Republic. Atmosphere 12, 175. Available from: https://doi.org/10.3390/atmos12020175.

Georgescu, M., Moustaoui, M., Mahalov, A., 2013. Summer-time climate impacts of projected megapolitan expansion in Arizona. Nature Climate Change 3, 37−41.

Georgescu, M., Morefield, P.E., Bierwagen, B.G., Weaver, C.P., 2014. Urban adaptation can roll back warming of emerging megapolitan regions. Proceedings of the National Academy of Sciences of the United States of America 111, 2909−2914. Available from: https://doi.org/10.1073/pnas.1322280111.

Giometto, M.G., Christen, A., Egli, P.E., Schmid, M.F., Tooke, R.T., Coops, N.C., et al., 2017. Effects of trees on mean wind, turbulence and momentum exchange within and above a real urban environment. Advances in Water Resources 106, 154−168. Available from: https://doi.org/10.1016/j.advwatres.2017.06.018.

Grimmond, C.S.B., Oke, T.R., 1999. Aerodynamic properties of urban areas derived from analysis of surface form. Journal of Appllied Meteorology and Climatology 38, 1262−1292. Available from: https://doi.org/10.1175/1520-0450(1999)038 < 1262: APOUAD > 2.0.CO;2.

Grimmond, C.S.B., Oke, T.R., Steyn, D.G., 1986. Urban water balance: 1. A model for daily totals. Water Resources Research 22, 1397−1403. Available from: https://doi.org/10.1029/wr022i010p01397.

Grimmond, C.S.B., Best, M., Barlow, J., Arnfield, A.J., Baik, J.-J., Baklanov, A., et al., 2009. Urban surface energy balance models: model characteristics and methodology for a comparison study. In: Baklanov, A., Sue, G., Alexander, M., Athanassiadou, M. (Eds.), Meteorological and Air Quality Models for Urban Areas. Springer Berlin Heidelberg, Berlin, Heidelberg, pp. 97−123. Available from: https://doi.org/10.1007/978-3-642-00298-4_11.

Grimmond, C.S.B., Blackett, M., Best, M.J., Baik, J.-J., Belcher, S.E., Beringer, J., et al., 2011. Initial results from Phase 2 of the international urban energy balance model comparison: results from international urban energy balance model comparison: Phase 2. International Journal of Climatology 31, 244−272. Available from: https://doi.org/10.1002/joc.2227.

Gu, K., Fang, Y., Qian, Z., Sun, Z., Wang, A., 2020. Spatial planning for urban ventilation corridors by urban climatology. Ecosystem Health and Sustainability 6, 1747946. Available from: https://doi.org/10.1080/20964129.2020.1747946.

Guo, X., Li, W., Iorio, F., 2016. Convolutional neural networks for steady flow approximation, In: Proceedings of the 22nd ACM SIGKDD International Conference on Knowledge Discovery and Data Mining, KDD '16. Association for Computing Machinery, New York, NY, USA, 481−490. <https://doi.org/10.1145/2939672.2939738>.

Haddad, S., Barker, A., Yang, J., Kumar, D.I.M., Garshasbi, S., Paolini, R., et al., 2020. On the potential of building adaptation measures to counterbalance the impact of climatic change in the tropics. Energy and Buildings 229, 110494. Available from: https://doi.org/10.1016/j.enbuild.2020.110494.

Hamdi, R., Kusaka, H., Doan, Q.-V., Cai, P., He, H., Luo, G., et al., 2020. The state-of-the-art of urban climate change modeling and observations. Earth Systems and Environment . Available from: https://doi.org/10.1007/s41748-020-00193-3.

Harlan, S.L., Brazel, A.J., Prashad, L., Stefanov, W.L., Larsen, L., 2006. Neighborhood microclimates and vulnerability to heat stress. Social Science & Medicine 63, 2847−2863. Available from: https://doi.org/10.1016/j.socscimed.2006.07.030.

Hertwig, D., Ng, M., Grimmond, S., Vidale, P.L., McGuire, P.C., 2021. High-resolution global climate simulations: representation of cities. International Journal of Climatology. Available from: https://doi.org/10.1002/joc.7018.

Hsieh, C.-M., Huang, H.-C., 2016. Mitigating urban heat islands: a method to identify poten-tial wind corridor for cooling and ventilation. Computers, Environment and Urban Systems 57, 130−143. Available from: https://doi.org/10.1016/j.compenvurbsys.2016.02.005.

Huttner, S., Bruse, M., 2009. Numerical modeling of the urban climate−a preview on ENVI-met 4.0, In: The Seventh International Conference on Urban Climate.

Hwang, M.-K., Bang, J.-H., Kim, S., Kim, Y.-K., Oh, I., 2019. Estimation of thermal comfort felt by human exposed to extreme heat wave in a complex urban area using a WRF-MENEX model. International Journal of Biometeorology 63, 927−938. Available from: https://doi.org/10.1007/s00484-019-01705-1.

Jacobson, M.Z., Ten Hoeve, J.E., 2012. Effects of urban surfaces and white roofs on global and regional climate. Journal of Climate 25, 1028−1044. Available from: https://doi.org/10.1175/JCLI-D-11-00032.1.

Jandaghian, Z., Berardi, U., 2020. Comparing urban canopy models for microclimate simulations in weather research and forecasting models. Sustainable Cities and Society 55, 102025. Available from: https://doi.org/10.1016/j.scs.2020.102025.

Järvi, L., Grimmond, C.S.B., Christen, A., 2011. The Surface Urban Energy and Water Balance Scheme (SUEWS): evaluation in Los Angeles and Vancouver. Journal of Hydrology 411, 219−237. Available from: https://doi.org/10.1016/j.jhydrol.2011.10.001.

Jasak, H., 2009. OpenFOAM: open source CFD in research and industry. International Journal of Naval Architecture And Ocean Engineering 1, 89−94. Available from: https://doi.org/10.2478/ijnaoe-2013-0011.

Jin, M., Zuo, W., Chen, Q., 2013. Simulating natural ventilation in and around buildings by fast fluid dynamics. Numerical Heat Transfer, Part A: Applications 64, 273−289. Available from: https://doi.org/10.1080/10407782.2013.784131.

Jin, L., Schubert, S., Fenner, D., Meier, F., Schneider, C., 2021. Integration of a building energy model in an urban climate model and its application. Boundary − Layer Meteorology 178, 249−281. Available from: https://doi.org/10.1007/s10546-020-00569-y.

Kanda, M., 2006. Progress in the scale modeling of urban climate: review. Theoretical and Applied Climatology 84, 23−33. Available from: https://doi.org/10.1007/s00704-005-0141-4.

Kanda, M., Kawai, T., Kanega, M., Moriwaki, R., Narita, K., Hagishima, A., 2005. A sim-ple energy balance model for regular building arrays. Boundary − Layer Meteorology 116, 423−443.

Kanda, M., Kawai, T., Moriwaki, R., Narita, K., Hagishima, A., Sugawara, H., 2006. Comprehensive outdoor scale model experiments for urban climate (COSMO), In: Proc., 6th Int. Conf. on Urban Climate. techno-office.com, 270−273.

Karagulian, F., Belis, C.A., Dora, C.F.C., Prüss-Ustün, A.M., Bonjour, S., Adair-Rohani, H., et al., 2015. Contributions to cities' ambient particulate matter (PM): a systematic review of local source contributions at global level. Atmospheric Environment 120, 475−483. Available from: https://doi.org/10.1016/j.atmosenv.2015.08.087.

Kastner, P., Dogan, T., 2019. Towards high-resolution annual outdoor thermal comfort mapping in urban design, In: Proceedings of Building Simulation 2019: 16th Conference of IBPSA. Presented at the Building Simulation 2019, IBPSA. <https://doi.org/10.26868/25222708.2019.210458>.

Katzfey, J., Schlünzen, H., Hoffmann, P., Thatcher, M., 2020. How an urban parameteri-zation affects a high-resolution global climate simulation. Quarterly Journal of the Royal Meteorological Society 146, 3808−3829. Available from: https://doi.org/10.1002/qj.3874.

Kikegawa, Y., Genchi, Y., Yoshikado, H., Kondo, H., 2003. Development of a numerical simulation system toward comprehensive assessments of urban warming countermeasures including their impacts upon the urban buildings' energy-demands. Applied Energy 76, 449−466. Available from: https://doi.org/10.1016/S0306-2619(03)00009-6.

Kikegawa, Y., Tanaka, A., Ohashi, Y., Ihara, T., 2014. Observed and simulated sensitivities of summertime urban surface air temperatures to anthropogenic heat in downtown areas of two Japanese major cities, Tokya and Osaka. Theoretical and Applied Climatology 117, 175−193.

Kondo, H., Genchi, Y., Kikegawa, Y., Ohashi, Y., Yoshikado, H., Komiyama, H., 2005. Development of a multi-layer urban canopy model for the analysis of energy consumption in a big city: structure of the urban canopy model and its basic performance. Boundary − Layer Meteorology 116, 395−421.

Krayenhoff, E.S., Voogt, J.A., 2007. A microscale three-dimensional urban energy balance model for studying surface temperatures. Boundary − Layer Meteorology 123, 433−461. Available from: https://doi.org/10.1007/s10546-006-9153-6.

Krayenhoff, E.S., Moustaoui, M., Broadbent, A.M., Gupta, V., Georgescu, M., 2018. Diurnal interaction between urban expansion, climate change and adaptation in US cities. Nature Climate Change 8, 1097−1103. Available from: https://doi.org/ 10.1038/s41558-018-0320-9.

Krayenhoff, E.S., Jiang, T., Christen, A., Martilli, A., Oke, T.R., Bailey, B.N., et al., 2020. A multi-layer urban canopy meteorological model with trees (BEP-Tree): street tree impacts on pedestrian-level climate. Urban Climate 32, 100590. Available from: https://doi.org/10.1016/j.uclim.2020.100590.

Krayenhoff, E.S., Broadbent, A M , Zhao, L., Georgescu, M., Middel, A., Voogt, J.A., et al., 2021. Cooling hot cities: a systematic and critical review of the numerical modelling literature. Environmental Research Letters. Available from: https://doi.org/ 10.1088/1748-9326/abdcf1.

Kusaka, H., Kondo, H., Kikegawa, Y., Kimura, F., 2001. A simple single-layer urban canopy model for atmospheric models: comparison with multi-layer and slab models. Boundary − Layer Meteorology 101, 329−358. Available from: https://doi.org/ 10.1023/A:1019207923078.

Kusaka, H., Chen, F., Tewari, M., Dudhia, J., Gill, D.O., Duda, M.G., et al., 2012. Numerical simulation of urban heat island effect by the WRF model with 4-km grid increment: an inter-comparison study between the urban canopy model and slab model. Journal of the Meteorological Society of Japan. Ser. II 90B, 33−45. Available from: https://doi.org/10.2151/jmsj.2012-B03.

Lachapelle, J., Menheere, N., Krayenhoff, S., Middel, A., Broadbent, A.M., 2020. TUF-pedestrian: aA three-dimensional microscale model for pedestrian thermal exposure in urban environments, In: 100th American Meteorological Society Annual Meeting. AMS.

Landsberg, H.E., 1981. The Urban Climate. Academic Press.

Lee, S.-H., Park, S.-U., 2008. A vegetated urban canopy model for meteorological and environmental modelling. Boundary − Layer Meteorology 126, 73−102. Available from: https://doi.org/10.1007/s10546-007-9221-6.

Lemonsu, A., Masson, V., Shashua-Bar, L., Erell, E., Pearlmutter, D., 2012. Inclusion of vegetation in the town energy balance model for modelling urban green areas. Geoscientific Model Development 5, 1377−1393. Available from: https://doi.org/ 10.5194/gmd-5-1377-2012.

Li, X.-X., Liu, C.-H., Leung, D.Y.C., Lam, K.M., 2006. Recent progress in CFD modelling of wind field and pollutant transport in street canyons. Atmospheric Environment (Oxford, England: 1994) 40, 5640−5658. Available from: https://doi.org/10.1016/j. atmosenv.2006.04.055.

Li, X.-X., Leung, D.Y.C., Liu, C.-H., Lam, K.M., 2008. Physical modeling of flow field inside urban street canyons. Journal of Appllied Meteorology and Climatology 47, 2058−2067. Available from: https://doi.org/10.1175/2007JAMC1815.1.

Li, Y., Fowler, H.J., Argüeso, D., Blenkinsop, S., Evans, J.P., Lenderink, G., et al., 2020. Strong intensification of hourly rainfall extremes by urbanization.

Geophysical Research Letters 47. Available from: https://doi.org/10.1029/2020gl088758.

Lindberg, F., Holmer, B., Thorsson, S., 2008. SOLWEIG 1.0—modelling spatial variations of 3D radiant fluxes and mean radiant temperature in complex urban settings. International Journal of Biometeorology 52, 697—713. Available from: https://doi.org/10.1007/s00484-008-0162-7.

Lindberg, F., Grimmond, C.S.B., Gabey, A., Huang, B., Kent, C.W., Sun, T., et al., 2018. Urban Multi-scale Environmental Predictor (UMEP): an integrated tool for city-based climate services. Environmental Modelling & Software 99, 70—87. Available from: https://doi.org/10.1016/j.envsoft.2017.09.020.

Lipson, M.J., Thatcher, M., Hart, M.A., Pitman, A., 2018. A building energy demand and urban land surface model. Quarterly Journal of the Royal Meteorological Society 144, 1572—1590. Available from: https://doi.org/10.1002/qj.3317.

Lipson, M.J., Thatcher, M., Hart, M.A., Pitman, A., 2019. Climate change impact on energy demand in building-urban-atmosphere simulations through the 21st century. Environmental Research Letters 14, 125014. Available from: https://doi.org/10.1088/1748-9326/ab5aa5.

Lu, J., Pal Arya, S., Snyder, W.H., Lawson, R.E., 1997. A laboratory study of the urban heat island in a calm and stably stratified environment. Part I: temperature field. Journal of Appllied Meteorology and Climatology 36, 1377—1391. Available from: https://doi.org/10.1175/1520-0450(1997)036 < 1377:ALSOTU > 2.0.CO;2.

Mackey, C., Galanos, T., Norford, L., Roudsari, M.S., 2017. Wind, sun, surface temperature, and heat island: critical variables for high-resolution outdoor thermal comfort, In: Proceedings of the 15th International Conference of Building Performance Simulation Association. San Francisco, USA. <https://doi.org/10.26868/25222708.2017.260>.

Maraun, D., Widmann, M., 2018. Statistical Downscaling and Bias Correction for Climate Research. Cambridge University Press.

Maronga, B., Banzhaf, S., Burmeister, C., Esch, T., Forkel, R., Fröhlich, D., et al., 2020. Overview of the PALM model system 6.0. Geoscientific Model Development 13, 1335—1372. Available from: https://doi.org/10.5194/gmd-13-1335-2020.

Martilli, A., 2007. Current research and future challenges in urban mesoscale modelling. International Journal of Climatology 27, 1909—1918. Available from: https://doi.org/10.1002/joc.1620.

Martilli, A., Clappier, A., Rotach, M.W., 2002. An urban surface exchange parameterisation for mesoscale models. Boundary — Layer Meteorology 104, 261—304. Available from: https://doi.org/10.1023/A:1016099921195.

Masson, V., 2000. A physically-based scheme for the urban energy budget in atmospheric models. Boundary — Layer Meteorology 94, 357—397.

Masson, V., Heldens, W., Bocher, E., Bonhomme, M., Bucher, B., Burmeister, C., et al., 2020a. City-descriptive input data for urban climate models: model requirements, data sources and challenges. Urban Climate 31, 100536. Available from: https://doi.org/10.1016/j.uclim.2019.100536.

Masson, V., Lemonsu, A., Hidalgo, J., Voogt, J., 2020b. Urban climates and climate change. Annual Review of Environment and Resources 45, 411—444. Available from: https://doi.org/10.1146/annurev-environ-012320-083623.

Matzarakis, A., Rutz, F., Mayer, H., 2010. Modelling radiation fluxes in simple and complex environments: basics of the RayMan model. International Journal of Biometeorology 54, 131—139. Available from: https://doi.org/10.1007/s00484-009-0261-0.

Meili, N., Manoli, G., Burlando, P., Bou-Zeid, E., Chow, W.T.L., Coutts, A.M., et al., 2020. An urban ecohydrological model to quantify the effect of vegetation on urban climate and hydrology (UT&C v1.0). Geoscientific Model Development 13, 335—362. Available from: https://doi.org/10.5194/gmd-13-335-2020.

Melnikov, V., Krzhizhanovskaya, V.V., Sloot, P.M.A., 2017. Models of pedestrian adaptive behaviour in hot outdoor public spaces. Procedia Computer Science 108, 185−194. Available from: https://doi.org/10.1016/j.procs.2017.05.006.

Melnikov, V., Krzhizhanovskaya, V.V., Lees, M.H., Sloot, P.M.A., 2018. System dynamics of human body thermal regulation in outdoor environments. Building and Environment 143, 760−769. Available from: https://doi.org/10.1016/j.buildenv.2018.07.024.

Mills, G., 1997. Building density and interior building temperatures: a physical modeling experiment. Physical Geography 18, 195−214. Available from: https://doi.org/10.1080/02723646.1997.10642616.

Mirzaei, P.A., 2015. Recent challenges in modeling of urban heat island. Sustainable Cities and Society 19, 200−206. Available from: https://doi.org/10.1016/j.scs.2015.04.001.

Moradi, M., Dyer, B., Nazem, A., Nambiar, M.K., Nahian, M.R., Bueno, B., et al., 2019. The Vertical City Weather Generator (VCWG v1.0.0). Geoscientific Model Development Discussions 1−42. Available from: https://doi.org/10.5194/gmd-2019-176.

Mussetti, G., Brunner, D., Allegrini, J., Wicki, A., Schubert, S., Carmeliet, J., 2020. Simulating urban climate at sub-kilometre scale for representing the intra-urban variability of Zurich, Switzerland. International Journal of Climatology 40, 458−476. Available from: https://doi.org/10.1002/joc.6221.

Myrup, L.O., 1969. A numerical model of the urban heat island. Journal of Appllied Meteorology and Climatology 8, 908−918 Available from: https://doi.org/10.1175/1520-0450(1969)008 < 0908:ANMOTU > 2.0.CO;2.

Nakajima, K., Takane, Y., Kikegawa, Y., Furuta, Y., Takamatsu, H., 2021. Human behaviour change and its impact on urban climate: restrictions with the G20 Osaka Summit and COVID-19 outbreak. Urban Climate 35, 100728. Available from: https://doi.org/10.1016/j.uclim.2020.100728.

Nathan Kutz, J., 2017. Deep learning in fluid dynamics. Journal of Fluid Mechanics 814, 1−4. Available from: https://doi.org/10.1017/jfm.2016.803.

Nazarian, N., Acero, J.A., Norford, L., 2019a. Outdoor thermal comfort autonomy: performance metrics for climate-conscious urban design. Building and Environment 155, 145−160. Available from: https://doi.org/10.1016/j.buildenv.2019.03.028.

Nazarian, N., Dumas, N., Kleissl, J., Norford, L., 2019b. Effectiveness of cool walls on cooling load and urban temperature in a tropical climate. Energy and Buildings. Available from: https://doi.org/10.1016/j.enbuild.2019.01.022.

Nazarian, N., Kleissl, J., 2016. Realistic solar heating in urban areas: air exchange and street-canyon ventilation. Building and Environment 95, 75−93. Available from: https://doi.org/10.1016/j.buildenv.2015.08.021.

Nazarian, N., Krayenhoff, E.S., Martilli, A., 2020a. A one-dimensional model of turbulent flow through "urban" canopies (MLUCM v2.0): updates based on large-eddy simulation. Geoscientific Model Development 13, 937−953. Available from: https://doi.org/10.5194/gmd-13-937-2020.

Nazarian, N., Krayenhoff, S., Martilli, A., 2020b. Developing an urban canopy Model for neighborhood-scale thermal exposure assessment, In: Proceedings of the 100th American Meteorological Society Annual Meeting. AMS.

Nazarian, N., Lee, J.K.W., 2021. Personal assessment of urban heat exposure: a systematic review. Environmental Research Letters 16, 033005. Available from: https://doi.org/10.1088/1748-9326/abd350.

Nazarian, N., Martilli, A., Kleissl, J., 2018a. Impacts of realistic urban heating, Part I: spatial variability of mean flow, turbulent exchange and pollutant dispersion. Boundary−Layer Meteorology. Available from: https://doi.org/10.1007/s10546-017-0311-9.

Nazarian, N., Martilli, A., Norford, L., Kleissl, J., 2018b. Impacts of realistic urban heating. Part II: air quality and city breathability. Boundary − Layer Meteorology 168, 321−341. Available from: https://doi.org/10.1007/s10546-018-0346-6.

Nazarian, N., Sin, T., Norford, L., 2018c. Numerical modeling of outdoor thermal comfort in 3D. Urban Climate 26, 212−230. Available from: https://doi.org/10.1016/j.uclim.2018.09.001.

Ng, E., Yuan, C., Chen, L., Ren, C., Fung, J.C.H., 2011. Improving the wind environment in high-density cities by understanding urban morphology and surface roughness: a study in Hong Kong. Landscape and Urban Planning 101, 59−74. Available from: https://doi.org/10.1016/j.landurbplan.2011.01.004.

Nice, K.A., Coutts, A.M., Tapper, N.J., 2018. Development of the VTUF-3D v1.0 urban micro-climate model to support assessment of urban vegetation influences on human thermal comfort. Urban Climate 24, 1052−1076. Available from: https://doi.org/10.1016/j.uclim.2017.12.008.

Oke, T.R., 1982. The energetic basis of the urban heat island. Quarterly Journal of the Royal Meteorological Society. Available from: https://doi.org/10.1002/qj.49710845502.

Oleson, K.W., Feddema, J., 2020. Parameterization and surface data improvements and new capabilities for the Community Land Model Urban (CLMU). Journal of Advances in Modelling Earth Systems 12, e2018MS001586. Available from: https://doi.org/10.1029/2018MS001586.

Oleson, K.W., Bonan, G.B., Feddema, J., Vertenstein, M., Grimmond, C.S.B., 2008. An urban parameterization for a global climate model. Part I: formulation and evaluation for two cities. Journal of Appllied Meteorology and Climatology 47, 1038−1060. Available from: https://doi.org/10.1175/2007JAMC1597.1.

Ortiz, L., González, J.E., Lin, W., 2018. Climate change impacts on peak building cooling energy demand in a coastal megacity. Environmental Research Letters 13, 094008. Available from: https://doi.org/10.1088/1748-9326/aad8d0.

Pantelic, J., Nazarian, N., Meggers, F., Lee, J.K.W., Miller, C., Licinia, D., 2021. Transformational IoT technologies for air quality and thermal comfort at the urban, building, and human scales. Submitted to Environment International.

Papangelis, G., Tombrou, M., Dandou, A., Kontos, T., 2012. An urban "green planning" approach utilizing the Weather Research and Forecasting (WRF) modeling system. A case study of Athens, Greece. Landscape and Urban Planning 105, 174−183. Available from: https://doi.org/10.1016/j.landurbplan.2011.12.014.

Plate, E.J., 1999. Methods of investigating urban wind fields—physical models. Atmospheric Environment (Oxford, England: 1994) 33, 3981−3989. Available from: https://doi.org/10.1016/S1352-2310(99)00140-5.

Porson, A., Harman, I.N., Bohnenstengel, S.I., Belcher, S.E., 2009. How many facets are needed to represent the surface energy balance of an urban area? Boundary − Layer Meteorology 132, 107−128. Available from: https://doi.org/10.1007/s10546-009-9392-4.

Porson, A., Clark, P.A., Harman, I.N., Best, M.J., Belcher, S.E., 2010. Implementation of a new urban energy budget scheme in the MetUM. Part I: description and idealized simulations. Quarterly Journal of the Royal Meteorological Society 136, 1514−1529. Available from: https://doi.org/10.1002/qj.668.

Redon, E., Lemonsu, A., Masson, V., 2020. An urban trees parameterization for modeling microclimatic variables and thermal comfort conditions at street level with the Town Energy Balance model (TEB-SURFEX v8.0). Geoscientific Model Development 13, 385−399. Available from: https://doi.org/10.5194/gmd-13-385-2020.

Reinhart, C.F., Cerezo Davila, C., 2016. Urban building energy modeling—a review of a nascent field. Building and Environment 97, 196−202. Available from: https://doi.org/10.1016/j.buildenv.2015.12.001.

Ribeiro, I., Martilli, A., Falls, M., Zonato, A., Villalba, G., 2021. Highly resolved WRF-BEP/BEM simulations over Barcelona urban area with LCZ. Atmospheric Research 248, 105220. Available from: https://doi.org/10.1016/j.atmosres.2020.105220.

Ridder, K.D., De Ridder, K., Lauwaet, D., Maiheu, B., 2015. UrbClim—a fast urban boundary layer climate model. Urban Climate. Available from: https://doi.org/10.1016/j.uclim.2015.01.001.

Roth, M., Lim, V.H., 2017. Evaluation of canopy-layer air and mean radiant temperature simulations by a microclimate model over a tropical residential neighbourhood. Building and Environment 112, 177−189. Available from: https://doi.org/10.1016/j.buildenv.2016.11.026.

Sailor, D.J., 2011. A review of methods for estimating anthropogenic heat and moisture emissions in the urban environment: estimating anthropogenic heat and moisture emissions. International Journal of Climatology 31, 189−199. Available from: https://doi.org/10.1002/joc.2106.

Salamanca, F., Martilli, A., 2010. A new building energy model coupled with an urban canopy parameterization for urban climate simulations—part II. Validation with one dimension off-line simulations. Theoretical and Applied Climatology. Available from: https://doi.org/10.1007/s00704-009-0143-8.

Salamanca, F., Krpo, A., Martilli, A., Clappier, A., 2010. A new building energy model coupled with an urban canopy parameterization for urban climate simulations—part I. formulation, verification, and sensitivity analysis of the model. Theoretical and Applied Climatology.

Salamanca, F., Georgescu, M., Mahalov, A., Moustaoui, M., Wang, M., 2014. Anthropogenic heating of the urban environment due to air conditioning: anthropogenic heating due to AC. Journal of Geophysical Research 119, 5949−5965. Available from: https://doi.org/10.1002/2013jd021225.

Sanaieian, H., Tenpierik, M., van den Linden, K., Mehdizadeh Seraj, F., Mofidi Shemrani, S.M., 2014. Review of the impact of urban block form on thermal performance, solar access and ventilation. Renewable and Sustainable Energy Reviews 38, 551−560. Available from: https://doi.org/10.1016/j.rser.2014.06.007.

Santamouris, M., 2014. On the energy impact of urban heat island and global warming on buildings. Energy and Buildings 82, 100−113. Available from: https://doi.org/10.1016/j.enbuild.2014.07.022.

Sarrat, C., Lemonsu, A., Masson, V., Guedalia, D., 2006. Impact of urban heat island on regional atmospheric pollution. Atmospheric Environment (Oxford, England: 1994).

Schoetter, R., Masson, V., Bourgeois, A., Pellegrino, M., Lévy, J.-P., 2017. Parametrisation of the variety of human behaviour related to building energy consumption in the Town Energy Balance (SURFEX-TEB v. 8.2). Geoscientific Model Development 10, 2801−2831. Available from: https://doi.org/10.5194/gmd-10-2801-2017.

Schoetter, R., Hidalgo, J., Jougla, R., Masson, V., Rega, M., Pergaud, J., 2020a. A statistical−dynamical downscaling for the urban heat island and building energy consumption—analysis of its uncertainties. Journal of Applied Meteorology and Climatology 59, 859−883. Available from: https://doi.org/10.1175/JAMC-D-19-0182.1.

Schoetter, R., Kwok, Y.T., de Munck, C., Lau, K.K.L., Wong, W.K., Masson, V., 2020b. Multi-layer coupling between SURFEX-TEB-v9. 0 and Meso-NH-v5. 3 for modelling the urban climate of high-rise cities. Geoscientific Model Development 13, 5609−5643.

Sharma, A., Conry, P., Fernando, H.J.S., Hamlet, A.F., Hellmann, J.J., Chen, F., 2016. Green and cool roofs to mitigate urban heat island effects in the Chicago metropolitan area: evaluation with a regional climate model. Environmental Research Letters 11, 064004. Available from: https://doi.org/10.1088/1748-9326/11/6/064004.

Sharma, A., Wuebbles, D.J., Kotamarthi, R., 2021. The need for urban-resolving climate modeling across scales. AGU Advances 2. Available from: https://doi.org/10.1029/2020av000271.

Stewart, I.D., Oke, T.R., 2012. Local climate zones for urban temperature studies. Bulletin of the American Meteorological Society 93, 1879−1900. Available from: https://doi.org/10.1175/BAMS-D-11-00019.1.

Sun, T., Grimmond, C.S.B., Ni, G.-H., 2016. How do green roofs mitigate urban thermal stress under heat waves?: green roofs reduce urban thermal stress. Journal of Geophysical Research 121, 5320−5335. Available from: https://doi.org/10.1002/2016jd024873.

Takane, Y., Kikegawa, Y., Hara, M., Ihara, T., Ohashi, Y., Adachi, S.A., et al., 2017. A climatological validation of urban air temperature and electricity demand simulated by a regional climate model coupled with an urban canopy model and a building energy model in an Asian megacity: a climatological validation of a RCM-UCM + BEM in an Asian megacity. International Journal of Climatology 37, 1035−1052. Available from: https://doi.org/10.1002/joc.5056.

Takane, Y., Kikegawa, Y., Hara, M., Grimmond, C.S.B., 2019. Urban warming and future air-conditioning use in an Asian megacity: importance of positive feedback. Npj Climate and Atmospheric Science 2, 39. Available from: https://doi.org/10.1038/s41612-019-0096-2.

Takane, Y., Ohashi, Y., Grimmond, C.S.B., Hara, M., Kikegawa, Y., 2020. Asian megacity heat stress under future climate scenarios: impact of air-conditioning feedback. Environmental Research Communications 2 (1), 015004.

Tartarini, F., Schiavon, S., 2020. pythermalcomfort: a Python package for thermal comfort research. SoftwareX 12, 100578. Available from: https://doi.org/10.1016/j.softx.2020.100578.

Tartarini, F., Schiavon, S., Cheung, T., Hoyt, T., 2020. CBE thermal comfort tool: online tool for thermal comfort calculations and visualizations. SoftwareX 12, 100563. Available from: https://doi.org/10.1016/j.softx.2020.100563.

Teixeira, J.C., Fallmann, J., Carvalho, A.C., Rocha, A., 2019. Surface to boundary layer coupling in the urban area of Lisbon comparing different urban canopy models in WRF. Urban Climate 28, 100454. Available from: https://doi.org/10.1016/j.uclim.2019.100454.

Thatcher, M., Hurley, P., 2012. Simulating Australian urban climate in a mesoscale atmospheric numerical model. Boundary − Layer Meteorology 142, 149−175. Available from: https://doi.org/10.1007/s10546-011-9663-8.

Theeuwes, N.E., Solcerová, A., Steeneveld, G.J., 2013. Modeling the influence of open water surfaces on the summertime temperature and thermal comfort in the city: surface water and urban temperatures. Journal of Geophysical Research 118, 8881−8896. Available from: https://doi.org/10.1002/jgrd.50704.

Tominaga, Y., Stathopoulos, T., 2013. CFD simulation of near-field pollutant dispersion in the urban environment: a review of current modeling techniques. Atmospheric Environment (Oxford, England: 1994) 79, 716−730. Available from: https://doi.org/10.1016/j.atmosenv.2013.07.028.

Tong, Z., Chen, Y., Malkawi, A., 2016. Defining the Influence Region in neighborhood-scale CFD simulations for natural ventilation design. Applied Energy 182, 625−633. Available from: https://doi.org/10.1016/j.apenergy.2016.08.098.

Toparlar, Y., Blocken, B., Maiheu, B., van Heijst, G.J.F., 2017. A review on the CFD analysis of urban microclimate. Renewable and Sustainable Energy Reviews 80, 1613−1640. Available from: https://doi.org/10.1016/j.rser.2017.05.248.

Trusilova, K., Früh, B., Brienen, S., Walter, A., Masson, V., Pigeon, G., et al., 2013. Implementation of an urban parameterization scheme into the regional climate model COSMO-CLM. Journal of Appllied Meteorology and Climatology 52, 2296−2311. Available from: https://doi.org/10.1175/JAMC-D-12-0209.1.

Uehara, K., Murakami, S., Oikawa, S., Wakamatsu, S., 2000. Wind tunnel experiments on how thermal stratification affects flow in and above urban street canyons. Atmospheric Environment (Oxford, England: 1994) 34, 1553−1562. Available from: https://doi.org/10.1016/S1352-2310(99)00410-0.

Vanos, J.K., Baldwin, J.W., Jay, O., Ebi, K.L., 2020. Simplicity lacks robustness when projecting heat-health outcomes in a changing climate. Nature Communications 11, 6079. Available from: https://doi.org/10.1038/s41467-020-19994-1.

Wang, K., Li, Y., Li, Y., Lin, B., 2018a. Stone forest as a small-scale field model for the study of urban climate. International Journal of Climatology 38, 3723–3731. Available from: https://doi.org/10.1002/joc.5536.

Wang, Y., Li, Y., Di Sabatino, S., Martilli, A., Chan, P.W., 2018b. Effects of anthropogenic heat due to air-conditioning systems on an extreme high temperature event in Hong Kong. Environmental Research Letters 13, 034015.

Wang, C., Wang, Z.-H., Ryu, Y.-H., 2021. A single-layer urban canopy model with transmissive radiation exchange between trees and street canyons. Building and Environment 191, 107593. Available from: https://doi.org/10.1016/j.buildenv.2021.107593.

Willems, P., 2012. Impacts of Climate Change on Rainfall Extremes and Urban Drainage Systems. IWA Publishing.

Willems, P., Arnbjerg-Nielsen, K., Olsson, J., Nguyen, V.T.V., 2012. Climate change impact assessment on urban rainfall extremes and urban drainage: methods and shortcomings. Atmospheric Research 103, 106–118. Available from: https://doi.org/10.1016/j.atmosres.2011.04.003.

Wise, D.J., Boppana, V.B.L., Li, K.W., Poh, H.J., 2018. Effects of minor changes in the mean inlet wind direction on urban flow simulations. Sustainable Cities and Society 37, 492–500. Available from: https://doi.org/10.1016/j.scs.2017.11.041.

Wong, M.S., Nichol, J.E., To, P.H., Wang, J., 2010. A simple method for designation of urban ventilation corridors and its application to urban heat island analysis. Building and Environment 45, 1880–1889. Available from: https://doi.org/10.1016/j.buildenv.2010.02.019.

World Meteorological Organization, 2020. 2020 State of Climate Services Report.

Wouters, H., Demuzere, M., Ridder, K.D., van Lipzig, N.P.M., 2015. The impact of impervious water-storage parametrization on urban climate modelling. Urban Climate 11, 24–50. Available from: https://doi.org/10.1016/j.uclim.2014.11.005.

Wouters, H., Demuzere, M., Blahak, U., Fortuniak, K., Maiheu, B., Camps, J., et al., 2016. The efficient urban canopy dependency parametrization (SURY) v1.0 for atmospheric modelling: description and application with the COSMO-CLM model for a Belgian summer. Geoscientific Model Development 9, 3027–3054. Available from: https://doi.org/10.5194/gmd-9-3027-2016.

Yaghoobian, N., Kleissl, J., 2012. An indoor–outdoor building energy simulator to study urban modification effects on building energy use—model description and validation. Energy and Buildings. Available from: https://doi.org/10.1016/j.enbuild.2012.07.019.

Yin, S., Li, Y., Sandberg, M., Lam, K., 2017. The effect of building spacing on near-field temporal evolution of triple building plumes. Building and Environment 122, 35–49. Available from: https://doi.org/10.1016/j.buildenv.2017.05.030.

Yuan, C., Ng, E., 2012. Building porosity for better urban ventilation in high-density cities—a computational parametric study. Building and Environment.

Zhang, Y., Gu, Z., Yu, C.W., 2020. Impact factors on airflow and pollutant dispersion in urban street canyons and comprehensive simulations: a review. Current Pollution Reports 6, 425–439. Available from: https://doi.org/10.1007/s40726-020-00166-0.

Zhao, L., Oppenheimer, M., Zhu, Q., Baldwin, J.W., Ebi, K.L., Bou-Zeid, E., et al., 2018. Interactions between urban heat islands and heat waves. Environmental Research Letters 13, 034003. Available from: https://doi.org/10.1088/1748-9326/aa9f73.

Zhao, L., Oleson, K., Bou-Zeid, E., Krayenhoff, E.S., Bray, A., Zhu, Q., et al., 2021. Global multi-model projections of local urban climates. Nature Climate Change. Available from: https://doi.org/10.1038/s41558-020-00958-8.

CHAPTER 5

Urban overheating—energy, environmental, and heat-health implications

Matthaios Santamouris
School of Built Environment, Faculty of Arts, Design and Architecture, University of New South Wales (UNSW), Sydney, NSW, Australia

5.1 Introduction

Urban overheating is the most documented phenomenon of climate change. In more than 450 major cities in the world, urban overheating is well-documented and quantified through detailed experiments (Oke et al., 1991; Santamouris, 2016a,b; Santamouris, 2015a,b; Santamouris, 2018). The amplitude of urban overheating may be as high as 10°C, with an average value around 5°C−6°C (Santamouris, 2015a,b). Higher urban ambient temperatures result from the positive thermal balance of cities caused by several reasons, as summarized in Chapter 1. Among the major reasons driving the temperature increase in cities are the absorption of solar radiation by the urban fabric, the canyon radiative geometry that traps the emitted infrared radiation, the released anthropogenic heat, the reduction of the evapotranspiration fluxes in cities, the urban heat island (UHI) effect that blocks the emitted infrared radiation, the reduced ventilation potential of urban places, and the potential advective heat fluxes from neighboring warm zones (Santamouris and Kolokotsa, 2016). As already described in Chapter 3, the recent strengthening of the magnitude and the increase of the frequency of extreme heat events like heat waves intensify the amplitude of urban overheating and extend the duration of the hot spells in cities (Founda and Santamouris, 2017; Khan et al., 2020, 2021; Pyrgou et al., 2020).

An increase in the urban ambient temperature causes a serious impact on the quality of life of urban citizens, affects the global environmental quality of cities, puts at risk the health of urban dwellers, and causes serious vulnerability and economic problems. As discussed in Chapter 1, urban overheating affects both the energy generation and energy demand sectors by increasing the energy consumption for cooling, raising the peak electricity demand during the warm period obliging utilities to build additional power plants, reduces the generation efficiency of the power plants, and lowers the capacity of the distribution systems. Apart from the impact on the energy sector, higher urban temperatures increase the concentration of harmful pollutants like ground-level ozone and particulate matter by accelerating the photochemical reaction in the atmosphere. At the same time, overheating affects the vulnerability of the low-income households living in deprived urban zones suffering from higher urban temperatures, while it seriously increases the levels of heat-related mortality and hospital admissions.

Extensive work carried out in the recent years has documented and quantified the specific impact of urban overheating. However, there is a need for a more global and holistic methodology considering the synergies and the trade-offs between the various sectors.

This chapter aims to present in a comprehensive way, document and quantify the specific impact of urban overheating, investigate the synergies between the affected sectors, and highlight the interdisciplinary nature of the regional and global climate change.

5.2 Impact of urban overheating on energy generation and energy supply systems

While the impact of the temperature increase in cities is quite evident, it is essential to classify, document, and quantify the different ways the energy sector is affected. A classification of the specific impacts of urban overheating on energy generation and supply systems is given in Table 5.1 (Chandramowli and Felder, 2014; Santamouris, 2020).

Using the analysis provided in Santamouris (2020), there are six main types of studies evaluating the impact of higher urban temperatures on the energy supply and demand systems. Most of the studies focus on how urban overheating affects the heating and cooling demand of buildings and the future energy implications of global climate change. In particular, the six types of studies focus on the following:

Type 1: Impact of urban overheating on the energy consumption of reference buildings. This type of study usually evaluates the energy consumption of a reference building under urban and rural climatic data. The energy consumption for heating and cooling is calculated through building energy simulation tools, and the increase or the

Table 5.1 Implications of ambient and urban overheating on demand and supply-side components of electricity.

Implications of ambient and urban overheating on demand-side components of electricity

Demand system component	Ambient and urban overheating effect	Implications
Cooling load of buildings	Higher ambient temperature in summer	Increase of the cooling demand of buildings
Peak electricity demand	Higher ambient temperature in summer	Increase of the peak electricity demand
Load duration curves	Important change of air conditioning profile	Higher demand curve peaks and much greater load variability may increase chances of breaching the market price cap
Nontemperature sensitive demand	Increased cooling water temperatures	Generation curtailments and potential interruption of power to avoid blackouts

Implications of ambient and urban overheating on supply-side components of electricity

Supply system component	Overheating effect	Implications
Thermal electricity generation plants and components	Increased ambient temperatures Increased water temperatures	Decreased efficiency of electricity generating equipment like gas turbines, coal power plants, etc
Transmission network	Higher ambient temperatures and longer spells of dry weather	Power disruptions, increased cost of adaptation designs. Reduced equipment lifetime. Reduced power carrying capacity of transmission lines may cause disruptions because of the power line sagging. Disruptions because of the power line sagging
Substations and transformers	High ambient temperatures	Increased losses within substations and transformers

(Continued)

Table 5.1 (Continued)

Fuel stock	High ambient temperatures	Coal stocks may spontaneously combust or self-ignite
Power plants	Increased ambient temperature and extreme events Increase of the peak electricity demand	Utilities must build additional power plants to cover the peaks Increased cost of electricity production during the peak hours

Source: Adapted from Santamouris, M., 2020. Recent progress on urban overheating and heat island research. Integrated assessment of the energy, environmental, vulnerability and health impact. Synergies with the global climate change. Energy and Buildings 207, 109482. https://doi.org/10.1016/j.enbuild.2019.109482. Chandramowli, S.N., Felder, F.A., 2014. Impact of climate change on electricity systems and markets—a review of models and forecasts. Sustainable Energy Technologies and Assessments 5, 62—74. https://doi.org/10.1016/j.seta.2013.11.003.

decrease of the energy load under urban overheating conditions is evaluated.

Type 2: Impact of urban overheating on the temporal variation of the energy consumption of buildings. These studies are based on the simulation of the energy consumption of buildings using multiyear climatic data. The aim is to evaluate the temporal variation of the energy consumption caused by the local overheating.

Type 3: Impact of overheating on the energy consumption of the total building stock of a city. These studies aim to evaluate the impact of local or global climate change on the total energy consumption of the building stock of a city.

Type 4: Impact of the future overheating on the energy consumption of buildings. These studies are using climatic data predicted by global climatic models to assess the future energy consumption of buildings and the potential impact of global and local climate change.

Type 5: Impact of overheating on the global electricity consumption of a city or a country. Usually, these studies correlate the electricity energy consumption against the corresponding levels of the ambient temperature and establish associations that are able to predict the impact of higher temperatures on the electricity demand.

Type 6: Impact of overheating on the peak electricity demand. These studies associate the demand for power with the corresponding ambient temperature and predict the variability of the power demand as a function of the temperature magnitude.

Apart from the aforementioned six types of studies, we propose a seventh category as in the following:

Type 7: Impact of overheating on the performance of the electricity generation and distribution systems. This type of study investigates the impact of higher ambient temperatures on the efficiency of the power plants and the potential effect on the electricity distribution systems.

In the following, the actual developments and the main conclusions drawn from each of the above seven types of studies are presented.

5.2.1 Impact of urban overheating on the energy consumption of reference buildings

Twenty-two studies have evaluated the impact of urban overheating on the energy consumption of reference buildings (Australia, 2012; Bagiorgas and Mihalakakou, 2016; Bründl and Höppe, 1984; Calice et al., 2015; Chandler, 1965; Chen et al., 2012; Fanchiotti et al., 2012; Hassid et al., 2000; Hwang et al., 2017; Ignatius et al., 2016; Kolokotroni et al., 2007, 2012; Magli et al., 2014; Palme et al., 2017; Papanastasiou et al., 2013; Radhi and Sharples, 2013; Rong, 2006; Santamouris et al., 2001; Skelhorn et al., n.d.; Street et al., n.d.; Zinzi et al., 2018; Zinzi and Carnielo, 2017). Studies are undertaken for Athens and Volos, Greece, London, UK, Munich, Germany, Rome, Italy, Boston, New York, California, Texas, USA, Melbourne, Australia, Bahrain, Lima, Peru, Valparaiso and Antofagasta, Chile, Guayaquil, Equator, Modena, Rome, Italy, Taichung, Taiwan, and Manchester, UK. The studies are analyzed and presented in Santamouris (2014, 2020). Sixteen of the case studies refer to residential buildings, and the rest to tertiary buildings. The calculated cooling demand of all the reference buildings without considering the energy impact of the urban overheating varies between 0.5 kWh/m^2/y and 210 kWh/m^2/y (Fig. 5.1B). When the influence of urban overheating is considered, then the cooling load varies between 2 and 230 kWh/m^2/y (Fig. 5.1C). It is clear that almost 90% of the data are below a consumption of 45 kWh/m^2/y for the reference case and 50 kWh/m^2/y for the urban scenario. In Fig. 5.1A, the cumulative frequency distribution of the cooling penalty per degree of temperature increase is given. As shown, it varies between 0.5 and 8 kWh/m^2/y/°C. The magnitude of the urban

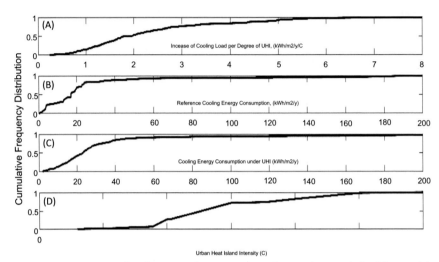

Figure 5.1 Impact of urban overheating on reference—typical buildings. (A) Cumulative frequency distribution of the increase of the cooling load of the buildings per degree of UHI. (B) Cumulative frequency distribution of the cooling energy consumption of the buildings under reference climatic conditions. (C) Cumulative frequency distribution of the cooling energy consumption of the buildings under urban—UHI conditions. (D) Cumulative frequency distribution of the UHI intensity for all considered cases.

Source: Santamouris, M., 2020. Recent progress on urban overheating and heat island research. Integrated assessment of the energy, environmental, vulnerability and health impact. Synergies with the global climate change. Energy and Buildings 207, 109482. https://doi.org/10.1016/j.enbuild.2019.109482.

overheating in all the considered cities varied between 0.5°C and 7°C, with almost 90% of the cases below 4.5°C (Fig. 5.1D) (Fig. 5.2).

The analysis of the 22 considered cases has shown that the average increase of the cooling load for all types of buildings is close to 12%, while the corresponding increase for the tertiary buildings is close to 18% (Santamouris, 2020). Urban overheating is found to cause an energy penalty between 0.1 and 20 kWh/m^2/y/°C, with an average value close to 2.3 kWh/m^2/y/C for the residential and 2.4 kWh/m^2/y/C for the tertiary buildings. As expected, higher cooling penalties are found for buildings presenting a high than average or low cooling load (Fig. 5.3).

The average decrease of the heating demand in buildings is found close to 19%, with the highest decrease observed in heating-dominated cities. It is evident that the specific increase/decrease of the energy load is

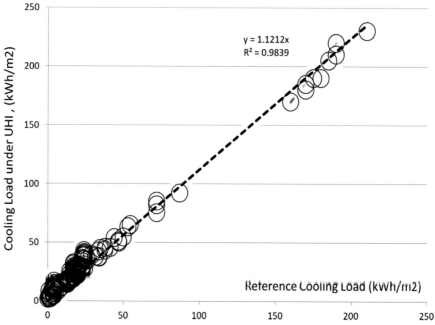

Figure 5.2 Correlation between the calculated cooling energy demand under reference, rural, and, urban overheating conditions.

Source: Santamouris, M., 2020. Recent progress on urban overheating and heat island research. Integrated assessment of the energy, environmental, vulnerability and health impact. Synergies with the global climate change. Energy and Buildings 207, 109482. https://doi.org/10.1016/j.enbuild.2019.109482.

a function of the local climate and the characteristics of the buildings and projects. As reported in Santamouris (2014), the increase of the cooling demand is significant in climatic zones presenting an average summer ambient temperature above 27°C.

While the characteristics of the local climate highly determine the magnitude of the urban overheating and the subsequent variability of the energy load, the thermal balance of the considered buildings leads the impact of the local climate conditions. Given that the balance point temperature of the tertiary buildings is low because of the high internal gains, cooling is needed even in heating-dominated zones. It is shown that in London, United Kingdom, office buildings with high internal gains present a higher increase of the cooling demand than the corresponding decrease of heating needs (Kolokotroni et al., 2012).

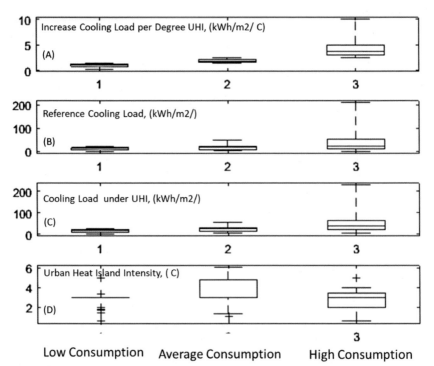

Low Consumption   Average Consumption   High Consumption

Figure 5.3 (A) Variability of the cooling penalty for low, average, and high cooling loads of the reference building. (B) Levels of the reference cooling load. (C) Levels of the cooling load under UHI conditions. (D) The corresponding urban heat island intensity.
Source: Santamouris, M., 2020. Recent progress on urban overheating and heat island research. Integrated assessment of the energy, environmental, vulnerability and health impact. Synergies with the global climate change. Energy and Buildings 207, 109482. https://doi.org/10.1016/j.enbuild.2019.109482.

5.2.2 Impact of urban overheating on the temporal variation of the energy consumption of buildings

Global and local climate change increases the ambient temperature in cities. The warming trend varies between cities (Fig. 5.4) and can be as high as 1°C per decade while the average is close to 0.4°C/decade. To assess the potential temporal increase of the cooling load and the possible decrease of the heating demand, building energy simulations are carried out using multiyear climatic data. Relevant simulation studies for both residential and commercial buildings are available for 18 different cities or regions, and in particular: Athens, Larisa, Corfu and Heraklion in Greece, Honk Kong, Nicosia, Paphos, Limassol, Larnaca, Famagusta, and Kerynia in Cyprus, Zurich, Geneva, Lugano and Davos in Switzerland, Phoenix,

GLOBAL OVERHEATING TREND

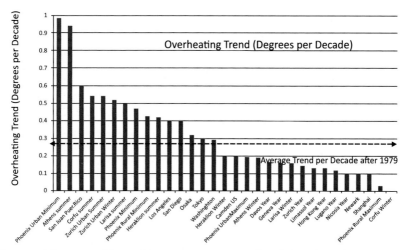

Figure 5.4 Increase of the ambient temperature per decade in selected cities.
Source: Santamouris, M., 2020. Recent progress on urban overheating and heat island research. Integrated assessment of the energy, environmental, vulnerability and health impact. Synergies with the global climate change. Energy and Buildings 207, 109482. https://doi.org/10.1016/j.enbuild.2019.109482.

Washington DC, Puerto Rico in the United States, and Resolute in Canada, (Christenson et al., 2006; Crawley, 2008; Doddaballapur et al., n.d.; Frank, 2005; Golden et al., 2006; Kapsomenakis et al., 2013; Lam et al., 2004; Lange et al., n.d.).

A full analysis of the available information is offered in Santamouris (2014). Fig. 5.5 reports the cumulative frequency distribution of the estimated cooling demand for two reference years, 1970 and 2010. It also reports the range of the overheating trend and the increase of the cooling demand per year and degree of overheating (Santamouris, 2020). As observed, for 1970, the cooling demand of the considered buildings varies between 14 and 100 $kWh/m^2/y$, while it increases to 20−125 $kWh/m^2/y$ for 2010. This corresponds to an overheating trend between 0.1°C and 0.94°C per decade. The mean rise of the cooling demand is close to 11 kWh/m^2y and corresponds to an increase close to 23% of the 1970 cooling demand. The estimated rise of the cooling demand per year and degree of temperature increase is found to vary between 67 and 161 $kWh/m^2/y/°C$, with a mean value close to 76 $kWh/m^2/y/°C$. As it concerns the possible decrease of the heating demand, it is estimated that

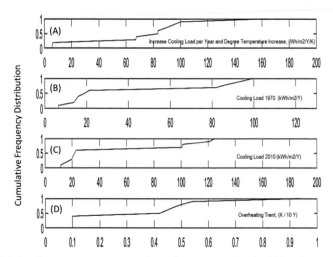

Figure 5.5 Cooling energy consumption of representative buildings in 18 cities for the period 1970–2019. (A) Cumulative frequency distribution of the increase of the cooling load per year and degree of temperature increase. (B) Cumulative frequency distribution of the calculated cooling load for 1970. (C) Cumulative frequency distribution of the calculated cooling load for 2010. (D) Cumulative frequency distribution of the overheating trent.
Source: Santamouris, M., 2020. Recent progress on urban overheating and heat island research. Integrated assessment of the energy, environmental, vulnerability and health impact. Synergies with the global climate change. Energy and Buildings 207, 109482. https://doi.org/10.1016/j.enbuild.2019.109482.

it is decreased by 19% during the considered period, while the corresponding sum of the heating and cooling needs increased by 11%.

5.2.3 Impact of overheating on the energy consumption of the total building stock of a city

Knowledge of the energy consumption as well as of the spatial and temporal variation of reference buildings is very useful; however, it does not reflect the global impact of urban and regional overheating on the energy consumption of cities. Several studies have assessed the energy penalty induced by urban overheating on the total energy consumption of cities (Ding et al., 2018; Hassid et al., 2000; Santamouris et al., 2001, 2007). The existing studies refer to: Athens, Greece, Western Athens, Greece, Tokyo, Japan, Beijing, and Qingdao, China. The existing results and data are analyzed in Khan et al. (2020) and Santamouris (2020), and it is proposed to assess the global impact of urban overheating using the four following energy indices:

GEPS: The global energy penalty per unit of city surface (kWh/m^2). This index is calculated to vary between 0.13 and $\pm$ 5.5 kWh/m^2, with a mean value around 2.3 (1.5) kWh/m^2.

GEPSI: The global energy penalty per unit of city area and per degree of the UHI intensity ($kWh/m^2/K$). This index is found to vary between 0.12 and 2.22 $kWh/m^2/°C$, with a mean value close to 0.73 $\pm$ (0.64) $kWh/m^2/°C$.

GEPP: The global energy penalty per person (kWh/p). This index varies between 104 and 305 kWh/p, with a mean value close to 230 $\pm$ (120) kWh/p.

GEPPI: The global energy penalty per person and per degree of the UHI intensity ($kWh/p/K$). This index varies between 20 and 154 $kWh/p/C$, with a mean value close to 78 $\pm$ (47) $kWh/p/°C$.

Given the low number of available data, it is evident that the above figures may change once more data are added.

5.2.4 Impact of the future overheating on the energy consumption of buildings

Numerous studies have attempted to evaluate the impact of the future overheating caused by global climate change on the energy consumption of buildings. Almost all studies are based on future climatic projections without having to consider the impact of local overheating. A complete analysis of 114 articles predicting the future energy consumption of reference office buildings is performed and reported in Santamouris (2016a,b). As noted, the reference cooling load at time zero differed between 12 and 360 $kWh/m^2/y$, varying as a function of the building characteristics, the local climate, and the set point temperature. In parallel, the estimated cooling penalty induced by the global overheating varied between 1 and 86 $kWh/m^2/y$. Figs. 5.6 and 5.7 present the relation and the variability between the expected increase of the cooling energy consumption ($kWh/m^2/y$) and the current cooling degree days and the corresponding ambient temperature for each of the considered locations. It is concluded that the lower the number of the cooling degree days, the lower the reference cooling load and the lower the calculated cooling penalty. In parallel, buildings presenting a high cooling consumption are located in places presenting a high number of cooling degree days, while the absolute value of the cooling penalty depends highly on the considered future climate

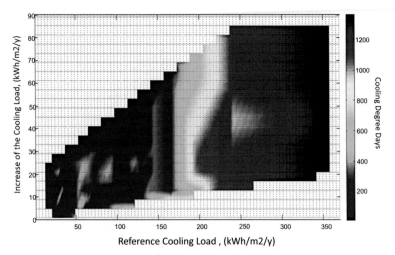

Figure 5.6 A 3D representation of the used current cooling energy consumption of office buildings (kWh/m²/y), the expected increase of the cooling energy consumption, (kWh/m²/y), and the current cooling degree days for each of the considered locations.
Source: From Santamouris, M., 2016a. Innovating to zero the building sector in Europe: minimising the energy consumption, eradication of the energy poverty and mitigating the local climate change. Solar Energy 128, 61–94. https://doi.org/10.1016/j.solener.2016.01.021. Santamouris, M., 2016b. Cooling the buildings—past, present and future. Energy and Buildings 128, 617–638. https://doi.org/10.1016/j.enbuild.2016.07.034.

scenario. A strong nonlinear correlation is also observed between the reference cooling load and the increase of the cooling demand per degree of temperature increase (Fig. 5.8). It is also found that the impact of the expected temperature increase is significantly lower in places with a high number of cooling degree days. Fig. 5.9 presents the relation of the reference cooling demand with the relative increase of the cooling consumption per degree of temperature rise. As shown, high reference cooling loads correspond to low relative increase of the cooling demand per degree of temperature rise.

5.2.5 Impact of overheating on the global electricity consumption of a city, or a country

High ambient temperatures increase the need for air conditioning and the corresponding electricity demand. The electricity penalty induced by the ambient overheating depends on the quality and the characteristics of the

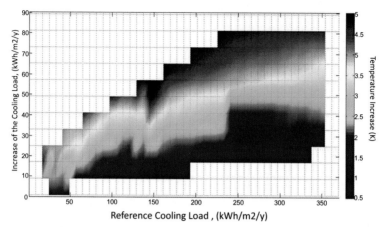

Figure 5.7 A 3D representation of the used current cooling energy consumption of office buildings (kWh/m²/y), the expected increase of the cooling energy consumption (kWh/m²/y), and the corresponding considered increase of the ambient temperature for each of the considered locations.

Source: From Santamouris, M., 2016a. Innovating to zero the building sector in Europe: minimising the energy consumption, eradication of the energy poverty and mitigating the local climate change. Solar Energy 128, 61—94. https://doi.org/10.1016/j.solener.2016.01.021. Santamouris, M., 2016b. Cooling the buildings—past, present and future. Energy and Buildings 128, 617—638. https://doi.org/10.1016/j.enbuild.2016.07.034.

building stock, the local climate, the degree of the penetration of air conditioning in the considered place, the used set point temperature, and the characteristics of the electricity network. Data evaluating the impact of the ambient temperature on the hourly, daily, and monthly electricity demand are available for 16 case studies, states or countries and in particular, Bangkok, Thailand, Spain, California, USA and part of the state, Athens, Greece, New Orleans, USA, Hong Kong, Ohio, Louisiana, Chicago, Maryland, Massachusetts, USA, Singapore, The Netherlands, and Delhi, India (Amato et al., 2005; Franco and Sanstad, 2007; Fung et al., 2006; Giannakopoulos and Psiloglou, 2006; Hayhoe et al., 2010; Hekkenberg et al., 2009; Joutz et al., 2013; Mirasgedis et al., 2006; Pardo et al., 2002; Rosenfeld et al., 1995; Ruth and Lin, 2006; Sailor and Muñoz, 1997; Wangpattarapong and Maneewan, 2008; Wong et al., 2010; Yadav et al., 2017).

The specific data are analyzed in detail in Santamouris et al. (2015). Fig. 5.10 presents the percentage increase of the electricity demand per degree of temperature increase for the above studies. It is found that the hourly, daily, or monthly electricity demand per degree of temperature

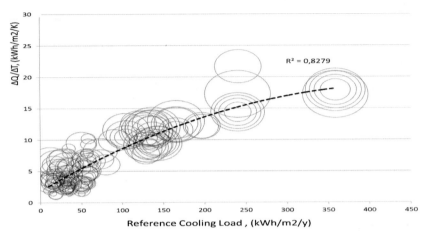

Figure 5.8 The relation between the reference cooling energy for the considered office buildings and of the increase of the cooling energy demand per degree of ambient temperature. The size of the circles presents the percentage increase of the reference cooling load.
Source: From Santamouris, M., 2016a. Innovating to zero the building sector in Europe: minimising the energy consumption, eradication of the energy poverty and mitigating the local climate change. Solar Energy 128, 61–94. https://doi.org/10.1016/j.solener.2016.01.021. Santamouris, M., 2016b. Cooling the buildings—past, present and future. Energy and Buildings 128, 617–638. https://doi.org/10.1016/j.enbuild.2016.07.034.

increase varies between 0.5% and 8.5%, with a mean value close to 4.2% (Wangpattarapong and Maneewan, 2008). The threshold ambient temperature where the electricity demand starts to increase varies between 11.7°C and 22°C (Santamouris et al., 2015). As expected, countries presenting a high penetration of air conditioning present a relatively higher elasticity value.

5.2.6 Impact of overheating on the peak electricity demand

Higher urban temperatures increase the need for air conditioning, and the peak electricity demand obliging utilities to build additional power plants to satisfy the electricity needs and increasing the cost of the electricity supply. The relation between the magnitude of the ambient temperature and the corresponding requirements of the peak electricity is studied for 13 cities or countries and particular: Tokyo, Japan, Thailand, Ontario, East Canada, Los Angeles, Washington, Dallas, Colorado Springs, Phoenix, and Tuscon, USA, Israel, part of Carolina USA, New South Wales, and

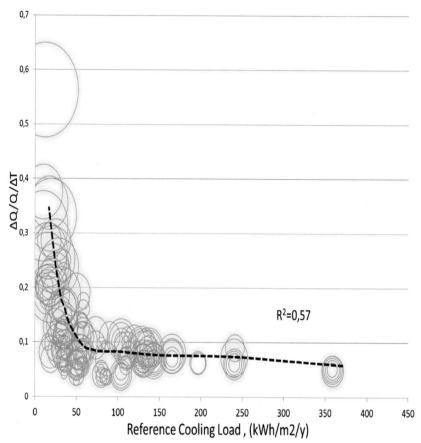

Figure 5.9. Relation between the reference cooling energy for the considered office buildings and of the ratio of the cooling energy increase and the reference cooling demand per degree of ambient temperature. The size of the circles presents the percentage increase of the reference cooling load.

Source: From Santamouris, M., 2016a. Innovating to zero the building sector in Europe: minimising the energy consumption, eradication of the energy poverty and mitigating the local climate change. Solar Energy 128, 61–94. https://doi.org/10.1016/j.solener.2016.01.021. Santamouris, M., 2016b. Cooling the buildings—past, present and future. Energy and Buildings 128, 617–638. https://doi.org/10.1016/j.enbuild.2016.07.034.

Darwin, Australia, (Akbari, 2009; Colombo et al., 1999; Franco and Sanstad, 2007; Haddad et al., 2018; Paolini et al., 2018; Segala et al., 1992; Synnefa et al., 2018; Yabe, 2005). The existing data have been analyzed in Santamouris et al. (2015) and Santamouris (2020). Fig. 5.11 reports the base electrical load and its percentage rise per degree of

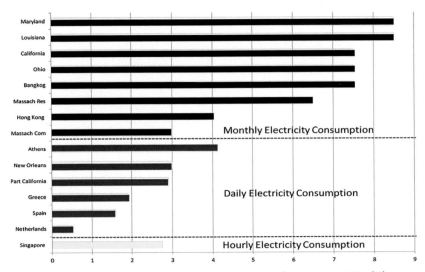

Figure 5.10 Increase of the electricity demand (%) per degree of ambient temperature rise for various countries.
Source: From Santamouris, M, 2015a. Regulating the damaged thermostat of the cities—status, impacts and mitigation challenges. Energy and Buildings 91, 43–56. https://doi.org/10.1016/j.enbuild.2015.01.027. Santamouris, M., 2015b. Analyzing the heat island magnitude and characteristics in one hundred Asian and Australian cities and regions. Science of the Total Environment 512, 582–598. https://doi.org/10.1016/j.scitotenv.2015.01.060.

ambient temperature increase (Santamouris et al., 2015). The analysis has shown that the rise in the peak electricity demand varies between 0.45% and 12.3% per degree of temperature increase. This is highly influenced by the local climatic conditions, the characteristics of the electricity network, the level of penetration of the air conditioning in the building stock, and the used setpoint temperatures. It is estimated that the penalty of the peak electricity demand per person and degree of temperature increase is around 21.9 ± (11.8) W/°C/person, while the average rise of the urban peak electricity demand is close to 3.7% or 215 MW per degree of temperature increase.

5.2.7 Impact of overheating on the performance of the electricity production and distribution systems

Three factors define the adequacy and the quality of the electricity supply systems (Bartos et al., 2016): (1) realizable generation capacity, (2) capacity of the transmission and distribution subsystems, and (3) expected peak

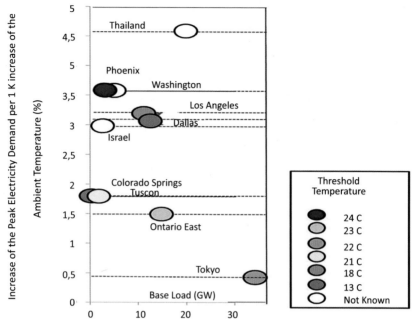

Figure 5.11 Increase of the peak electricity demand (%) per degree of ambient temperature rise.

Source: From Santamouris, M, 2015a. Regulating the damaged thermostat of the cities—status, impacts and mitigation challenges. Energy and Buildings 91, 43—56. https://doi.org/10.1016/j.enbuild.2015.01.027. Santamouris, M., 2015b. Analyzing the heat island magnitude and characteristics in one hundred Asian and Australian cities and regions. Science of the Total Environment 512, 582—598. https://doi.org/10.1016/j.scitotenv.2015.01.060.

load. Several studies have demonstrated that higher ambient temperatures affect in a negative way the efficiency of both the thermal and nuclear power plants. Table 5.2 summarizes the estimated current and future impact of ambient overheating on the demand and supply-side components of electricity.

It is reported that the efficiency of nuclear power plants is reduced by 0.8% per degree of temperature increase, while the expected increase of the water temperature may decrease the available capacity of the nuclear power plants by 6 GW (Rübbelke and Vögele, 2011). In a similar way, ambient overheating is decreasing the power capacity of the coal and gas power plants operating under Rankine and Brayton cycles (Arrieta and Lora, 2005; Paeth et al., 2008). It is estimated that increased ambient

Table 5.2 Quantified implications of the ambient and urban overheating on demand and supply-side components of energy and electricity demand.

Demand-side component	Overheating effect	Average implications
Cooling demand of reference buildings	Urban heat island	Analysis of 22 studies has shown: The average increase of the cooling demand induced by UHI is 12% compared to the demand under reference–rural climatic conditions The average cooling penalty induced by UHI, is close to 2.3 $kWh/m^2/y/$and per degree of temperature increase
Cooling demand of reference buildings	Combined impact of urban heat island and global climate change	Analysis of 18 studies comparing the heating and cooling needs of reference buildings between 1970 and 2010 has shown: The average increase of the cooling demand is 23% or 11 $kWh/m^2/y$ The average decrease of the heating demand is 19% The average increase of the total heating and cooling demand is 11%
Cooling demand of the total city stock	Combined impact of urban heat island and global climate change	Analysis of six studies evaluating the energy impact of regional and global overeating in cities has shown: The Global Energy Penalty per unit of city surface, GEPS varies between 0.13 and 5.5 kWh/m^2, with an average value close to 2.3 $\pm$ (1.5) kWh/m^2

(Continued)

Table 5.2 (Continued)

Demand-side component	Overheating effect	Average implications
		The Global Energy Penalty per city surface and degree of temperature increase, GEPSI, varies between 0.12 and 2.22 kWh/m2/°C, with an average value close to 0.73 $\pm$ (0.64) kWh/m^2/°C
		The Global Energy Penalty per Person, GEPP, varies between 104 and 305 kWh/p, with an average value close to 230 $\pm$ (120) kWh/p
		The Global Energy Penalty per person and degree of temperature increase, GEPPI, varies between 20 and 154 kWh/p/°C, with an average value close to 78 $\pm$ (47) kWh/p/°C
Future cooling load of reference buildings	Global climate change	Analysis of 114 studies forecasting the future cooling demand of buildings has shown:
		For actual reference cooling loads close to 50 kWh/m^2/y, the average expected future increase of the cooling demand per degree of temperature increase is close to 6 kWh/m^2/°C
		For actual reference cooling loads close to 150 kWh/m^2/y, the average expected future increase of the cooling demand is close to 12 kWh/m^2/y

(*Continued*)

Table 5.2 (Continued)

Demand-side component	Overheating effect	Average implications
Peak electricity demand	Combined impact of urban heat island and global climate change	Analysis of 11 studies evaluating the impact of temperature on peak electricity demand has shown: The additional peak electricity demand per degree of temperature increase varies between 0.45% and 12.3% The average additional electricity demand per person and degree temperature increase is close to 21.9 ± (11.8) W/°C/person The average increase of the peak electricity demand is close to 3.7% or 215 MW per degree of temperature increase
Supply-side component	Overheating effect	Average implications on electricity capacity
Power output of coil and gas power plants	Combined impact of urban heat island and global climate change	
Power output of nuclear plants	A temperature increase of 1°C reduces the power output of coal and gas power stations by 0.6%	
Carrying capacity of electric transmission network	A temperature increase of 1°C, decreases their power output by 0.8%	
Additional investments	Rise of the ambient temperature by 2040−60 may decrease the mean summertime transmission	

(Continued)

Table 5.2 (Continued)

Demand-side component	Overheating effect	Average implications
	capacity in the United States by 1.9%—5.8% relative to the 1990—2010 period. Increased ambient temperatures and extreme phenomena, may cause 14%—23% additional investments on electricity capacity in the United States, relative to a nonclimate change scenario for the years between 2010 and 2055	

Source: Santamouris, M., 2020. Recent progress on urban overheating and heat island research. Integrated assessment of the energy, environmental, vulnerability and health impact. Synergies with the global climate change. Energy and Buildings 207, 109482. https://doi.org/10.1016/j.enbuild.2019.109482.

temperatures raise the thermal losses of the power plants and may reduce the electricity output of the coal and gas power plants by 0.6% per degree of temperature increase (Linnerud et al., 2009).

Apart from the important reduction of the efficiency of the thermal and nuclear power plants, the increase of the ambient temperature limits the carrying capacity of the transmission networks considerably because of the important power line sagging (Bartos et al., 2016). It is predicted that ambient overheating may decrease the average transmission capacity in the United States during the period between 2040 and 2060 between 1.9% and 5.8% relative to the actual situation (Bartos et al., 2016), while increased losses within substations and transformers should be additionally considered (Mikellidou et al., 2018). Forecasting of the impact of ambient overheating on the electricity capacity of the United States found that global and local climate change may increase the required investments between 14% and 23% in the period between 2010 and 2055 (Mikellidou et al., 2018).

5.3 Impact of urban overheating of urban vulnerability

It is well known that biophysical and socioeconomic vulnerability in cities presents an important spatial distribution with low-income households living in the most deprived urban zones (Harlan et al., 2006; Harlan and Ruddell, 2011; Leal Filho et al., 2018; Reid et al., 2009; Ruddell et al., 2009; Weber et al., 2015). Several studies have analyzed the association between heat exposure and neighborhoods' quality (Table 5.3). In Philadelphia, USA, it is found that during heat waves, there is a significant overlap between the heat exposure and social sensitivity (Weber et al., 2015), while in Georgetown, NC, USA (Cutter et al., 2000), it is observed that districts combining medium levels of biophysical with medium to high socioeconomic vulnerability were the most vulnerable ones.

Table 5.3 Quantified implications of ambient and urban overheating on low-income and vulnerable population.

Component	Overheating ambient effect	Implication
Cooling energy cost	Combined impact of global and regional climate change	The cost of air conditioning of low-income households may increase up to 100% relative to the average conditions (4)
Indoor temperature	Combined impact of global and regional climate change	Indoor peak temperatures in low-income houses during heat waves may exceed 40°C
Cooling needs	Combined impact of global and regional climate change	Low-income households cover a very small part (even 3%) of their cooling needs
Urban heat island and vulnerability	Combined impact of global and regional climate change	Low-income population is living in deprived neighborhoods characterized by excess heat stress and high urban heat island intensity. Districts of high vulnerability levels are usually associated with a higher risk of heat-related mortality

Numerous studies have concluded that there is a significant association between socioeconomic parameters like the quality of housing, education levels, income etc., with the exposure to high ambient temperature in cities. As already mentioned, most of the studies concluded that low-income households used to live in deprived urban zones suffering from serious overheating problems and a high risk of heat-related mortality (Goggins et al., 2013; Rosenthal et al., 2014; Rydman et al., 1999; Schinasi et al., 2018; Smargiassi et al., 2009; Taylor et al., 2015). As reported in Xu et al. (2013), in Barcelona, Spain, deprived urban zones presented almost a double heat-related mortality rate during heat waves than in the rest of the city. Apart from the health issues, low-income households living in low-quality buildings in overheated precincts face a serious energy penalty to keep their dwelling at appropriate indoor temperatures (Asimakopoulos et al., 2012; Gouveia et al., 2018; Santamouris et al., 2014). The World Health Organization has found that in eight European countries, almost 9% of the population lives in dwellings presenting serious overheating problems in summer. As reported in Santamouris (2016a,b), the vulnerable and low-income population in Greece pays the double cost of air conditioning than the average population. The serious lack of resources to satisfy the energy needs obliges low-income households to live under extreme indoor ambient temperatures. It is characteristic that in Portugal vulnerable population covers just 2% of their cooling electricity needs (Gouveia et al., 2018).

The specific indoor climatic conditions in low-income houses located in deprived urban zones are extensively studied in the United Kingdom and elsewhere (Beizaee et al., 2013; Firth, Wright, n.d.; Lomas and Kane, 2013; Mavrogianni et al., 2010; Sakka et al., 2012; Summerfield et al., 2007; Wingfield et al., 2011; Wright et al., 2005; Yohanis and Mondol, 2010). There is a general conclusion that serious overheating occurs in most of the houses during the summer period, with indoor temperatures reaching 40°C while thermal comfort conditions are out of the accepted levels exceeding the health thresholds for most of the time.

5.4 Impact of urban overheating on air quality

Ambient overheating affects the outdoor air quality in cities in three main ways:

1. High temperatures increase the speed of photochemical reactions in the atmosphere resulting in the formation of ground-level ozone

through interaction with nitrogen oxides and hydrocarbons (Coates et al., 2016).

2. Overheating affects the turbulent exchange, the airflow rate and characteristics in cities, and the height of the planetary boundary layer resulting in increased concentration of pollutants (Sakka et al., 2012). In coastal cities overheating may slow down the sea breeze and block pollutants (Meier et al., 2017).

3. The extended operation of power plants, required to satisfy the increased summer air conditioning demand, results in a serious increase in the emitted pollutants (Lo and Quattrochi, 2003).

Table 5.4 reports and summarizes the various impacts of ambient overheating on air quality.

It is well known that ozone is a high oxidant and toxic pollutant to humans and plants. It irritates the human lungs and affects the cardiovascular and respiratory system of humans. Several studies have associated the ground-level ozone with elevated hospital admissions and heat-related mortality, and it is ranked as a causal intermediate in the heat-related mortality association (Reid et al., 2012; Wang et al., 2017). The formation of the ground-level ozone depends on several parameters like the intensity of the solar radiation, the magnitude of the ambient temperature, as well as from the concentration of NO_x and VOCs and the ratio of VOCs and NO_x (Stathopoulou et al., 2008).

Recent strict policies aiming to decrease the concentration of atmospheric pollutants have resulted in a serious reduction of their concentration. Unfortunately, this is not the case with ground-level ozone, presenting continuously increased concentrations. As mentioned in Wang et al. (2013), the average ozone concentration in 74 Chinese cities has increased between 2013 and 2015 from 60 to 75 ppbv, while future projections show a much higher increase in the near future.

Urban overheating and urban heat island is strongly associated with high concentrations of the ground-level ozone during the day or night time (Jenkin and Clemitshaw, 2000; Ryu et al., 2013; Sarrat et al., 2006; Swamy et al., 2017). High solar radiation combined with high ambient temperatures speeds up the photochemical processes in the atmosphere, while the thermally induced airflow in cities caused by the change of the height of the boundary layer and the corresponding dilution of NOx results in an increased ozone concentration during the daytime (Li et al., 2016). During the night period, the high surface temperature results in an increased boundary layer that favors the vertical transport of pollutants, decreases the NOx concentration at the

Table 5.4 Quantified implications of ambient and urban overheating on ambient air quality.

Air quality component	Overheating effect	Implication
Ozone concentration	Global overheating—heat waves	Increase of the ozone concentration between 9.6% and 20% during heat waves
Ozone concentration	Combined impact of urban heat island and global climate change	Increase of the ambient temperature by 1% increases the number of days exceeding the threshold of ozone concentration by 10%. Urban heat island increases the number of days exceeding the threshold ozone concentration by 18%. Urban overheating increases the ozone concentration between 10−30 ppb and 1−13 ppb during the night and day time. About 67%−84% of the annual variability of the ozone concentration is due to the change of temperature and other meteorological parameters. Predicted future concentration of the ozone ranges between 6 and 12 ppb. Predicted percentage future increase of the concentration between 20% and 60% in 2050, 80% in 2080, and 400% by 2100
Emissions of pollutants by power plants	Combined impact of urban heat island and global climate change	Increased emission of pollutants by power plants per degree of temperature increase: $3.32\%/°C \pm 0.36\%/°C$ increase in CO_2

(Continued)

Table 5.4 (Continued)

Air quality component	Overheating effect	Implication
		emissions, 3.35%/°C ± 0.50%/°C increase in SO_2 emissions, and a 3.60%/°C ± 0.49%/°C increase in NO_X emissions
Future emissions of pollutants by power plants	Combined impact of urban heat island and global climate change	Estimated increase of emissions by power plants: Plus 16% NOx emissions and plus 18% SO_2 emissions by 2050

lower atmospheric level, reduces the titration phenomena, and finally increases the ozone concentration (Li et al., 2016).

The association between urban overheating and ground-level ozone is very well-documented by several theoretical and experimental studies (Akbari, 2009; Civerolo et al., 2007; Fallmann et al., 2016; Kovač-Andrić et al., 2009; Li et al., 2014; Ooka et al., 2011; Sillman and Samson, 1995; Tarasova and Karpetchko, 2003; Wang et al., 2007). Studies performed for Eastern and North-Eastern United States, California, Athens, Greece, Seoul, Korea, Kanto province, Japan, Croatia, Stuttgart, Germany, Phoenix, Arizona and New York, USA, Pearl River area, China and Cyprus, show a strong correlation between the magnitude of overheating and the corresponding ozone concentration. Detailed information is provided in Santamouris (2020). The expected increase of the ambient temperature caused by global climate change may have a serious impact on the future concentration of the ground-level ozone, as the chemical reaction rates will increase (Hogrefe et al., 2004). It is estimated that in the North-Eastern USA, the concentration of ozone will increase by 10%–30% by 2020 and 100% by 2050 (Lin et al., 2007), while a similar study for four Canadian cities predicts an increase of the concentration close to 50% by 2050 and 80% in 2080 (Cheng et al., 2007).

While the association of urban overheating with the ground-level ozone is quite strong, the impact of higher urban temperatures on the concentration of particulate matter (PM) is not so obvious and mainly depends on the specific

PM component (Aw and Kleeman, 2003; Dawson et al., 2007; Kleeman, 2008). In particular, it is found that while the concentration of sulfate particles is increasing with temperature because of the faster SO_2 oxidation, this is not the case of organic volatile and nitrate particles. Not a significant correlation between the PM concentration and the ambient temperature is found in Jacob and Winner (2009).

The extended use of power plants, because of the increased demand for air conditioning, raises the primary pollutant emissions like sulfur dioxide and nitrogen oxides while it increases the concentration of secondary pollutants like PMs and ground-level ozone (Abel et al., 2017). Meier et al. (2017), found that in the Eastern United States, the electricity generation between the summers of 2007 and 2012, the concentration of the pollutant released by the power plants increased by 3.87% $\pm$ 0.41% for each degree of temperature rise. In particular, it resulted in a 3.32%/°C $\pm$ 0.36%/°C rise in CO_2 emissions, a 3.60%/°C $\pm$ 0.49%/°C increase in NO_x emissions, and a 3.35%/°C $\pm$ 0.50%/°C in SO_2 emissions. Given the increased rate of the electricity demand in the Eastern United States and the corresponding increase of the ambient temperature, it is estimated that by 2050, NO_x emissions will increase by 16%, and SO_2 emissions by 18%.

5.5 Impact of urban overheating on health

Exposure to high ambient temperatures is a serious health hazard for human beings (Anderson and Bell, 2011; Gasparrini et al., 2012; Kovats et al., 2004). When exposed above a certain temperature threshold, the human body and its thermoregulation system cannot offset the impact resulting in serious health problems or even mortality. Thousands of epidemiological studies are performed examining the impact of extreme heat on human beings. Most of the studies concluded that the elderly population presents the highest vulnerability, followed by the population with preexisting cardiovascular, respiratory, or mental problems (Chow et al., 2012; Johnson et al., 2005; Page et al., 2012; Stafoggia et al., 2006). In parallel, it is well-demonstrated that "those 'lacking in economic assets and access to public support systems, with diminished physical or cognitive capacities to respond to warnings and missing strong and enduring social support systems,' present a very high vulnerability to extreme heat" (Moghadamnia et al., 2017). Metanalysis of a high number of epidemiological studies shows that exposure to high temperatures increases by 1.3% and 8.1% the risk of cardiovascular mortality in the total and elderly population, respectively (Reid et al., 2009) Heat-related health problems are very much associated with demographic, socioeconomic, and biophysical factors (Reid et al., 2009), while it is much higher at urban than rural areas (Ho et al., 2017).

Numerous studies are carried out to analyze the characteristics of the intraurban heat-related health problems, investigate the impact of place and urban precincts, and design efficient prevention policies (Bélanger et al., 2016; Benmarhnia et al., 2017; Burkart et al., 2016; Chien et al., 2016; Ellen et al., 2001; Goggins et al., 2012; Gronlund et al., 2015; Harlan et al., 2013; Ho et al., 2017; Jenerette et al., 2016; Johnson et al., 2009; Madrigano et al., 2013; Milojevic et al., 2016; Rosenthal et al., 2014; Son et al., 2016; Tan et al., 2010; Taylor et al., 2015; Uejio et al., 2011; Vandentorren et al., 2006; Vaneckova et al., 2010). It is evident that urban vulnerability and the levels of deprivation may vary considerably in a city, mainly because of the high spatial heterogeneity of urban overheating. Most of the existing studies conclude that the thermal quality of urban neighborhoods affects heat-related health issues through three main pathways (Jesdale et al., 2013; Santamouris, 2020): (1) The potential stresses in the physical and social environment, (2) the availability of neighborhood institutions and resources, and (3) the relative influence and impact of the local networks. As a result of the above conclusions and findings, urban heat-related health issues should be investigated as the integrated and holistic impact of thermal, social, economic, and demographic risk factors.

5.5.1 Impact of urban overheating on heat-related morbidity

The impact of urban overheating on hospital admissions and morbidity on the economics of public health and the well-being of the urban population is of extreme importance (Li et al., 2019). However, the association between urban overheating and heat-related morbidity is not well-investigated, and quite a few studies are available. Most of the studies have concluded that there is a considerable association between high urban temperatures and hospital admissions (Santamouris, 2020); however, not all studies have agreed on this finding (Table 5.5). In particular, several studies have reported a strong association of higher urban temperatures with respiratory problems (Knowlton et al., 2009; Mastrangelo et al., 2006; Ye et al., 2012), cardiovascular diseases (Dawson et al., 2008; Ebi et al., 2004; Li et al., 2019; Lin et al., 2009; Michelozzi et al., 2009; Phung et al., 2016; Schwartz et al., 2004), acute renal failure, mental illnesses, diabetes, dehydration and electrolyte imbalance, and heat stroke. According to Knowlton et al. (2009), the observed heterogeneity of the

Table 5.5 Main characteristics and results of existing studies on heat-related morbidity.

Location	Population concerned	Exposure period	Climate data to assess spatial variability	Outcome	Temperature definition	Covariates	Results
Brisbane, Australia	All population	October – March 2007–2011	Local ambient T from two stations	All types of morbidity	Daily maximum temperature	Socioeconomic factors	During summer, increase in daily maximum temperature by 10°C was associated with a 7.2% increase in hospital admissions on the following day. A significant variability of morbidity with neighborhood ranging from a 55% decrease in admissions per 10° C increase in temperature to 102% increase
Toronto, Canada	All population	Summer 2002–2005	Local ambient T	All types of morbidity	Mean average T	Socioeconomic factors	There is clear geospatial heterogeneity in the burden of HRI in Toronto. Areas within the downtown core experienced high rates of HRI medical dispatch calls. While

(Continued)

Table 5.5 (Continued)

Location	Population concerned	Exposure period	Climate data to assess spatial variability	Outcome	Temperature definition	Covariates	Results
							the reasons for proportionately higher rates are unclear, possible explanations may include spatial risk factors like poorer housing type, lack of air conditioning, and particular local heat islands. It also appears that areas with high rates of HRI include low-income inner-city neighborhoods, areas with high rates of street-involved individuals such as the homeless. Further analysis using indicators of socioeconomic status would provide added information to explore the possible ecological association between

Location	Population	Period	Data	Outcome	Temperature		Comments
							socioeconomic factors and rate of HRI, as has been suggested in other studies. Areas along the waterfront also have a particularly high rate of HRI as compared with other neighborhoods. A plausible explanation for this is that these areas have a high rate of outdoor activities and therefore include a large temporary population exposed to hot weather. This is consistent with previous work that considered increases in all 911 medical dispatch calls in Toronto on heat alert days
NY, USA	Old population	Extreme hot days Summer 2005–2013	Local climate data	Cardiovascular diseases	Average T	None	A 7% increased risk of ischemic heart disease on the day of extreme heat, and increased risks of hypertension (4%) and cardiac dysrhythmias (6%) occurred on lag days 5

(Continued)

Table 5.5 (Continued)

Location	Population concerned	Exposure period	Climate data to assess spatial variability	Outcome	Temperature definition	Covariates	Results
							and 6, respectively. Important increase in some neighborhood
Mekong Delta River, Vietnam	All population	January 2002–December 2014 period	Local ambient T	All causes, and for infectious, cardiovascular, and respiratory diseases	Average temperature	Socioeconomic factors	For 1°C increase in average temperature, the risk of hospital admissions increased by 1.3% (95% CI, 0.9–1.8) for all causes, 2.2% (95% CI, 1.4–3.1) for infectious diseases, and 1.1% (95% CI, 0.5–1.7) for respiratory diseases. However, the result was inconsistent for cardiovascular diseases
Mekong River Delta, Vietnam	All population	2010–2013	Local ambient T from 12 local stations	All types of heat-related morbidity	Daily average T	Sociodemographic factors	The result of the first model indicated that an increase of 5°C in average temperature was associated with a statistically significant 6.1% increase (95%CI: 5.9, 6.2) in total

| Mekong River Delta, Vietnam | All population | Heat waves 2–12 years | Local ambient temperature | All-cause hospital admissions | Average ambient T | Socioeconomic factors | hospital admissions across districts. High variability of morbidity per district. 55.2% decrease (95% CI: −54, −56) to 24.4% The risk of admissions was higher in the North (5.4%, 95%CI: _0.1e11.5; 11.2%, 95%CI: 3.1e19.9; 7.5%, 95%CI: 1.1e14.4: 2.7%, 95% CI: _5.4e11.5) than in the South (1.3%; 95% CI: 0.1e2.6; 3.2%, 95%CI: 0.7e5.7; _1.2%, 95% CI: _2.6e2.3; 2.1%, 95% CI: _0.8e1.2) for all causes, infectious diseases, cardiovascular, and respiratory diseases, respectively | Morbidity |
| 114 US cities | Older than 65 years | 1992–2006 | Local ambient temperature | All cause hospital admissions | Apparent T | Socioeconomic factors | When comparing the effects of extreme heat by climate zone, we found significant | Morbidity |

(Continued)

Table 5.5 (Continued)

Location	Population concerned	Exposure period	Climate data to assess spatial variability	Outcome	Temperature definition	Covariates	Results
							heterogeneity between climate zones in the cumulative 8-day effect estimates for admissions for respiratory diseases, with pooled increases of 11.8% (95% CI: 2.3%, 22.2%) in climate zone 1 and −0.5% (95% CI: −4.0%, 3.1%) in climate zone 5 (Table 5.3). For admissions for both respiratory and renal diseases, we found significant heterogeneity in effect estimates within climate zones 1, 2, 3, and/or 4. Climate zone 5 is the hot one

12 European cities	All ages, 65–74 age group, and 751 age group	Heat waves: 1990–2001	Local ambient temperature	Cardiov-ascular, cerebrovascular, and respiratory causes	Apparent temperature	None	Morbidity
							For respiratory admissions, there was a positive association that was heterogeneous between cities. For an 18°C increase in maximum apparent temperature above a threshold, respiratory admissions increased by 14.5% (95% CI, 1.9–7.3) and13.1% (95%CI, 0.8–5.5) in the 751age group in Mediterranean and North-Continental cities, respectively. In contrast, the association between temperature and cardiovascular and cerebrovascular admissions was not statistical significant

reported results is attributed to four main factors (Knowlton et al., 2009): (1) The important demographic differences and the variability in acclimatization and adaptation between the investigated places, (2) the diversity of the overheating indicators and proxies used to describe heat exposure, like the average, the minimum of the maximum ambient temperature, the diurnal temperature range, and the apparent temperature, (3) the consideration of different measures and types of heat-related morbidity like hospital admissions, practitioner's visits, and emergency department visits, and (4) the methodological differences between the studies around the design and the execution of the research. Some of the reported results and conclusions are presented below.

Under the climatic conditions of Vietnam, Madrid, Spain, and Brisbane, Australia, it is found that all-cause heat-related morbidity increases between 0.05% and 3.6%, 0.46% and 0.72% per degree of temperature rise correspondingly (Lin et al., 2009; Santamouris, 2020). In parallel, data collected during heat waves in England, UK, Vietnam, the United States, France, and Australia, show that heat-related morbidity increases between 1% and 11% (Santamouris, 2020).

The threshold temperature above which heat-related morbidity starts to increase varies considerably between cities and the type of heat-related illness. The analysis of the existing studies shows clearly that local adaptation and acclimatization determines at large the temperature threshold. A large-scale European research has shown that the threshold morbidity temperature decreases as a function of the geographic latitude (Gronlund et al., 2016). It is found that the threshold morbidity temperature varies between 17.7°C in Dublin, Ireland, and 36.4°C in Valencia, Spain, while it is 22.8°C in Stockholm, Sweden, 24.6°C in London, UK, 27.3°C in Ljubljana, Slovenia, 27.8°C in Paris, France, 28.9°C in Budapest, Hungary, 30.8°C in Barcelona, Spain, 31.2°C in Turin, Italy, and 33.8°C in Milan, Italy. In parallel, it is reported that the morbidity threshold for respiratory diseases is close to 23°C in London, UK, 28.6°C in San Diego, California, 28.9°C in NY, and 44.4°C in Phoenix, Arizona (Santamouris, 2020).

Comparison of the heat-related morbidity in rural and urban areas show that cities present a much higher morbidity rate (Bassil et al., 2009). For example, while the relative risk of heat-related morbidity in the suburban zones of Chicago is close to 1.89, it increases up to 3.86 in the urban zones of the city (Rydman et al., 1999). In parallel, intraurban morbidity may be important because of the big differences in vulnerability levels between the various neighborhoods. In particular, in Brisbane, Australia, during the warm period, morbidity levels in the various neighborhoods were found to range between a decrease of 55% and a 102% increase per 10°C of temperature rise (Hondula and Barnett, 2014).

5.5.2 Impact of urban overheating on heat-related mortality

The impact of urban overheating and the intraurban temperature variability on heat-related mortality is a well-studied subject. Several climatic proxies are used to characterize the association between the intraurban thermal quality and heat-related mortality (Santamouris, 2020). Some of the most used climatic parameters and proxies are: (1) The intraurban distribution of the near-surface ambient temperature or a combination of it with other climatic parameters. (2) The distribution of the surface temperature in the city as obtained by remote sensing monitoring. (3) The density of vegetation in the neighborhood. (4) The percentage of impervious surfaces. (5) The proximity to water surfaces (Santamouris, 2020).

A total of 28 relevant studies are analyzed in Santamouris (2020). The main characteristics of the considered studies are given in Tables 5.6A and 5.6B (Santamouris, 2020). As concluded by most of the investigations, there is a significant correlation between the magnitude of overheating in the various neighborhoods of a city and the relative risk of heat-related mortality. The association between urban overheating and heat-related mortality is found to depend on the magnitude of the urban overheating in the different parts of the cities and also from demographic, socioeconomic, and factors related to the local health infrastructure. Results of systematic review analyzing the impact of intraurban heat stress on heat-related mortality concluded that those living in the warmer neighborhoods of the cities have in average almost 6% higher risk of heat-related mortality than those living in the cooler parts of the cities, while the relative risk varied between 0 and 13% (Schinasi et al., 2018). The same study also concluded that living in nonvegetated urban areas may increase by 5% on average the risk of heat-related mortality compared to those living in green urban zones.

The main conclusions drawn from the analysis of the existing studies associating intraurban overheating and heat-related mortality are summarized in the following (Santamouris, 2020).

1. Heat-related mortality in a city increases considerably above a temperature threshold. The threshold temperature varies between cities as a

Table 5.6A Main characteristics and results of existing studies associating intracities heat-related mortality with ambient and surface temperature.

Location	Population concerned	Exposure period	Climate data to assess spatial variability	Outcome	Temperature proxy	Considered covariates	Results
London, UK	All population	Hot spell: 26 of May to 19 of July 2006	Twelve stations in London + mesoscale simulations	All-cause mortality	A composite parameter with the mean maximum T, the UHI temperature anomalies, and the dwelling thermal quality	Quality of housing	Spatial variability of HRM follows the background mortality rates due to population age. Estimated UHI attributable deaths between 6.1 and 8.14 deaths per million of population. Dwelling characteristics cause a larger variation in temperature exposure and risk, than UHI
Hong Kong China	All population	June–September 2001–2009	Ambient T from several stations	All-cause mortality	Mean ambient T and calculated UHI index	Meteoro- logical parameters and concentration of PM10	Average T above 29°C and low wind speeds associated with higher mortality. Stronger impact in areas with high UHI index. A 1°C rise above 29°C caused a 4.1% and 0.7% increase in mortality in high and low UHI zones correspondingly
London, UK	All population	June–August 1993–2006	Simulated ambient T data	All-cause mortality	Mean ambient T	Age, socio- economic deprivation score	1°C UHI anomaly multiplied the risk of heat death by 1.004 and 1.070 when acclimatization or no acclimatization is considered
St Louis, Missouri, USA	Older than 65 years	Heat wave days 1980, 1983,1988, and 1995	Measured ambient temperature in one station	All-cause mortality	Mean ambient T	Socioeconomic, demographic	The cooler suburbs presented much lower heat-related mortality than the warmer ones. Heat wave mortality rates were higher in warmer, less stable IN socioeconomically

Location	Population	Period	Temperature source	Mortality	Temperature metric	Confounders	Results
Sydney, Australia	Older than 65 years	Warm months 1993–2004	Measured ambient T	All-cause mortality	Mean and maximum temperature	Socioeconomic factors	Zones of higher temp had a higher mortality risk, 0.8%–30% for 10°C temperature increase. No statistically significant correlation between socioeconomic factors and HRM
Vancouver, Canada	All population	Selected days 1998–2014	Ambient and surface T	All-cause mortality	Humidex, ambient, and surface temp	Socioeconomic, demographic, health	The OR for a 1°C increase in daily mean humidex was 1.13 and 1.04 for data above and below 34.2°C, respectively
Kaohsiung, Taiwan	All population	May – October 1999–2008	Measured ambient T	All-cause mortality	A composite thermal load index	Socioeconomic factors	Mortality increases per 1°C rise, increases by 2.8%, 2.3%, and −1.3% for the very warm, less warm, and cool urban zones, respectively. Above 29.0°C increase rates were 4.2%, 5.0%, and 0.3%, respectively
West Midlands, UK	All population	Heat wave: 1–10 August 2003	Mesoscale simulations for urban and rural zones	All-cause mortality	Mean daily temperature above 17.7° C	None	The relative risk of death above17.7 average daily temperature was 1.023 per degree of temperature increase. UHI contributes 50% of the total heat-related mortality in the city
Sao Paolo, Brazil	All population	1993–1994	Ambient T from three stations	Cardiovascular and respiratory	Urban heat island intensity	None	Significant correlation of the intensity of the UHI and the annual mortality rates. HRM 30% higher in zones of 2°C–4°C higher temp

(Continued)

Table 5.6A (Continued)

Location	Population concerned	Exposure period	Climate data to assess spatial variability	Outcome	Temperature proxy	Considered covariates	Results
Shanghai, China	All population	Heat waves 1975–2004	Measured ambient T data	All-cause mortality	Urban heat island intensity	None	HRM between 1 and 27 deaths/million following the intensity and length of the heat wave
Several US cities	All population	2006–2010	Measured temperatures	All-cause mortality	Not clear	None	The impact of the UHI in the death rate is 1.1 people per million
Texas, USA	Older than 65 years	2006–2001	A high number of metro-stations	All-cause mortality	Heat index, HI, combining temp and humidity	None	Risk of mortality is increasing for higher HI. Significantly higher RR of heat on mortality for HImax over 90th percentile
Paris, France	Older than 65 years	Heat wave 11–13 August 2003	61 images from NOAA-AVHRR	All-cause mortality	Min, max, mean and the diurnal daily surface temperatures	Age, gender	Mortality risk very well-associated with the min and max surface temp. Elderly population exposed to high night surface temp, has a double risk of death than less exposed people
Maricopa, Arizona, USA	All population	July of 2000–2008	Satellite-data surface T	Heat-related mortality	Mean daily surface temperature	Socioeconomic factors	Increase of the mean surface temperature by 1°C caused a 32% increase in the odds of death from heat exposure
Montreal, Canada	All population	June – August 1990–2003	Satellite-derived surface T plus one ambient T station	All-cause mortality	Surface temperature to classify urban zones. Mean ambient T	Ozone, residential property values	The odd ratios comparing mortality on days with a mean ambient T of 26°C against that on days at 20°C, was 1.21 in the "'hot'" urban zone and 1.11 in the cooler zone
New York, USA	Older than 65 years	Extremely hot days, May–Sept1997–2006	LANDSAT surface T data	All-cause mortality	Surface T to classify urban zones	Socioeconomic, health, and other parameters	The mean Mortality Rate Ratios, MRR, in zones with high and low surface T were 1.223 and 1, respectively

Location	Population	Period	Data source	Health outcome	Temperature metric	Confounding factors	Results
Paris, France	Older than 65 years	Heat wave: 8–13 August 2003	LANDSAT surface T data	All-cause mortality	Mean daily surface temp	Social factors + housing	The odd ratio concerning the surface temperature around the dwelling was high, 1.82
Philadelphia, USA	All population	Heat wave July 3 to July 14	LANDSAT surface T data	All-cause mortality	Mean and maximum daily surface T	Socioeconomic and demographic factors	For each degree increase in the mean LST, risk of death increases by a factor of 6, while for each degree decrease in maximum LST the odds of death increase by a factor of 2.84
11 cities	Various	Various	Measured ambient T and surface T	All-cause mortality + cardiovascular and respiratory	Daily mean T, mean apparent T, max T, UTCI, Min T, daily mean Humidex	Many	Population living in warmer areas within cities have almost 6% higher risk of mortality/morbidity compared to those living in cooler areas of cities. The estimated risk varied between the 11 studies from 0% to 13%

Table 5.6B Main characteristics and results of existing studies associating intracities heat-related mortality with landscape parameters.

Location	Population concerned	Exposure period	Outcome	Considered covariates	Results
10 cities[a]	Various	Various	All-cause mortality + cardiovascular, respiratory	Many[c]	Those living in less vegetated areas had 5% higher risk compared to those living in more vegetated areas. The risk varied between 0% and 30%
Barcelona, Spain	All population	Warm seasons 1999–2006	All-cause mortality	Socioeconomic housing quality	No modifying effects related to Percent Tree Cover HRM was higher in zones where residents perceiving little surrounding greenness (RR = 1.29)
Seoul, Korea	All population	2000–2009	All-cause mortality	None	Mortality risk increase per degree of temp rise above 25.1°C was 4.1%, 3.0%, and 2.2% in low, medium, and high NDVI group, respectively
Lisbon, Portugal	Above 65 years old	1998–2008	All cause mortality	None	For UTCI > 19.9°C, mortality was higher in areas of the lowest quartile of NDVI by 3.9%/deg. In the 2nd, 3rd, and 4th quartiles were 2.2%, 2.2%, and 1.2%, respectively. Above 24.8°C mortality was 14.7, 5.4, 5.1%, and 3.0%/deg

Location	Population	Time period	Mortality	Factor	Findings
Maritopa, Arizona	All population	July of 2000–08	Heat-related mortality	Socioeconomic factors	Lack of vegetation had a weak but significant positive association with the odds of at least one heat death in a census block (1.19)
Worcester, MA, USA	Patients > 25 years	April October 1995, 1997, 1999, 2001, 2003	Mortality from acute MI	Socioeconomic factors	No relation between greenery levels and mortality
Michigan, USA	All population	May – September 1990–2007	Cardiovascular mortality	Socioeconomic factors	The odds of cardiovascular mortality during extreme heat events were higher by 39% in urban zones without green spaces
Paris, France	Older than 65 years	Heat wave 8–13 August 2003	All-cause mortality	Socioeconomic factors	A strong negative correlation between mortality and Vegetative index, VI. Mortality increases by a factor of 2.8 per unit of decrease of the VI
Sydney, Australia	Older than 65 years old	Warm months 1993–2004	All-cause mortality	Socioeconomic factors	No correlation between mortality and level of vegetation
England	Nonretired population	2001–2005	All-cause mortality, circulatory, lung cancer	Socioeconomic factors	The incidence rate ratio for all-cause mortality for the most income-deprived quartile compared with the least deprived was 1 · 93 and 1.43 in the least and most

(Continued)

Table 5.6B (Continued)

Location	Population concerned	Exposure period	Outcome	Considered covariates	Results
					vegetated zones, respectively. For circulatory diseases, it was $2 \cdot 19$ and 1.54. respectively
New York, USA	Older than 65 years	Hot days, May–September 1997–2006	All-cause mortality	Socioeconomic + health factors	Nonimportant correlation between the percentage of vegetated surfaces and mortality (r = 0.3)
Paris France	Older than 65 years	Heat wave 2003	All-cause mortality	Socioeconomic factors	The density of vegetation has a protective effect with a beta coefficient of −0.005
109 US cities	Older than 65 years old	May – September 1992–2006	Renal heat respiratory	Socioeconomic factors	A weak correlation between HRM and vegetation levels
Berlin, Germany	All population	Heat waves 1990–2006			Significant correlation between the number of deaths and the proportion of land covered by impervious surfaces during very high temperature periods
New York, USA	Older than 65 years	Hot days, May–Sept1997–2006	All-cause mortality	Socioeconomic + health factors	Significant correlation between the area of impervious surfaces and mortality is found (r = 0.3)

Berlin, Germany	All population	Heat waves 1990–2006		Socioeconomic factors	Important relationship between the mortality rate and the density of urban structures within the city area
Worcester, MA, USA	Patients > 25 years	Summer 1995, 1997 1999, 2001, 2003	Mortality from acute MI	Socioeconomic factors	Population living in more urban areas were more likely to suffer from an acute MI on hot period than those living in less dense zones
7 US cities[b]	All population	22 years data	All-cause mortality	Socioeconomic factors	In most of the cities higher density was associated with higher risk of mortality
Paris, France	Older than 65 years	Heat wave 2003	All-cause mortality	Socioeconomic factors	Urban density increased the risk of dying
St Louis, Missouri	Older than 65 years old	Heat wave 1980, 1983, 1988 1995	All-cause mortality	Socioeconomic, demographic	Densities were higher, rather than lower, in tracts with low heat wave mortality rates
Lisbon, Portugal	Above 65 years old	1998–2008	All-cause mortality	None	In zones > 4 km from water, rise of UTCI by 1 C above the 99th percentile increased mortality by 7.1%. In zones ≤ 4 immortality increased by 2.1%

[a]Montreal, Canada, Hong Kong China, Kaohsing City Taiwan, Worcester MA, USA, Barcelona Spain, Lisbon, Spain, 8 cities in Michigan, USA, Vancouver Canada, London, UK, Seoul Korea.

[b]Atlanta, Georgia; Boston, Massachusetts; Minneapolis-St. Paul, Minnesota Philadelphia, Pennsylvania; Phoenix, Arizona; Seattle, Washington; St. Louis, Missouri.

[c]Residential property values, Climate, Concentration of pollutants, percentage of old population, density, education level, reception of social benefits, proximity to water, family status, age, race, age of dwelling increased mortality by 7.1%.

function of adaptation and acclimatization of the local population. It is considerably higher in warm cities than in cities of cold climate. Fig. 5.12A presents the threshold mortality temperature as a function of the geographic latitude for the Northern Hemisphere, while Fig. 5.12B plots the apparent threshold mortality temperatures (Santamouris, 2020). As shown, there is a clear decreasing trend of the daily maximum, daily average, and maximum apparent threshold mortality temperature with increasing latitudes, despite the variability of the socioeconomic and demographic factors affecting the levels of heat-related mortality.

2. To assess the impact of intraurban variability of overheating on heat-related mortality, several proxies are proposed and used. It is well-understood that the application of various proxies in the same city may result in conflicting conclusions and results. Thus the choice of

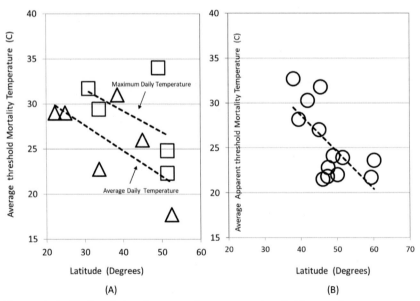

Figure 5.12 (A) Correlation between the average threshold mortality temperature and latitude. (B) Correlation between the average apparent threshold mortality temperature and latitude. Squares refer to the maximum daily temperature and triangles to the average daily temperature.
Santamouris, M., 2020. Recent progress on urban overheating and heat island research. Integrated assessment of the energy, environmental, vulnerability and health impact. Synergies with the global climate change. Energy and Buildings 207, 109482. https://doi.org/10.1016/j.enbuild.2019.109482.

the proper proxy depends highly on the characteristics of the city, the local climate, and the local demographic and socioeconomic issues. Among the most commonly used proxies are the average and the maximum ambient temperature. Aggregate and composite proxies like the operative temperature, Universal Thermal Climate Index (UTCI), Standard Effective Temperature, Humidex index, or proxies integrating socioeconomic and demographic factors may be a more suitable option to describe the dynamics of the heat-related mortality in a city.

3. A strong correlation between the intensity of the urban heat island and heat-related mortality in a city is found in most of the studies. As already mentioned, higher ambient temperature combined with negative demographic and socioeconomic factors increases the levels of heat-related mortality considerably. For example, in Hong Kong, it is found that increased UHI intensities are rising significantly the risk of heat-related mortality (Goggins et al., 2012), while in Sao Paolo, Brazil, the levels of heat-related mortality were found to be about 30% higher in urban zones with 2°C−4°C higher ambient temperature (Lowe, 2016). In a similar way, in Texas, USA, the relative mortality risk in the neighborhoods belonging to the upper and higher quartiles of the Heat Index were 1.12 and 0, respectively, while in Kaohsiung, Taiwan, in the warmer and cooler neighborhoods, the additional mortality above 29°C was 4.2% and 0.3%, respectively (Goggins et al., 2013).

4. The excess mortality caused by the UHI is found to vary between 1 and 27 additional deaths per million people (Santamouris, 2020). The highest value, 27 deaths, is observed in Shanghai, China. During the 1998 heatwave, the figures for London, UK varied between 6.1 and 8.1 additional deaths (Tan et al., 2010; Taylor et al., 2015). The corresponding additional mortality caused by the urban heat island during the nonheat waves period is found to be much lower. In particular, it is found that the additional HRM in the center of several US cities is close to 1.1 death per million (Lowe, 2016).

5. Very few epidemiological studies investigating the impact of urban overheating on heat-related mortality in developing countries are available. While most of the existing studies are from developed countries, because of the high intensity of urban overheating in large cities of the developing world, the lack of appropriate health infrastructures, the poor quality of housing, and the urban poverty, it is expected that

heat-related mortality may reach very significant levels. It is evident that there is an urgent need for additional information on the health burden caused by urban overheating in the developing world.

6. In most of the studied cities, a significant association between socio-economic status and heat-related mortality is identified. In Sydney, Australia, Kaohsiung, Taiwan, and London, UK, the warmer neighborhoods, despite the high deprivation levels, do not experience high heat-related mortality. This is mainly attributed to the extensive use of air conditioning and relatively good quality of buildings. In Hong Kong, heat-related mortality per degree of temperature above 29°C is found to increase by 5.6% for the population of low socioeconomic status living in warm neighborhoods, 3.0% for the population of high status living in warm urban zones, and 2.6% for those of low status living in cooler neighborhoods. In parallel, in Philadelphia, USA, the probability of heat-related mortality for the elderly population increased by a factor of 1.82, when poverty increased by one standard unit (Johnson et al., 2009).

7. The quality and the type of dwellings have a high impact on exposure risk than the UHI. A study in London, UK, shown that UHI and dwelling-attributable deaths were 6.1 and 23.5 deaths per million of population, respectively (Taylor et al., 2015).

8. The level of urban vegetation is used by many studies to characterize the association between urban overheating and heat-related mortality. However, as pointed out in Santamouris (2020), when microclimate is determined by other processes, like advection of heat, the vegetation fails to describe the quality of urban zones. It is characteristic that in Western Sydney, Australia, despite the high levels of vegetation, advection of warm air from the desert determines the local climatic conditions, largely. However, contrary to the above findings, for the rest of the studies, vegetation was found to have a strong association with heat-related mortality levels.

5.6 Conclusion

Urban overheating is the result of the positive thermal balance of cities. Global climate change increases the magnitude and the frequency of heat waves that acts synergistically and intensifies the levels of urban overheating and asks for an integrated and holistic analysis combining both regional and global climate issues in cities.

Higher urban temperatures affect the quality of life seriously in cities increasing the energy consumption of buildings for cooling purposes, raising the peak electricity demand obliging utilities to build additional power plants, affect the efficiency of the nuclear and thermal power plants, decrease the capacity of the distribution networks, increase the concentration of harmful pollutants like the ground-level ozone, affect heat-related mortality and morbidity, and create serious vulnerability problems on low-income population. Important quantitative and qualitative information and data are collected in the recent years to document the impact of urban overheating. However, the existing information is highly fragmented among the different scientific disciplines and fails to consider the problem of ambient overheating in an integrated and holistic way. It is now widely understood that the various types of impacts caused by the local overheating is strongly interconnected with important synergies between energy, health, pollution, and vulnerability as presented in Santamouris (2020):

- The provision of energy to operate air conditioning keeps indoor temperatures at appropriate levels and may decrease the levels of heat-related mortality.
- Lack of energy resources in low-income households increases dramatically indoor temperatures during heat waves raises vulnerability levels and skyrockets heat-related mortality.
- The prolonged operation of power plants to satisfy the additional electricity demand during heat waves increases significantly the emission of harmful pollutants and may exacerbate respirational and other health problems.
- The significant increase of the peak electricity demand obliges utilities to build additional power plants operating just for a fraction of time, resulting in a significant increase of the electricity prices and a serious increase of vulnerability in cities.
- Serious disruptions in electricity generation result in blackouts, limiting the potential of air conditioning to provide protection from extreme indoor temperatures and resulting in a significant increase in heat-related mortality and morbidity The serious decrease of the efficiency of power plants during the warm season increases the cost of electricity and puts an additional burden to the vulnerable population.
- Low-income population facing health problems need to maintain reasonably low indoor temperatures for health reasons. This results in increasing their energy bill.

- The thermal quality of housing determines the level of indoor temperature during the warm period and affects highly heat-related mortality, more than the increase of the outdoor temperature.
- High ground-level ozone concentrations affect health directly and increase the cost of ventilation and filtration systems in tertiary buildings.
- Increased needs for comfort and health protection indoors skyrockets the penetration of air conditioning, raising the global energy consumption and the peak electricity demand.

It is evident that forecasts on the future impact of ambient overheating are characterized by severe uncertainty. The important fuzziness on the future greenhouse gas emission scenarios, adaptation mechanisms, and technological developments is still very significant despite the recent improvements. The current forecasts of the future energy and generation requirements, pollutants concentration, vulnerability levels, and mortality and morbidity magnitude offer a very significant information background asking for a more proactive climatic policy.

References

Abel, D., Holloway, T., Kladar, R.M., Meier, P., Ahl, D., Harkey, M., et al., 2017. Response of power plant emissions to ambient temperature in the eastern United States. Environmental Science & Technology 51, 5838–5846.

Akbari, H., 2009. Cooling our Communities. A Guidebook on Tree Planting and Light-Colored Surfacing. LBNL Publications.

Amato, A.D., Ruth, M., Kirshen, P., Horwitz, J., 2005. Regional energy demand responses to climate change: methodology and application to the Commonwealth of Massachusetts. Climatic Change 71, 175–201. Available from: https://doi.org/10.1007/s10584-005-5931-2.

Anderson, G.B., Bell, M.L., 2011. Heat waves in the United States: mortality risk during heat waves and effect modification by heat wave characteristics in 43 US communities. Environmental Health Perspectives 119, 210–218. Available from: https://doi.org/10.1289/ehp.1002313.

Arrieta, F.R.P., Lora, E.E.S., 2005. Influence of ambient temperature on combined-cycle power-plant performance. Applied Energy 80, 261–272. Available from: https://doi.org/10.1016/j.apenergy.2004.04.007.

Asimakopoulos, D.A., Santamouris, M., Farrou, I., Laskari, M., Saliari, M., Zanis, G., et al., 2012. Modelling the energy demand projection of the building sector in Greece in the 21st century. Energy and Buildings 49, 488–498. Available from: https://doi.org/10.1016/j.enbuild.2012.02.043.

Australia, A., 2012. Economic Assessment of the Urban Heat Island Effect. https://www.melbourne.vic.gov.au/SiteCollectionDocuments/eco-assessment-of-urban-heat-island-effect.pdf.

Aw, J., Kleeman, M.J., 2003. Evaluating the first-order effect of intraannual temperature variability on urban air pollution. Journal of Geophysical Research: Atmospheres 108.

Bagiorgas, H.S., Mihalakakou, G., 2016. On the influence of the urban heat island on the cooling load of a school building in Athens, Greece. Journal of Atmospheric and Solar-Terrestrial Physics 138, 179−186.

Bartos, M., Chester, M., Johnson, N., Gorman, B., Eisenberg, D., Linkov, I., et al., 2016. Impacts of rising air temperatures on electric transmission ampacity and peak electricity load in the United States. Environmental Research Letters 11, 114008. Available from: https://doi.org/10.1088/1748-9326/11/11/114008.

Bassil, K.L., Cole, D.C., Moineddin, R., Craig, A.M., Lou, W.Y.W., Schwartz, B., et al., 2009. Temporal and spatial variation of heat-related illness using 911 medical dispatch data. Environmental Research 109, 600−606.

Beizaee, A., Lomas, K.J., Firth, S.K., 2013. National survey of summertime temperatures and overheating risk in English homes. Building and Environment 65, 1−17.

Bélanger, D., Abdous, B., Valois, P., Gosselin, P., Sidi, E.A.L., 2016. A multilevel analysis to explain self-reported adverse health effects and adaptation to urban heat: a cross-sectional survey in the deprived areas of 9 Canadian cities. BMC Public Health 16, 1−11.

Benmarhnia, T., Kihal-Talantikite, W., Ragettli, M.S., Deguen, S., 2017. Small-area spatiotemporal analysis of heatwave impacts on elderly mortality in Paris: A cluster analysis approach. Science of the Total Environment 592, 288−294.

Bründl, W., Höppe, P., 1984. Advantages and disadvantages of the urban heat island—an evaluation according to the hygro-thennic effects. Archives for Meteorology, Geophysics, and Bioclimatology, Series B 35, 55−66.

Burkart, K., Meier, F., Schneider, A., Breitner, S., Canário, P., Alcoforado, M.J., et al., 2016. Modification of heat-related mortality in an elderly urban population by vegetation (urban green) and proximity to water (urban blue): evidence from Lisbon, Portugal. Environmental Health Perspectives 124, 927−934.

Calice, C., Clemente, C., Salvati, A., Palme, M., Inostroza, L., 2015 Urban Heat Island effect on the energy consumption of institutional buildings in Rome, In: IOP Conference Series: Materials Science and Engineering. IOP Publishing, 8.

Chandler, T.J., 1965. The climate of London. Hutchinson and Co. Ltd, London, Urban temperature fields 91.

Chandramowli, S.N., Felder, F.A., 2014. Impact of climate change on electricity systems and markets—a review of models and forecasts. Sustainable Energy Technologies and Assessments 5, 62−74. Available from: https://doi.org/10.1016/j.seta.2013.11.003.

Chen, D., Ren, Z., Wang, C.-H., Thatcher, M., Wang, X., 2012. Urban heat island on Australian housing energy consumption. Healthy Buildings, Brisbane, Queensland.

Cheng, C.S., Campbell, M., Li, Q., Li, G., Auld, H., Day, N., et al., 2007. A synoptic climatological approach to assess climatic impact on air quality in south-central Canada. Part II: future estimates. Water, Air, and Soil Pollution 182, 117−130.

Chien, L.-C., Guo, Y., Zhang, K., 2016. Spatiotemporal analysis of heat and heat wave effects on elderly mortality in Texas, 2006−2011. Science of the Total Environment 562, 845−851.

Chow, W.T.L., Chuang, W.C., Gober, P., 2012. Vulnerability to extreme heat in metropolitan Phoenix: spatial, temporal, and demographic dimensions. Professional Geographer 64, 286−302. Available from: https://doi.org/10.1080/00330124.2011.600225.

Christenson, M., Manz, H., Gyalistras, D., 2006. Climate warming impact on degree-days and building energy demand in Switzerland. Energy Conversion and Management 47, 671−686.

Civerolo, K., Hogrefe, C., Lynn, B., Rosenthal, J., Ku, J.-Y., Solecki, W., et al., 2007. Estimating the effects of increased urbanization on surface meteorology and ozone concentrations in the New York City metropolitan region. Atmospheric Environment 41, 1803−1818.

Coates, J., Mar, K.A., Ojha, N., Butler, T.M., 2016. The influence of temperature on ozone production under varying NOx conditions—a modelling study. Atmospheric Chemistry and Physics 16, 11601—11615.

Colombo, A., Etkin, D., Karney, B., 1999. Climate variability and the frequency of extreme temperature events for nine sites across Canada: implications for power usage. Journal of Climate 12, 2490—2502.

Crawley, D.B., 2008. Estimating the impacts of climate change and urbanization on building performance. Journal of Building Performance Simulation 1, 91—115.

Cutter, S.L., Mitchell, J.T., Scott, M.S., 2000. Revealing the vulnerability of people and places: a case study of Georgetown County, South Carolina. Annals of the Association of American Geographers 90, 713—737.

Dawson, J.P., Adams, P.J., Pandis, S.N., 2007. Sensitivity of PM2.5 to climate in the Eastern US: a modeling case study. Atmospheric Chemistry and Physics 7, 4295—4309. Available from: https://doi.org/10.5194/acp-7-4295-2007.

Dawson, J., Weir, C., Wright, F., Bryden, C., Aslanyan, S., Lees, K., et al., 2008. Associations between meteorological variables and acute stroke hospital admissions in the west of Scotland. Acta Neurologica Scandinavica 117, 85—89.

Ding, F., Pang, H., Guo, W., 2018. Impact of the urban heat island on residents' energy consumption: a case study of Qingdao, In: IOP Conference Series: Earth and Environmental Science Impact of the Urban Heat Island on Residents' Energy Consumption: a case study of Qingdao, 32026. <https://doi.org/10.1088/1755-1315/121/3/032026>.

Doddaballapur, S., Bryan, H., Faia, F., n.d. Analysis of the impact of urban heat island on building energy consumption, In: Proceedings of the 2012 ACES Solar Conference, World Renewable Energy Forum (WREF), Denver, CO, USA. 13—17.

Ebi, K.L., Exuzides, K.A., Lau, E., Kelsh, M., Barnston, A., 2004. Weather changes associated with hospitalizations for cardiovascular diseases and stroke in California, 1983—1998. International Journal of Biometeorology 49, 48—58. Available from: https://doi.org/10.1007/s00484-004-0207-5.

Ellen, I.G., Mijanovich, T., Dillman, K.-N., 2001. Neighborhood effects on health: exploring the links and assessing the evidence. Journal of Urban Affairs 23, 391—408.

Fallmann, J., Forkel, R., Emeis, S., 2016. Cool cities—clean cities? Secondary impacts of urban heat island mitigation strategies on urban air quality. Air Pollution Modeling and Its Application XXIV. Springer, pp. 371—375.

Fanchiotti, A., Carnielo, E., Zinzi, M., 2012. Impact of cool materials on urban heat islands and on buildings comfort and energy consumption, In: Proc. ASES Conference.

Firth, S.K., Wright, A.J., n.d. Investigating the thermal characteristics of English dwellings: summer temperatures, In: Proceedings of Windsor Conference, UK, 27—29.

Founda, D., Santamouris, M., 2017. Synergies between urban heat island and heat waves in Athens (Greece), during an extremely hot summer (2012). Scientific Reports 7, 10973. Available from: https://doi.org/10.1038/s41598-017-11407-6.

Franco, G., Sanstad, A.H., 2007. Climate change and electricity demand in California. Springer, p. 87. Available from: https://doi.org/10.1007/s10584-007-9364-y.

Frank, T., 2005. Climate change impacts on building heating and cooling energy demand in Switzerland. Energy and Buildings 37, 1175—1185.

Fung, W., Lam, K., Hung, W., Pang, S., Energy, Y.L., 2006. Impact of urban temperature on energy consumption of Hong Kong. Energy 31, 2623—2637.

Gasparrini, A., Armstrong, B., Kovats, S., Wilkinson, P., 2012. The effect of high temperatures on cause-specific mortality in England and Wales. Occupational and Environmental Medicine 69, 56−61.

Giannakopoulos, C., Psiloglou, B.E., 2006. Trends in energy load demand for Athens, Greece: weather and non-weather related factors, int-res.com.

Goggins, W.B., Chan, E.Y.Y., Ng, E., Ren, C., Chen, L., Braga, A.L.F., et al., 2012. Effect modification of the association between short-term meteorological factors and mortality by urban heat islands in Hong Kong. PLoS One 7, e38551. Available from: https://doi.org/10.1371/journal.pone.0038551.

Goggins, W.B., Ren, C., Ng, E., Yang, C., Chan, E.Y.Y., 2013. Effect modification of the association between meteorological variables and mortality by urban climatic conditions in the tropical city of Kaohsiung, Taiwan. Geospatial Health 8, 37−44.

Golden, J.S., Brazel, A., Salmond, J., Laws, D., 2006. Energy and water sustainability: the role of urban climate change from metropolitan infrastructure. Journal of Green Building 1, 124−138.

Gouveia, J.P., Seixas, J., Long, G., 2018. Mining households' energy data to disclose fuel poverty: lessons for Southern Europe. Journal of Cleaner Production 178, 534−550.

Gronlund, C.J., Berrocal, V.J., White-Newsome, J.L., Conlon, K.C., O'Neill, M.S., 2015. Vulnerability to extreme heat by socio-demographic characteristics and area green space among the elderly in Michigan, 1990−2007. Environmental Research 136, 449−461.

Gronlund, C.J., Zanobetti, A., Wellenius, G.A., Schwartz, J.D., O'Neill, M.S., 2016. Vulnerability to renal, heat and respiratory hospitalizations during extreme heat among US elderly. Climatic Change 136, 631−645.

Haddad, S., Paolini, R., Synnefa, A., Santamouris, M., 2018. Mitigation of urban overheating in three Australian cities (Darwin, Alice Springs and Western Sydney).

Harlan, S.L., Brazel, A.J., Prashad, L., Stefanov, W.L., Larsen, L., 2006. Neighborhood microclimates and vulnerability to heat stress. Social Science & Medicine 63, 2847−2863.

Harlan, S.L., Declet-Barreto, J.H., Stefanov, W.L., Petitti, D.B., 2013. Neighborhood effects on heat deaths: social and environmental predictors of vulnerability in Maricopa County, Arizona. Environmental Health Perspectives 121, 197−204.

Harlan, S.L., Ruddell, D.M., 2011. Climate change and health in cities: impacts of heat and air pollution and potential co-benefits from mitigation and adaptation. Current Opinion in Environmental Sustainability 3, 126−134. Available from: https://doi.org/10.1016/J.COSUST.2011.01.001.

Hassid, S., Santamouris, M., Papanikolaou, N., Linardi, A., Klitsikas, N., Georgakis, C., et al., 2000. Effect of the Athens heat island on air conditioning load. Energy and Buildings 32, 131−141. Available from: https://doi.org/10.1016/S0378-7788(99)00045-6.

Hayhoe, K., Robson, M., Rogula, J., Auffhammer, M., Miller, N., VanDorn, J., et al., 2010. An integrated framework for quantifying and valuing climate change impacts on urban energy and infrastructure: a Chicago case study. Journal of Great Lakes Research 36, 94−105. Available from: https://doi.org/10.1016/j.jglr.2010.03.011.

Hekkenberg, M., Benders, R.M.J., Moll, H.C., Schoot Uiterkamp, A.J.M., 2009. Indications for a changing electricity demand pattern: the temperature dependence of electricity demand in the Netherlands. Energy Policy 37, 1542−1551. Available from: https://doi.org/10.1016/j.enpol.2008.12.030.

Ho, H.C., Knudby, A., Walker, B.B., Henderson, S.B., 2017. Delineation of spatial variability in the temperature−mortality relationship on extremely hot days in greater Vancouver, Canada. Environmental Health Perspectives 125, 66−75. Available from: https://doi.org/10.1289/EHP224.

Hogrefe, C., Lynn, B., Civerolo, K., Ku, J., Rosenthal, J., Rosenzweig, C., et al., 2004. Simulating changes in regional air pollution over the eastern United States due to changes in global and regional climate and emissions. Journal of Geophysical Research: Atmospheres 109.

Hondula, D.M., Barnett, A.G., 2014. Heat-related morbidity in Brisbane, Australia: spatial variation and area-level predictors. Environmental Health Perspectives 122, 831–836. Available from: https://doi.org/10.1289/ehp.1307496.

Hwang, R.-L., Lin, C.-Y., Huang, K.-T., 2017. Spatial and temporal analysis of urban heat island and global warming on residential thermal comfort and cooling energy in Taiwan. Energy and Buildings 152, 804–812.

Ignatius, M., Wong, N.H., Jusuf, S.K., 2016. The significance of using local predicted temperature for cooling load simulation in the tropics. Energy and Buildings 118, 57–69.

Jacob, D.J., Winner, D.A., 2009. Effect of climate change on air quality. Atmospheric Environment 43, 51–63.

Jenerette, G.D., Harlan, S.L., Buyantuev, A., Stefanov, W.L., Declet-Barreto, J., Ruddell, B.L., et al., 2016. Micro-scale urban surface temperatures are related to land-cover features and residential heat related health impacts in Phoenix, AZ USA. Landscape Ecology 31, 745–760.

Jenkin, M.E., Clemitshaw, K.C., 2000. Ozone and other secondary photochemical pollutants: chemical processes governing their formation in the planetary boundary layer. Atmospheric Environment 34, 2499–2527.

Jesdale, B.M., Morello-Frosch, R., Cushing, L., 2013. The racial/ethnic distribution of heat risk-related land cover in relation to residential segregation. Environmental Health Perspectives 121, 811–817.

Johnson, H., Kovats, R.S., Mcgregor, G., Stedman, J., Gibbs, M., Walton, H., et al., 2005. The impact of the 2003 heat wave on mortality and hospital admissions in England. Health Statistics Quarterly 6–11.

Johnson, D.P., Wilson, J.S., Luber, G.C., 2009. Socioeconomic indicators of heat-related health risk supplemented with remotely sensed data. International Journal of Health Geographics 8, 1–13.

Joutz, F., Rohatgi, A., Doshi, N.B.Z.T., 2013. Impact of climate change on electricity demand of Singapore. Available from: https://esi.nus.edu.sg/publications/esi-publications/publication/2015/12/18/impact-of-climate-change-on-electricity-demand-of-singapore.

Kapsomenakis, J., Kolokotsa, D., Nikolaou, T., Santamouris, M., Zerefos, S.C., 2013. Forty years increase of the air ambient temperature in Greece: the impact on buildings. Energy Conversion and Management 74, 353–365.

Khan, H.S., Paolini, R., Santamouris, M., Caccetta, P., 2020. Exploring the synergies between urban overheating and heatwaves (HWs) in Western Sydney. Energies 13, 470. Available from: https://doi.org/10.3390/en13020470.

Khan, H.S., Santamouris, M., Paolini, R., Caccetta, P., Kassomenos, P., 2021. Analyzing the local and climatic conditions affecting the urban overheating magnitude during the heatwaves (HWs) in a coastal city: a case study of the greater Sydney region. Science of the Total Environment 755, 142515. Available from: https://doi.org/10.1016/j.scitotenv.2020.142515.

Kleeman, M.J., 2008. A preliminary assessment of the sensitivity of air quality in California to global change. Climatic Change 87, 273–292.

Knowlton, K., Rotkin-Ellman, M., King, G., Margolis, H.G., Smith, D., Solomon, G., et al., 2009. The 2006 California heat wave: impacts on hospitalizations and emergency department visits. Environmental Health Perspectives 117, 61–67.

Kolokotroni, M., Ren, X., Davies, M., Mavrogianni, A., 2012. London's urban heat island: impact on current and future energy consumption in office buildings. Energy and Buildings 47, 302—311. Available from: https://doi.org/10.1016/j.enbuild.2011.12.019.

Kolokotroni, M., Zhang, Y., Watkins, R., 2007. The London Heat Island and building cooling design. Solar Energy 81, 102—110.

Kovač-Andrić, E., Brana, J., Gvozdić, V., 2009. Impact of meteorological factors on ozone concentrations modelled by time series analysis and multivariate statistical methods. Ecological Informatics 4, 117—122.

Kovats, R.S., Hajat, S., Wilkinson, P., 2004. Contrasting patterns of mortality and hospital admissions during hot weather and heat waves in Greater London, UK. Occupational and Environmental Medicine 61, 893—898.

Lam, J.C., Tsang, C.L., Li, D.H.W., 2004. Long term ambient temperature analysis and energy use implications in Hong Kong. Energy Conversion and Management 45, 315—327.

Lange, M.A., Hadjinicolaou, P., Santamouris, M., Zittis, G., n.d. Assessment of increasing electricity demand for space cooling under climate change condition in the Eastern Mediterranean, In: 2nd Conference on Power Options for the Eastern Mediterranean Region (POEM 2013).

Leal Filho, W., Echevarria Icaza, L., Neht, A., Klavins, M., Morgan, E.A., 2018. Coping with the impacts of urban heat islands. A literature based study on understanding urban heat vulnerability and the need for resilience in cities in a global climate change context. Journal of Cleaner Production 171, 1140—1149. Available from: https://doi.org/10.1016/j.jclepro.2017.10.086.

Li, J., Georgescu, M., Hyde, P., Mahalov, A., Moustaoui, M., 2014. Achieving accurate simulations of urban impacts on ozone at high resolution. Environmental Research Letters 9, 114019.

Li, M., Shaw, B.A., Zhang, W., Vásquez, E., Lin, S., 2019. Impact of extremely hot days on emergency department visits for cardiovascular disease among older adults in New York State. International Journal of Environmental Research and Public Health 16, 2119.

Li, M., Song, Y., Mao, Z., Liu, M., Huang, X., 2016. Impacts of thermal circulations induced by urbanization on ozone formation in the Pearl River Delta region, China. Atmospheric Environment 127, 382—392.

Lin, S., Luo, M., Walker, R.J., Liu, X., Hwang, S.-A., Chinery, R., 2009. Extreme high temperatures and hospital admissions for respiratory and cardiovascular diseases. Epidemiology (Cambridge, Mass.) 738—746.

Lin, C.Y.C., Mickley, L.J., Hayhoe, K., Maurer, E.P., Hogrefe, C., 2007. Rapid calculation of future trends in ozone exceedances over the northeast United States: results from three models and two scenarios. Consequences of global change for air quality festival, EPA, Research Triangle Park, NC.

Linnerud, K., Mideksa, T.K., Eskeland, G.S., 2009. The impact of climate change on thermal power supply. Manuscript, CICERO.

Lo, C.P., Quattrochi, D.A., 2003. Land-use and land-cover change, urban heat island phenomenon, and health implications. Photogrammetric Engineering & Remote Sensing 69, 1053—1063.

Lomas, K.J., Kane, T., 2013. Summertime temperatures and thermal comfort in UK homes. Building Research & Information 41, 259—280.

Lowe, S.A., 2016. An energy and mortality impact assessment of the urban heat island in the US. Environmental Impact Assessment Review 56, 139—144.

Madrigano, J., Mittleman, M.A., Baccarelli, A., Goldberg, R., Melly, S., Von Klot, S., et al., 2013. Temperature, myocardial infarction, and mortality: effect modification by individual and area-level characteristics. Epidemiology (Cambridge, Mass.) 24, 439.

Magli, S., Lodi, C., Lombroso, L., Muscio, A., Teggi, S., 2014. Analysis of the urban heat island effects on building energy consumption. International Journal of Energy and Environmental Engineering 6, 91−99. Available from: https://doi.org/10.1007/s40095-014-0154-9.

Mastrangelo, G., Hajat, S., Fadda, E., Buja, A., Fedeli, U., Spolaore, P., 2006. Contrasting patterns of hospital admissions and mortality during heat waves: are deaths from circulatory disease a real excess or an artifact? Medical Hypotheses 66, 1025−1028.

Mavrogianni, A., Davies, M., Wilkinson, P., Pathan, A., 2010. London housing and climate change: impact on comfort and health-preliminary results of a summer overheating study. Open House International 35 (2), 49−59.

Meier, P., Holloway, T., Patz, J., Harkey, M., Ahl, D., Abel, D., et al., 2017. Impact of warmer weather on electricity sector emissions due to building energy use. Environmental Research Letters 12, 64014.

Michelozzi, P., Accetta, G., De Sario, M., D'Ippoliti, D., Marino, C., Baccini, M., et al., 2009. High temperature and hospitalizations for cardiovascular and respiratory causes in 12 European cities. American Journal of Respiratory and Critical Care Medicine 179, 383−389. Available from: https://doi.org/10.1164/rccm.200802-217OC.

Mikellidou, C.V., Shakou, L.M., Boustras, G., Dimopoulos, C., 2018. Energy critical infrastructures at risk from climate change: a state of the art review. Safety Science 110, 110−120.

Milojevic, A., Armstrong, B.G., Gasparrini, A., Bohnenstengel, S.I., Barratt, B., Wilkinson, P., 2016. Methods to estimate acclimatization to urban heat island effects on heat-and cold-related mortality. Environmental Health Perspectives 124, 1016−1022. Available from: https://doi.org/10.1289/ehp.1510109.

Mirasgedis, S., Sarafidis, Y., Georgopoulou, E., Lalas, D.P., Moschovits, M., Karagiannis, F., et al., 2006. Models for mid-term electricity demand forecasting incorporating weather influences. Energy 31, 208−227. Available from: https://doi.org/10.1016/j.energy.2005.02.016.

Moghadamnia, M.T., Ardalan, A., Mesdaghinia, A., Keshtkar, A., Naddafi, K., Yekaninejad, M.S., 2017. Ambient temperature and cardiovascular mortality: a systematic review and meta-analysis. PeerJ 2017, 3574. Available from: https://doi.org/10.7717/peerj.3574.

Oke, T.R., Johnson, G.T., Steyn, D.G., Watson, I.D., 1991. Simulation of surface urban heat islands under "ideal" conditions at night part 2: diagnosis of causation. Boundary − Layer Meteorology 56, 339−358. Available from: https://doi.org/10.1007/BF00119211.

Ooka, R., Khiem, M., Hayami, H., Yoshikado, H., Huang, H., Kawamoto, Y., 2011. Influence of meteorological conditions on summer ozone levels in the central Kanto area of Japan. Procedia Environmental Sciences 4, 138−150.

Paeth, H., Scholten, A., Friederichs, P., Hense, A., 2008. Uncertainties in climate change prediction: El Niño-Southern oscillation and monsoons. Global and Planetary Change 60, 265−288. Available from: https://doi.org/10.1016/j.gloplacha.2007.03.002.

Page, L.A., Hajat, S., Kovats, R.S., Howard, L.M., 2012. Temperature-related deaths in people with psychosis, dementia and substance misuse. The British Journal of Psychiatry 200, 485−490.

Palme, M., Inostroza, L., Villacreses, G., Lobato-Cordero, A., Carrasco, C., 2017. From urban climate to energy consumption. Enhancing building performance simulation by including the urban heat island effect. Energy and Buildings 145, 107−120.

Paolini, R., Haddad, S., Synnefa, A., Garshasbi, S., Santamouris, M., 2018. Electricity demand reduction in Sydney and Darwin with local climate mitigation. n Engaging Architectural Science: Meeting the Challenges of Higher Density, The Architectural Science Asso-ciation (ANZAScA), Melbourne, Australia, 285−294.

Papanastasiou, D.K., Fidaros, D., Bartzanas, T., Kittas, C., 2013. Impact of urban heat island development on buildings' energy consumption. Fresenius Environmental Bulletin 22, 2087−2092.

Pardo, A., Meneu, V., Valor, E., 2002. Temperature and seasonality influences on Spanish electricity load. Energy Economics 24, 55−70. Available from: https://doi.org/10.1016/S0140-9883(01)00082-2.

Phung, D., Guo, Y., Nguyen, H.T.L., Rutherford, S., Baum, S., Chu, C., 2016. High temperature and risk of hospitalizations, and effect modifying potential of socio-economic conditions: a multi-province study in the tropical Mekong Delta Region. Environment International 92, 77−86.

Pyrgou, A., Hadjinicolaou, P., Santamouris, M., 2020. Urban-rural moisture contrast: Regulator of the urban heat island and heatwaves' synergy over a Mediterranean city. Environmental Research 182. Available from: https://doi.org/10.1016/j.envres.2019.109102.

Radhi, H., Sharples, S., 2013. Quantifying the domestic electricity consumption for air-conditioning due to urban heat islands in hot arid regions. Applied Energy 112, 371−380.

Rajagopalan, P., Andamon, M.M., eds., 2018. Engaging Architectural Science: Meeting the Challenges of Higher Density: 52nd International Conference of the Architectural Science Association, pp. 577−583. The Architectural Science Association and RMIT University, Australia.

Reid, C.E., O'neill, M.S., Gronlund, C.J., Brines, S.J., Brown, D.G., Diez-Roux, A.V., et al., 2009. Mapping community determinants of heat vulnerability. Environmental Health Perspectives 117, 1730−1736.

Reid, C.E., Snowden, J.M., Kontgis, C., Tager, I.B., 2012. The role of ambient ozone in epidemiologic studies of heat-related mortality. Environmental Health Perspectives 120, 1627−1630.

Rong, F., 2006. Impact of Urban Sprawl on US Residential Energy Use. http://hdl.handle.net/1903/3848.

Rosenfeld, A.H., Akbari, H., Bretz, S., Fishman, B.L., Kurn, D.M., Sailor, D., et al., 1995. Mitigation of urban heat islands: materials, utility programs, updates. Energy and Buildings 22, 255−265. Available from: https://doi.org/10.1016/0378-7788(95)00927-P.

Rosenthal, J.K., Kinney, P.L., Metzger, K.B., 2014. Intra-urban vulnerability to heat-related mortality in New York City, 1997−2006. Health & Place 30, 45−60.

Rübbelke, D., Vögele, S., 2011. Impacts of climate change on European critical infrastructures: the case of the power sector. Environmental Science and Policy 14, 53−63. Available from: https://doi.org/10.1016/j.envsci.2010.10.007.

Ruddell, D.M., Harlan, S.L., Grossman-Clarke, S., Buyantuyev, A., 2009. Risk and exposure to extreme heat in microclimates of Phoenix, AZ. Geospatial techniques in urban hazard and disaster analysis. Springer, Netherlands, Dordrecht, pp. 179−202. Available from: https://doi.org/10.1007/978-90-481-2238-7_9.

Ruth, M., Lin, A.C., 2006. Regional energy demand and adaptations to climate change: methodology and application to the state of Maryland, USA. Energy Policy 34, 2820−2833. Available from: https://doi.org/10.1016/j.enpol.2005.04.016.

Rydman, R.J., Rumoro, D.P., Silva, J.C., Hogan, T.M., Kampe, L.M., 1999. The rate and risk of heat-related illness in hospital emergency departments during the 1995

Chicago heat disaster. Journal of Medical Systems 23, 41−56. Available from: https://doi.org/10.1023/A:1020871528086.

Ryu, Y.-H., Baik, J.-J., Lee, S.-H., 2013. Effects of anthropogenic heat on ozone air quality in a megacity. Atmospheric Environment 80, 20−30.

Sailor, D.J., Muñoz, J.R., 1997. Sensitivity of electricity and natural gas consumption to climate in the U.S.A.—methodology and results for eight states. Energy 22, 987−998. Available from: https://doi.org/10.1016/S0360-5442(97)00034-0.

Sakka, A., Santamouris, M., Livada, I., Nicol, F., Wilson, M., 2012. On the thermal performance of low income housing during heat waves. Energy and Buildings 49, 69−77. Available from: https://doi.org/10.1016/j.enbuild.2012.01.023.

Santamouris, M., 2014. On the energy impact of urban heat island and global warming on buildings. Energy and Buildings 82, 100−113. Available from: https://doi.org/10.1016/j.enbuild.2014.07.022.

Santamouris, M., 2015a. Regulating the damaged thermostat of the cities—status, impacts and mitigation challenges. Energy and Buildings 91, 43−56. Available from: https://doi.org/10.1016/j.enbuild.2015.01.027.

Santamouris, M., 2015b. Analyzing the heat island magnitude and characteristics in one hundred Asian and Australian cities and regions. Science of the Total Environment 512, 582−598. Available from: https://doi.org/10.1016/j.scitotenv.2015.01.060.

Santamouris, M., 2016a. Innovating to zero the building sector in Europe: minimising the energy consumption, eradication of the energy poverty and mitigating the local climate change. Solar Energy 128, 61−94. Available from: https://doi.org/10.1016/j.solener.2016.01.021.

Santamouris, M., 2016b. Cooling the buildings—past, present and future. Energy and Buildings 128, 617−638. Available from: https://doi.org/10.1016/j.enbuild.2016.07.034.

Santamouris, M., 2018. Minimizing energy consumption, energy poverty and global and local climate change in the built environment: innovating to zero. Elsevier. https://doi.org/10.1016/C2016-0-01024-0.

Santamouris, M., 2020. Recent progress on urban overheating and heat island research. Integrated assessment of the energy, environmental, vulnerability and health impact. Synergies with the global climate change. Energy and Buildings 207, 109482. Available from: https://doi.org/10.1016/j.enbuild.2019.109482.

Santamouris, M., Alevizos, S.M., Aslanoglou, L., Mantzios, D., Milonas, P., Sarelli, I., et al., 2014. Freezing the poor—indoor environmental quality in low and very low income households during the winter period in Athens. Energy and Buildings 70, 61−70. Available from: https://doi.org/10.1016/j.enbuild.2013.11.074.

Santamouris, M., Cartalis, C., Synnefa, A., Kolokotsa, D., 2015. On the impact of urban heat island and global warming on the power demand and electricity consumption of buildings—a review. Energy and Buildings 98, 119−124. Available from: https://doi.org/10.1016/j.enbuild.2014.09.052.

Santamouris, M., Kolokotsa, D., 2016. Urban climate mitigation techniques. Urban climate mitigation techniques. Routledge, London, UK. Available from: https://doi.org/10.4324/9781315765839.

Santamouris, M., Papanikolaou, N., Livada, I., Koronakis, I., Georgakis, C., Argiriou, A., et al., 2001. On the impact of urban climate on the energy consumption of buildings. Solar Energy 70, 201−216. Available from: https://doi.org/10.1016/S0038-092X(00)00095-5.

Santamouris, M., Paraponiaris, K., Mihalakakou, G., 2007. Estimating the ecological footprint of the heat island effect over Athens, Greece. Climatic Change 80, 265−276. Available from: https://doi.org/10.1007/s10584-006-9128-0.

Sarrat, C., Lemonsu, A., Masson, V., Guedalia, D., 2006. Impact of urban heat island on regional atmospheric pollution. Atmospheric Environment 40, 1743−1758.

Schinasi, L.H., Benmarhnia, T., De Roos, A.J., 2018. Modification of the association between high ambient temperature and health by urban microclimate indicators: a systematic review and *meta*-analysis. Environmental Research . Available from: https://doi.org/10.1016/j.envres.2017.11.004.

Schwartz, J., Samet, J.M., Patz, J.A., 2004. Hospital admissions for heart disease: the effects of temperature and humidity. Epidemiology (Cambridge, Mass.) 15, 755−761.

Segala, M., Shafirb, H., Mandel, M., Alpert, P., Balmor, Y., 1992. Climatic-related evaluations of the summer peak-hours' electric load in Israel. Journal of Applied Meteorology and Climatology 31 (12), 1492−1498.

Sillman, S., Samson, P.J., 1995. Impact of temperature on oxidant photochemistry in urban, polluted rural and remote environments. Journal of Geophysical Research: Atmospheres 100, 11497−11508.

Skelhorn, C., Levermore, G., Buildings, S.L.-E. 2016, undefined, n.d. Impacts on cooling energy consumption due to the UHI and vegetation changes in Manchester, Energy and Buildings, UK. Elsevier. Volume 122, pp. 150−159.

Smargiassi, A., Goldberg, M.S., Plante, C., Fournier, M., Baudouin, Y., Kosatsky, T., 2009. Variation of daily warm season mortality as a function of micro-urban heat islands. Journal of Epidemiology and Community Health 63, 659−664. Available from: https://doi.org/10.1136/jech.2008.078147.

Son, J.-Y., Lane, K.J., Lee, J.-T., Bell, M.L., 2016. Urban vegetation and heat-related mortality in Seoul, Korea. Environmental Research 151, 728−733.

Stafoggia, M., Forastiere, F., Agostini, D., Biggeri, A., Bisanti, L., Cadum, E., et al., 2006. Vulnerability to heat-related mortality: a multicity, population-based, case-crossover analysis. Epidemiology (Cambridge, Mass.) 315−323.

Stathopoulou, E., Mihalakakou, G., Santamouris, M., Bagiorgas, H.S., 2008. On the impact of temperature on tropospheric ozone concentration levels in urban environments. Journal of Earth System Science 117, 227−236. Available from: https://doi.org/10.1007/s12040-008-0027-9.

Street, M., Reinhart, C., Norford, L., Ochsendorf, J., n.d. Urban heat island in Boston—an evaluation of urban air-temperature models for predicting building energy use, In: Proceedings of the BS2013: 13th Conference of International Building Performance Simulation Association, Chambéry, France, 26−28.

Summerfield, A.J., Lowe, R.J., Bruhns, H.R., Caeiro, J.A., Steadman, J.P., Oreszczyn, T., 2007. Milton Keynes Energy Park revisited: changes in internal temperatures and energy usage. Energy and Buildings 39, 783−791.

Swamy, G., Nagendra, S.M.S., Schlink, U., 2017. Urban heat island (UHI) influence on secondary pollutant formation in a tropical humid environment. Journal of the Air & Waste Management Association 67, 1080−1091.

Synnefa, A., Garshasbi, S., Haddad, S., Paolini, R, Santamouris, M., 2018. Impact of the mitigation of the local climate on building energy needs in Australian cities, In: Proceedings of the Engaging Architectural Science: Meeting the Challenges of Higher Density. 52nd International Conference of the Architectural Science Association, 277−284.

Tan, J., Zheng, Y., Tang, X., Guo, C., Li, L., Song, G., et al., 2010. The urban heat island and its impact on heat waves and human health in Shanghai. International Journal of Biometeorology 54, 75−84.

Tarasova, O.A., Karpetchko, A.Y., 2003. Accounting for local meteorological effects in the ozone time-series of Lovozero (Kola Peninsula). Atmospheric Chemistry and Physics 3, 941−949.

224 Urban Climate Change and Heat Islands

Taylor, J., Wilkinson, P., Davies, M., Armstrong, B., Chalabi, Z., Mavrogianni, A., et al., 2015. Mapping the effects of urban heat island, housing, and age on excess heat-related mortality in London. Urban Climate 14, 517–528. Available from: https://doi.org/10.1016/j.uclim.2015.08.001.

Uejio, C.K., Wilhelmi, O.V., Golden, J.S., Mills, D.M., Gulino, S.P., Samenow, J.P., 2011. Intra-urban societal vulnerability to extreme heat: the role of heat exposure and the built environment, socioeconomics, and neighborhood stability. Health & Place 17, 498–507.

Vandentorren, S., Bretin, P., Zeghnoun, A., Mandereau-Bruno, L., Croisier, A., Cochet, C., et al., 2006. August 2003 heat wave in France: risk factors for death of elderly people living at home. European Journal of Public Health . Available from: https://doi.org/10.1093/eurpub/ckl063.

Vaneckova, P., Beggs, P.J., Jacobson, C.R., 2010. Spatial analysis of heat-related mortality among the elderly between 1993 and 2004 in Sydney, Australia. Social Science & Medicine 70, 293–304. Available from: https://doi.org/10.1016/J.SOCSCIMED.2009.09.058.

Wang, X.M., Lin, W.S., Yang, L.M., Deng, R.R., Lin, H., 2007. A numerical study of influences of urban land-use change on ozone distribution over the Pearl River Delta region, China. Tellus, Series B: Chemical and Physical Meteorology. Blackwell, Munksgaard, pp. 633–641. Available from: https://doi.org/10.1111/j.1600-0889.2007.00271.x.

Wang, Y., Shen, L., Wu, S., Mickley, L., He, J., Hao, J., 2013. Sensitivity of surface ozone over China to 2000–2050 global changes of climate and emissions. Atmospheric Environment 75, 374–382.

Wang, T., Xue, L., Brimblecombe, P., Lam, Y.F., Li, L., Zhang, L., 2017. Ozone pollution in China: a review of concentrations, meteorological influences, chemical precursors, and effects. Science of the Total Environment 575, 1582–1596.

Wangpattarapong, K., Maneewan, S., Ketjoy, N., Rakwichian, W., 2008. The impacts of climatic and economic factors on residential electricity consumption of Bangkok Metropolis. Energy and Buildings 40(8), 1419–1425.

Weber, S., Sadoff, N., Zell, E., de Sherbinin, A., 2015. Policy-relevant indicators for mapping the vulnerability of urban populations to extreme heat events: a case study of Philadelphia. Applied Geography 63, 231–243.

Wingfield, J., Bell, M., Miles-Shenton, D., South, T., Lowe, R., 2011. Evaluating the impact of an enhanced energy performance standard on load-bearing masonry domestic construction: understanding the gap between designed and real performance: lessons from Stamford Brook.

Wong, S., Wan, K., Li, D., Lam, J., 2010. Impact of climate change on residential building envelope cooling loads in subtropical climates. Energy and Buildings 42, 2098–2103.

Wright, A.J., Young, A.N., Natarajan, S., 2005. Dwelling temperatures and comfort during the August 2003 heat wave. Building Services Engineering Research and Technology 26, 285–300.

Xu, Y., Dadvand, P., Barrera-Gómez, J., Sartini, C., Marí-Dell'Olmo, M., Borrell, C., et al., 2013. Differences on the effect of heat waves on mortality by sociodemographic and urban landscape characteristics. Journal of Epidemiology and Community Health 67, 519–525.

Yabe, K., 2005. Evaluation of energy saving effect for the long-term maximum power forecast (title only in original language). In: Proceedings of National Convention of the Institute of Electrical Engineers of Japan (IEE J).

Yadav, N., Sharma, C., Peshin, S.K., Masiwal, R., 2017. Study of intra-city urban heat island intensity and its influence on atmospheric chemistry and energy consumption in

Delhi. Sustainable Cities and Society 32, 202—211. Available from: https://doi.org/10.1016/j.scs.2017.04.003.

Ye, X., Wolff, R., Yu, W., Vaneckova, P., Pan, X., Tong, S., 2012. Ambient temperature and morbidity: a review of epidemiological evidence. Environmental Health Perspectives 120, 19—28.

Yohanis, Y.G., Mondol, J.D., 2010. Annual variations of temperature in a sample of UK dwellings. Applied Energy 87, 681—690.

Zinzi, M., Carnielo, E., 2017. Impact of urban temperatures on energy performance and thermal comfort in residential buildings. The case of Rome, Italy. Energy and Buildings 157, 20—29.

Zinzi, M., Carnielo, E., Mattoni, B., 2018. On the relation between urban climate and energy performance of buildings. A three-year experience in Rome, Italy. Applied Energy 221, 148—160. Available from: https://doi.org/10.1016/J.APENERGY.2018.03.192.

CHAPTER 6

Fighting urban climate change—state of the art of mitigation technologies

Jie Feng[1], Shamila Haddad[1], Kai Gao[1], Samira Garshasbi[1], Giulia Ulpiani[1], Matthaios Santamouris[1], Gianluca Ranzi[2] and Carlos Bartesaghi-Koc[3]

[1]School of Built Environment, Faculty of Arts, Design and Architecture, University of New South Wales (UNSW), Sydney, NSW, Australia
[2]School of Civil Engineering, The University of Sydney, Sydney, NSW, Australia
[3]School of Architecture and Built Environment, ECMS, The University of Adelaide, Adelaide, SA, Australia

6.1 Introduction

The urban heat island (UHI) and urban overheating are the most documented phenomena of climate change. Higher ambient temperatures increase the cooling energy consumption of buildings, affect human health, raise the concentration of urban pollutants, and affect the quality of life of urban citizens. The phenomenon is experimentally documented in more than 450 large cities around the world (Santamouris, 2015a, 2016). The intensity of urban overheating depends on several parameters like the characteristics of the local climate, the landscape, and the features of the city, such as the materials used and the strength of the local sinks and sources. According to the existing data, the average maximum magnitude of urban overheating is close to 5°C, but it may vary up to 10°C. The highest intensities are observed during anticyclonic climatic conditions, low wind speed, and clear sky conditions. In parallel, precipitation tends to decrease the strength of the urban overheating as it increases the thermal admittance of the rural areas (Santamouris, 2015a).

Urban overheating causes a serious increase in energy consumption and peak electricity demand in cities. According to recent studies, the average energy penalty is estimated close to 2.4 ($\pm$ 1.5) kWh per unit of city surface, or 0.74 ($\pm$ 0.67) kWh per square meter and degree of temperature increase, equivalent to 237 ($\pm$ 130) kWh per person while the energy penalty close to 70 ($\pm$ 45) kWh per person and degree of UHI intensity (Santamouris, 2014a,b). The corresponding increase of the peak electricity requirements is close to 21 ($\pm$ 10.4) W per person and degree (Santamouris et al., 2015).

Apart from the energy impact, urban overheating is increasing the ecological footprint of cities (Santamouris et al., 2007), raises the concentration of urban pollutants—mainly of the ground-level ozone (Santamouris, 2015b)—and increases the heat-related mortality and morbidity as higher urban temperatures limit the capacity of the human body to adapt (Paravantis et al., 2017; Pyrgou and Santamouris, 2018; Santamouris, 2015b).

To fight UHIs and urban overheating and counterbalance their impact, mitigation systems and technologies have been proposed, developed, and implemented (Akbari et al., 2016; Pisello et al., 2018). Mitigation strategies involve the use of additional green infrastructure (GI) in cities, advanced materials for buildings and the urban fabric, evaporative technologies, solar control systems, and heat dissipation systems to reject the excess heat in atmospheric sinks of low temperature. Existing performance data around hundreds of large-scale mitigation projects show a very significant cooling potential to drop the peak urban temperature up to 3°C (Santamouris et al., 2017).

6.2 Mitigating the urban heat using advanced materials for the urban fabric

6.2.1 Introduction to mitigation materials

The thermal balance of cities is highly influenced by the properties and the characteristics of the materials used in the building envelope and the urban fabric. While solar absorbing materials (usually dark colored) exhibit high surface temperatures and high release of sensible heat and contribute to increasing the urban ambient temperature, high-albedo materials reflect the solar radiation, present lower surface temperature, and do not increase the ambient temperature. The recent development of artificial materials of very high solar reflectance permits to easily increase the albedo of cities and reduce the magnitude of urban overheating. However, technical issues related to the optical aging of the materials have to be solved, while issues related to visual annoyance have to be confronted (Santamouris et al., 2017, 2008).

Recent research has succeeded in developing the so-called super cool materials presenting a very high reflectance in the solar spectrum and a high emissivity value in the atmospheric window wavelengths. In parallel, more advanced chromic materials as well as fluorescent materials are developed and are available for mitigation purposes.

The present chapter aims to review and analyze the most recent development in the field of innovative materials for urban cooling and mitigation.

6.2.2 High reflectance white coatings

White materials presenting a very high reflectance to solar spectrum based on acrylic, elastomeric, silicone, and fluoropolymer technologies have been available for the last 20 years (Akbari and Matthews, 2012). The initial reflectance of the materials may exceed 0.90; however, the 3-year value is substantially lower and varies between 0.65 and 0.85 because of aging.

Extensive experimental testing of the available high reflectance white materials shows that they present low surface temperature but above the ambient one. In contrast, engineered coatings present a substantially lower temperature than natural white materials (Synnefa et al., 2006). As expected, the surface temperature of reflective materials depends on the materials' specific optical and thermal characteristics and the corresponding climatic conditions. Several experiments have shown substantial temperature differences up to 17°C, even between white materials tested under the same conditions. In particular, materials covered with metal-based paints presented the highest surface temperature mainly because of their low emissivity values, while the lowest surface temperature corresponds to materials with high reflectance and emissivity values. As shown in Fig. 6.1 (Santamouris and Yun, 2020), the surface temperature of the coolest and warmest materials was 3°C and 12°C above the ambient temperature, respectively.

As already mentioned, manufactured high reflectance white coatings may present an important problem of weatherization and optical aging. Testing has shown that acrylic elastomeric coatings exposed to solar radiation for 60 days increased their surface temperature considerably because of the important loss of their reflectance caused by the deposition of salts, clay, black carbon soot particles and particulate organic matter, and the significant growth of bacteria on the surfaces (Mastrapostoli et al., 2016).

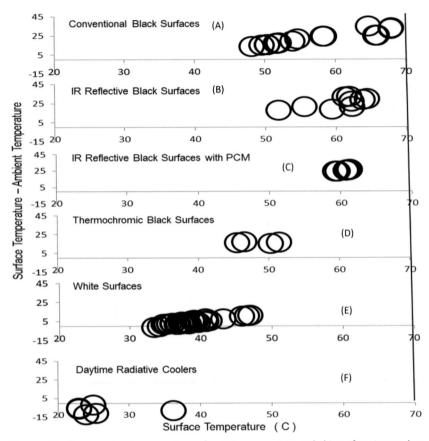

Figure 6.1 Comparative presentation of existing experimental data of various urban mitigation technologies. Relating surface temperature decrease against the corresponding surface temperature of the materials (Santamouris and Yun, 2020).

Intensive cleaning of the roofs contributes to reasonably restoring the initial reflectance, but as reported, it is reasonable to lose almost 20% of the initial albedo during the first $3 - 5e$ years (Ferrari et al., 2014; Takebayashi et al., 2016).

Research aiming to improve the optical durability of white coatings considers parameters related to the composition and treatment of the coatings, as well as on the specific environmental conditions (Khataee et al., 2016; Ohama and Van Gemert, 2011). Research has shown that the addition of inorganic components like titanium dioxide with Al, Li, or K in the paints is found to increase their optical stability as well as their solar reflectivity and emissivity compared to organic binder paints (Pal et al., 2016), while it helps to degrade the deposited organic substances into sulfate, water, nitrate, and carbon dioxide that are removed from the surface of the material by rain or water cleaning (Zapata et al., 2014). In parallel, it is found that a very significant decrease of the bacteria concentration on the surface of the coating may occur (Hwang et al., 2017).

6.2.3 Colored infrared reflective coatings

Colored infrared-reflective materials are characterized by high near-infrared (NIR)-solar reflection and high thermal emission (Levinson et al., 2007; Synnefa et al., 2007b). Colored infrared materials have the same color/visible-range solar reflection as conventional coatings but present significantly lower surface temperature due to their high solar reflection in the NIR range. A comparative study on the optical and thermal properties of infrared reflective materials of various colors revealed a significant solar reflection increase of 22%, 19%, 18%, 15%, 14%, 12%, 11%, 10%, and 7% for black, anthracite, chocolate brown, blue, light brown, light blue, brown, orange, and green colored cool coatings compared to their corresponding conventional coatings (Synnefa et al., 2007b). The emissivity measurements showed a high infrared emittance of 0.88 for both infrared reflective and conventional samples. The thermal assessment of the samples under the sunlight demonstrated that black, chocolate brown, blue, and anthracite are the best performing cool colored coatings with a mean daily surface temperature difference of up to 5.2°C, 4.7°C, 4.7°C, and 2.8°C, respectively. The black infrared reflective and its corresponding conventional sample recorded the highest surface temperature difference of 10.2°C. In the meantime, the lowest temperature difference of 1.6°C was observed between green infrared reflective and standard green colored coatings. Another study on various mitigation technologies in western Sydney showed that the large-scale implementation of infrared reflective colored coatings can decrease the ambient temperature by up to 3°C when increasing the global reflection by 0.6 (Santamouris et al., 2018c). The application of cool colored coatings also minimizes the heat transfer into the building by reflecting the solar radiation, and therefore, decreases the cooling load of buildings during warm periods. A study on the annual energy impact of cool colored coatings shows cooling loads reduction of $9-48 \, kWh/m^2$ (Synnefa et al., 2007a). The study also reported an annual heating penalty of $0.2-17 \, kWh/m^2$ during the cold periods, which is less important when comparing with the huge cooling load reduction. Results also show that cooling loads reduction of infrared reflective materials is more significant for buildings with minimal or no building insulation (Synnefa et al., 2007a).

Infrared reflective coatings can be made using organic and inorganic NIR-reflective pigments (e.g., metal oxides) (Jeevanandam et al., 2007). The size of dispersed metal oxides is a critical factor through which the wavelength range with the highest solar reflection could be determined (Fang et al., 2013). In general, the particle size should be more than half of the wavelength to achieve a very high solar reflection in the desired wavelength range (e.g., particle size should be more than $0.4 - 0.6$ microns for a very high solar reflection between 800 and 1200 nm). Several studies reported the limitations on the amount of metal oxides, especially for dark-colored infrared-reflective materials. The main reason behind this is that a large amount of metal oxides can change the visible range reflection/color (e.g., from black to gray). A US patent proposed two-layered cool coatings to address the limitation for the addition of metal oxides (Genjima and Mochizuki, 2002). The proposed infrared reflective coating comprises an NIR-permeable top layer to attain the desired color and an NIR-reflective base layer to reflect the NIR light. According to the US patent, the NIR-reflective coating should have a solar reflectance of 60% or more and an NIR-permeability of 25% or less; and the NIR-permeable layer should have a reflectance of less than 60%, an absorbance of 50% or less, and a permeability of 30% or more in the NIR range.

6.2.4 Reflecting materials of high thermal capacity

The use of phase change materials added in the mass helps to increase the apparent thermal capacity of the coatings and contribute to decreasing their peak surface temperature. Phase change materials have been extremely well-developed and are widely used for storage purposes in the built environment (Konstantinidou et al., 2019). Phase change materials are usually compatible with other building materials like concrete or asphalt and can be easily used in buildings and open spaces.

Important research has been carried out to investigate the mitigation potential of phase change materials. As reported, the incorporation of phase change materials in the mass of pavements can considerably reduce their surface temperature and release the stored energy at a later time. The potential for temperature decrease depends on many parameters of the volume of the Phase Change Materials (PCMs) in the pavement mass, its melting temperature, the way PCMs are incorporated into the pavement, the local climatic conditions, and the thermal and optical characteristics of the pavement materials. Experiments have shown that PCM-doped pavements' potential surface temperature drop may vary between 2.5°C and 19.7°C, as a function of the parameters mentioned above (Guan et al., 2011; Kim et al., 2015; Sharifi and Mahboub, 2018) (Fig. 6.1C).

Current research on PCM-doped pavements for mitigation purposes is quite limited and focus mainly on the most appropriate methods to integrate them in the mass of the pavements, improve aging issues and increase the thermal cycling of phase change materials.

6.2.5 Temperature-sensitive/color changing materials

Temperature-sensitive coatings technologies are advanced coatings capable of integrating different optical and fluorescent properties into one system for the best functionality during both warm and cold periods (Garshasbi and Santamouris, 2019; Karlessi et al., 2009). Temperature-sensitive materials can be divided into temperature-sensitive materials for transparent and opaque surfaces. Temperature-sensitive/thermochromic materials for the transparent surfaces (e.g., building glazing) modulate the incoming solar radiation by changing from opaque to transmissive by a decrease in temperature (Nazemiyan and Jalili, 2013; Warwick and Binions, 2014). Temperature-sensitive coatings for opaque surfaces can be categorized into two major groups: (1) Thermochromic materials with temperature-sensitive solar reflection. (2) Temperature-sensitive fluorescent materials/quantum dots (QDs) with temperature-sensitive photoluminescence (PL)-intensity and/or temperature-sensitive absorption edge wavelength (λ_{AE})/PL peak wavelength (Garshasbi and Santamouris, 2019). Temperature-sensitive materials for opaque surfaces can be also categorized into two groups, including dye-based and nondye thermochromics.

Dye-based thermochromic materials, as common types of thermochromic materials for opaque surfaces, have been extensively investigated by several research groups (Berardi et al., 2020; Karlessi et al., 2009; Perez et al., 2020; Zhang and Zhai, 2019). Dye-based thermochromic materials are smart materials composed of three main components, including dye, color developer, and solvent, which undergo a transition from color to colorless state at temperatures above their solvent melting temperature.

A thermal and optical evaluation of 11 different dye-based thermochromic materials with various colors has been performed to assess their outdoor thermal performance. As reported, the thermochromic materials demonstrated up to 43% solar reflection variation between colored and colorless states at their thermochromic transition temperature at around 30°C. The energy performance evaluation of dye-based thermochromic materials has also shown very promising results, especially in low-insulated buildings. Using dye-based thermochromic coatings with a transition temperature at 31°C on a low-insulated two-storey building, for instance, showed up to $4.2-5 \text{ kWh/m}^2$ and $0.7-1.4 \text{ kWh/m}^2$ compared to the reference building covered by conventional and cool coatings, respectively (Yuxuan et al., 2020).

While dye-based thermochromics demonstrate a high energy saving potential due to their interesting temperature-sensitive features, including transition temperature and optical properties variations, photodegradation is a serious issue for the outdoor application of dye-based thermochromics. A study has been performed to detect the parts of solar radiation contributing to photodegradation using several UV and optical filters (Karlessi and Santamouris, 2015). As reported, in addition to the UV portion of the solar radiation, the visible-range wavelengths, especially those closer to the blue end of the visible range, contribute significantly to the photodegradation process. According to the study, unprotected thermochromic samples demonstrate up to 8.8% solar reflection change in colored state and 3.4% solar reflection change in colorless state when exposed to the outdoor environment. However, red filters covering all wavelengths below 600 nm fully protected the thermochromic material in the colored state and decreased the solar reflection change to 0.8% in the colorless state. All other filters are reported to provide less protection against the outdoor environment than the red filter. Several studies have also been conducted to understand the underlying mechanism of photodegradation in thermochromics. Dealkylation of the alkylamino groups, oxidative cleavage, and reduction of an excited state dye cation to a colorless leuco dye form are reported to be the three major mechanisms contributing to the photodegradation process (Seeboth and Lötzsch, 2008). Several stabilizers have been shown to be able to inhibit the photodegradation process completely. However, none of the dye-based thermochromics-stabilizer systems demonstrate a reversible thermochromic effect.

Given the serious problem of photodegradation in dye-based thermochromics, nondye thermochromic materials working based on intriguing nanoscale optical effect, including quantum confinement, surface plasmon resonance, and photonics effects, could be used as great alternatives to their conventional counterparts. There are three main types of nanoscale thermochromics: (1) QDs, (2) plasmonics, and (3) photonic crystals.

QDs may demonstrate two different temperature-sensitive fluorescent properties: (1) Temperature-sensitive PL intensity (i.e., thermal quenching or antiquenching effect), (2) λ_{AE}/PL peak red-shift. For QDs with PL intensity variation by temperature, thermal quenching is the commonly observed temperature-sensitive mechanism in which QDs demonstrate low PL intensity at high temperatures and high PL intensity at low temperatures. However, research on QDs shows PL enhancement by temperature/thermal antiquenching effect. The PL enhancement is believed to be mainly because of the thermal rectification of surface traps in QDs having specific capping agents. An important study reported the interesting observation of the antiquenching effect for liposome encapsulated CdSe QDs in the temperature range between 20°C and 80°C (Garshasbi et al., 2020a). Another study reported the thermal quenching effect for CdSe capped with TOP/TOPO/HDA for temperatures above a transition temperature at around -23°C. As reported, for CdSe capped with alkylamines, the transition temperature could be controlled by changing the alkyl chain length (Wuister et al., 2004). Another study on CdSe QDs with different shell structures showed that shell thickness can significantly decrease the surface-defect-related thermal quenching effect (Jing et al., 2009). As reported, when increasing temperature from -19.3°C to 27°C, the PL intensity of CdSe/CdS QDs decreases at a much slower rate than that of CdSe core-only QDs. More interestingly, CdSe/CdS/CdZnS/ZnS core/multishell QDs show almost no thermal quenching when the same temperature variation is applied. QDs may also exhibit a λ_{AE}/PL peak red-shift by temperature variation, which usually occurs along with the thermal quenching effect. The λ_{AE}/PL peak red-shift effect can be explained by temperature-dependant bandgap shrinkage (Li et al., 2018a). An important study on CdSe cluster solution in n-butanol revealed a constant reversible PL peak red-shift of 10 nm and PL intensity drop of 93% by temperature increase from 15°C to 80°C after four to five heating – cooling cycles (Biju et al., 2005). Another study reported 0.34 nm/°C red-shift of CdTe QDs in the temperature range between 27°C and 170°C (Li and Li, 2014). Given the complex interactions between the PL effect and other heat rejection mechanisms, a comprehensive experimental and theoretical study is required to explore the optimal temperature-sensitive properties of QDs for the building application.

Plasmonics, as another type of nanoscale engineerable thermochromic materials, demonstrate intriguing tuneable thermochromic properties. Surface plasmon resonance, the nanoscale optical effect responsible for coloration in plasmonic materials, refers to the collective oscillation of conduction electrons of metal nanoparticles with incident light. The thermochromic mechanism of plasmonics is based on adjusting the surface plasmon wavelength through changing several factors, including matrix refractive index and interparticle distance by temperature. A study on silver nanoparticles capped by long-chain alkyl-thiolate chains reported a significant color change from dark brown to yellow (Carotenuto and Nicolais, 2009). The temperature-sensitive characteristic of the material is mainly because of the interparticle distance modulation at the melting point of thiolate-capping crystallites. Embedding metal nanoparticles in a temperature-sensitive polymer matrix is another common technique that modulates interparticle distance for the fabrication of thermochromic plasmonics (Seeboth and Lötzsch, 2013). In this case, the melting point of the polymer determines the thermochromic transition temperature.

Photonic crystals are another promising nanoscale thermochromic materials. They are periodic dielectric or metallic-dielectric nanostructures that are designed to form bandgap structures to control propagation of incident light. The layer thickness and refractive index are the two main factors determining the reflected wavelength in photonic crystals, and therefore, thermochromic photonic crystals can be produced using materials having temperature-sensitive thickness/volume and/or refractive index variation feature. An interesting study fabricated thermochromic photonic crystal through embedment of polystyrene spheres in PNIPAM (Seeboth and Lötzsch, 2013; Weissman et al., 1996). As reported, the developed 3D-photonic crystal displayed a reversible color change between 10°C and 35°C, which is mainly due to the temperature-sensitive volume change of the PNIPAM matrix.

Overall, while thermochromic materials demonstrate a very high energy conservation potential during the whole year, the real application of thermochromic materials for buildings' application needs further investigation. The two main areas for future research on thermochromic materials for buildings' applications include: (1) Gain a deep understanding of photodegradation in dye-based thermochromics. (2) Development and thermal performance evaluation of bespoke dye-based and nanoscale thermochromic materials with optimal temperature-sensitive transition temperature and optical/fluorescent.

6.2.6 Fluorescent materials for mitigation

Fluorescent materials are advanced coatings capable of rejecting the absorbed light through nonthermal emission of the absorbed light, a phenomenon known as the PL cooling effect (Berdahl et al., 2016; Garshasbi et al., 2020b). PL effect refers to the nonthermal radiation of bandgap energy for photons having an energy level equal to or higher than the bandgap energy. In addition to the nonthermal radiation, several thermal radiation mechanisms may occur when excited electrons release their absorbed energy. Thermalization loss as one of the major thermal radiation mechanisms refers to the thermal relaxation of the excess energy from higher energy bands to the conduction band (Garshasbi et al., 2020b). Also, fluorescent materials absorb shorter wavelength/higher energy light and reemit at a longer wavelength (Heidarzadeh et al., 2020). The difference between absorption peak and reemission peak is known as Stokes shift. The Stokes shift as another thermal radiation mechanism is a critical factor determining the fluorescent cooling potential. Unlike the commonly observed Stokes-shift effect, some fluorescent materials may reemit shorter wavelength/higher energy light than the originally absorbed light under specific conditions. This effect, known as antiStokes shift, is typically initiated by a laser light, and not sunlight. However, a very recent US patent proposed a method to create the antiStokes shift fluorescent effect using sunlight as the light source (Akey et al., 2012). Other nonradiant mechanisms, including thermal radiation caused by surface defects and reabsorption, affect the radiative relaxation efficiency of excited electrons when moving from conduction to valence band. Quantum yield is a key fluorescent property through which the surface defect and reabsorption-related thermal relaxation could be precisely measured.

There are three major fluorescent properties determining the fluorescent cooling potential: (1) Absorption peak/absorption edge wavelength (λ_{AE}), (2i) quantum yield, (3) Stokes/antiStokes shift. Fluorescent materials can be categorized into two groups: (1) Mineral-based fluorescent materials with fixed fluorescent properties and (2) man-made nanoscale light-emitting materials with adjustable fluorescent properties, known as QDs (Shenhav and Grottas, 2019). The tuneable fluorescent properties of

QDs are attributed to the intriguing quantum confinement effect at the nanoscale. Absorption peak/absorption edge wavelength (λ_{AE}) is the wavelength where the energy of absorbed photon corresponds to the bandgap energy. More precisely, only photons with a wavelength equal to or shorter than the λ_{AE} participate in the fluorescent cooling effect, while those with a wavelength longer than λ_{AE} pass through the QDs layer. λ_{AE} as a major fluorescent cooling potential can be precisely adjusted by simply changing the QDs size. The number of reemitted photons divided by the number of absorbed photons, known as quantum yield, is another fluorescent key property determining the fluorescent cooling potential. Nonradiant relaxation of excited photons caused by surface defects, reabsorption, and augur recombination are the major thermal radiation mechanisms resulting in a decrease in the quantum efficiency of the fluorescent materials (Babentsov and Sizov, 2008; Cho et al., 2018; Garshasbi and Santamouris, 2019). Several techniques, including the introduction of surface ligands and/or shell of other semiconductor materials around QDs core and embedment of QDs particles in a polymeric matrix, could be employed as effective measures to enhance the quantum efficiency of QDs coatings (G. Li et al., 2018b; Park et al., 2015; Vasudevan et al., 2015).

Some research has shown that the Stokes shift of QDs can also be controlled by changing the QDs size (Xu et al., 2015). The Stokes shift of QDs' coatings can be as large as 300−400 nm. As a general rule, the smaller the Stokes shift, the higher the fluorescent cooling potential. In contrast, antiStokes fluorescent cooling is defined by nonthermal radiation of higher energy photons than the originally absorbed light. The higher fluorescent cooling potential of antiStokes fluorescent materials is highly important, especially for fluorescent radiative materials with a λ_{AE} at the blue end of the spectrum where Stokes-shift fluorescent cooling potential is at its minimum value. Two-photon absorption (TPA) and triplet − triplet annihilation (TTA) are the two main processes to convert low-energy light to high-energy light (Zhu et al., 2017). TPA refers to simultaneous absorption of two lower-energy photons and reemission of one higher energy photon. The laser excitation light requirement of TPA considerably limits its application. A recent US patent designed a novel two-layered coating capable of demonstrating TPA antiStokes shift under natural sunlight irradiation (Shenhav and Grottas, 2019). The two-layered coating is composed of a highly reflective top layer capable of filtering selected wavelengths to create the laser effect and a fluorescent material as the base layer. As for TTA antiStokes shift fluorescent materials, the antiStokes mechanism is based on a sequential process beginning with initial absorption of light by a sensitizer and ending with emission from an acceptor (Ye et al., 2016). More precisely, TTA antiStokes shift consists of three stages: First, the sensitizer molecules are excited from ground state to a singlet state and return to a triplet step. Second, the energy transfers from sensitizer molecules to acceptor molecules in their triplet states; and third, when two acceptor molecules are close to each other, one molecule transfers its excited state to the other acceptor molecules. The higher energy photon is produced when one of excited photons returns to the ground state while the other one is promoted to a singlet excited state. The TTA antiStokes emission efficiency can be enhanced by varying sensitizer and acceptor concentration for an efficient cross-relaxation process while avoiding nonradiative quenching by the acceptor.

In addition to fluorescent properties, solar absorption is a key factor determining the fluorescent cooling and net heat-rejecting potential of fluorescent materials. As a general rule, the higher the solar absorption, the higher the PL effect cooling potential. Considering the positive correlation between solar reflection and PL effect as two major heat rejection mechanisms, a detailed analysis of the contribution of each heat rejection mechanism is required. Effective solar reflection, defined by the portion of rejected light through reflection and PL effect, is a reliable indicator to explore materials with the highest heat rejection potential through both reflection and PL effect.

Several studies have been performed to study the fluorescent cooling of potential of several mineral-based fluorescent materials and QDs' coatings. An experimental study on the fluorescent cooling of ruby samples with a PL peak wavelength at 694 nm and quantum yield of 0.83 revealed up to 6.5°C lower surface temperature than the corresponding nonfluorescent reference sample (Berdahl et al., 2016). The experimental approach in this study is based on the comparative thermal performance evaluation of the fluorescent sample and its corresponding nonfluorescent sample. Given the complexity and inaccuracy of the experimental approach, a very recent study proposed a theoretical model for the precise calculation of fluorescent cooling. The developed theoretical framework is a very useful tool for the precise computation of fluorescent cooling potential using a hypothetical reference sample. More importantly, the proposed deterministic model could be employed as a reliable tool to determine the optimal optical and fluorescent properties of QDs to design best performing fluorescent/heat-rejecting materials. In addition to the intriguing tuneable fluorescent properties, QDs also possess a very high transmission at wavelengths longer than their λ_{AE}. A recent study combined heat rejection through NIR reflection with that of PL effect by taking advantage of the high NIR transmission feature of QDs' coatings (Garshasbi et al., 2020c). As reported, the fluorescent cooling potential of CdSe/ZnS QDs is about 2.5°C, which is improved by another 8.1°C when a highly reflective base coat (silver-coated polyethylene terephthalate film) is employed.

6.2.7 Photonic and materials of daytime radiative cooling

Daytime radiative cooling surface is one of the passive cooling innovations and is attracting increasing attention due to its game-changing cooling potential. Eqs. 6.1 and 6.2 describe the energy balance of a surface exposed to the sun. P_{out} is the energy emitted from the material; e is the emittance of the surface; σ is Stephan − Boltzmann constant; $T_{ambient}$ and T_{roof} are the ambient air temperature and surface temperature. Two major parts of the energy that comes into the material are the solar radiation and longwave radiation it absorbs. r and t are the reflectance and transmissivity of the material. R_{solar} and $R_{ambient}$ are the incoming radiation in $0.25-3\ \mu m$ (short wave range) and $4.5-40\ \mu m$ (long wave range), respectively. $\alpha_{cond+conv}$ is the combined nonradiative heat transfer coefficient.

$$Pout = e \cdot \sigma \cdot T_{roof}^4 \tag{6.1}$$

$$Pin = R_{solar} \cdot \left(1 - (r+t)_{short}\right) + \alpha_{cond+conv} \cdot (T_{ambient} - T_{roof})$$
$$+ R_{ambient} \cdot \left(1 - (r+t)_{long}\right) \tag{6.2}$$

To achieve a lower surface temperature, the absorption of solar radiation and long-wave radiation needs to be minimized. The material needs to have high reflectivity within the solar wavelength to reduce the absorption of solar radiation. At most of the long-wave wavelengths, as the emission curve of the material almost overlaps the incoming longwave radiation, no matter what the emissivity is, there will not be a massive difference in the net cooling power. But between 8 and 13 microns, because the atmosphere is transparent, if the material has a high emissivity, it can emit radiation normally but absorbs much less radiation. This difference provides a chance to reduce the absorption of longwave radiation and lower the surface temperature. In summary, in order to have effective radiation cooling, the material needs to have a very high solar reflectivity and a very high emissivity in the range of 8−13 μm.

The development of effective radiation cooling devices originated from identifying natural materials or creating composite materials with satisfactory optical properties and then modifying the structure of the material at the nanometer level, thereby further improving solar reflectance and infrared emissivity. A thorough review of the previously presented radiative cooling devices and their performance was summarized in Santamouris et al. (2018c). Specifically, the following major sections were reviewed: multilayer planar photonic structures (Bao et al., 2017; Chen et al., 2016; Hervé et al., 2018; Kecebas et al., 2017; Kou et al., 2017; Narayanaswamy et al., 2014; Raman et al., 2014); 2D−3D photonic materials and metamaterials (Hossain et al., 2015; Rephaeli et al., 2013; Wu et al., 2018; Yang et al., 2017; Zou et al., 2017); polymers (Gentle and Smith, 2015, 2010; Goldstein et al., 2017; Huang and Ruan, 2017; Zhai et al., 2017), and paints (Atiganyanun et al., 2018; Mandal et al., 2018).

Recent advances on cheap and scalable structures, like the use of common polymers, spray, or paints, are more competitive in large-scale applications and save tremendously the original investment. Polymers having functional groups, such as $C-N$, $C-Cl$, $C--F$, and $C--O$, can be effective radiative cooling emitters as these bonds strongly emit within 8−13 μm (Aili et al., 2019). For example, a polyvinyl fluoride (PVF) film containing $C-H$ and $C-F$ bonds was used as a radiative cooler. In 1975, PVF film produced by Dupont was combined with polished aluminum and a maximum temperature reduction from the ambient of 15°C during the day was reported in Napoli, Italy (Catalanotti et al., 1975). In 2020, a PVF film with a silver coating was reported to provide a subambient temperature of 2°C in Tempe, Arizona. Another polymer, polymethylpentene, which contains a $C-H$ bond, was employed as an emitter for radiative coolers (Grenier, 1979). It was also the substrate for distributing silica spheres (Zhai et al., 2017). Two other common polymers, polyvinylidene fluoride (PVDF), having $C-H$ and $C-F$ groups, and poly (methyl methacrylate) (PMMA), with $C-H$, and $C-O$ groups, both of 50 μm thickness, were placed on a silver layer and tested under the direct sun in Aili et al. (2019). A subambient cooling degree of 6°C and 4°C during solar noon were observed for PVDF and PMMA, respectively. Other polymers that can be used as emitters are polytetra-fluoroethylene (Teflon) with $C - F$ group (Yang et al., 2018) and acrylic with $C - O$ and $C - H$ bonds (Liu et al., 2020). Spray-based cooling material is another choice that is convenient to use, and the production cost is usually cheap. It can easily adapt to existing building surfaces of various shapes and formats. The disadvantage is that the application cannot be quantitatively controlled, and the material itself is easy to settle and agglomerate if the formulation is not accurately controlled. A radiative cooling surface was produced in Ao et al. (2019) by spraying zinc phosphate sodium ($NaZnPO_4$) particles onto an aluminum substrate. A 7.3°C subambient temperature when solar radiation was 430 W/m^2 at noontime in Beijing was reported.

Maintaining the color or diffuse property of the radiative surfaces can benefit the building esthetics and lighting environment. The various colors of the cooler come from solar absorption, surface plasmon resonance structural, or Bragg diffraction. A new type of colloidal photonic opal, assembled from 300-nm silica colloidal nanospheres, was reported for the first time as a colorful radiative cooler in 2020 (Kim et al., 2020). The color preservation was achieved via structural Bragg diffraction (Arsenault et al., 2007) instead of selective absorption. This colorful radiative cooler with soft fluidity of colloidal suspensions is easy to craft and suitable for large-scale application. Another innovative colored radiative cooling structure was proposed by Yalçın et al. (2020). It included plasmonic core-shell nanospheres and silver nanoparticles distributed in a silica and polydimethylsiloxane matrix. By varying the radii of the core and shell, various colors were obtained as they altered the spectral position of the surface plasmon resonance.

It should be noted that climatic conditions are nonnegligible factors affecting the cooling performance. For example, the photonic structure presented in Raman et al. (2014) was able to reach 4.9°C subambient temperature when exposed to direct solar radiation exceeding 850 W/m^2. When the same structure built by the same company was tested in Hong Kong (Tso et al., 2017), it was reported that under any weather conditions in Hong Kong, the cooler could not produce a cooling effect during the day. In Okayama, Japan, under warm and humid conditions, a cooler with an average solar reflectance of 0.89 and an average emissivity of 0.72 in 8–13 μm was tested (Suichi et al., 2018). When the precipitable water vapor (PWV) was assumed to be 1 mm, it was expected to reach a surface temperature of 1.3°C subambient temperature. In the real experiment, despite having a convection shield, the measured cooler was 2.8°C higher than the ambient temperature. In Japan, the average value of PWV in early summer is about 20 mm, and this high humidity mainly caused the difference between simulated and experimental results. In a recent study, identical radiative coolers were tested in Phoenix, USA (arid, mid-latitude); New York, USA (mid-latitude, coastal); and Chattogram, Bangladesh (tropical, coastal) (Mandal et al., 2018). In a warm and dry location like Phoenix, the temperature at solar noon was 6°C below ambient temperature; and in a cold afternoon in New York, the subambient temperature became 5°C. When testing the surface in a foggy winter in Chattogram, the subambient temperature dropped to 3°C.

To improve the cooling performance under adverse climatic conditions, the use of an asymmetric electromagnetic transmission (AEMT) window with different transmittances for incoming and outgoing light was proposed and analyzed in Wong et al. (2018). In addition to controlling the optical spectra of the material, the AEMT grating window may be a completely innovative pathway for achieving radiative cooling. On the other hand, to avoid overcooling in the winter period, self-adaptive radiative cooling has been proposed, like thermochromic materials and VO_2 or photonic thermal management systems (Ulpiani et al., 2020c).

The theoretically predicted energy saving using daytime radiative cooling material is tremendous, and its correspondent benefit to the environment is significant. To investigate the energy-saving potential of radiative cooling technology, Fernandez et al. (2015) simulated the energy saving for a medium-sized office building with a total area of $5000\,m^2$ after implementing the photonic radiative cooling structure reported in Raman et al. (2014). The energy-saving analysis was conducted for five locations with four different climates. A heat transfer model was customized in EnergyPlus's energy management system framework, having overwritten some predefined points of intervention in the EnergyPlus model. As the calculation of photonic materials' radiative cooling power required mathematical functions not supported by EnergyPlus's calculation engine, a regression equation based on the calculation in Matlab was employed. The simulated results showed that the employed radiative cooling system could save electricity for 103 MWh per annum in Miami (hot and humid), approximately 50 MWh in Las Vegas (hot and dry), Los Angeles (hot and dry), and Chicago (cold climate), and around 24 MWh in San Francisco (marine climate). A large-scale application should be based on understanding and evaluation of the suitability of climate and morphology of a specific area.

This section has provided a picture of the state-of-the-art in daytime radiative cooling materials and the interaction between the environment and daytime radiative coolers. While there has been remarkable progress and development in this area, more investigation is necessary to move to the next stage of development and application of practical and commercially available radiative coolers.

6.2.8 Cooling with elastocaloric materials

Urban heat mitigation by conventional means (cool materials, greenery, solar shading, water-based technologies) is well-documented. Merging the results from a variety of geographical, climatological, and city-specific contexts, it is clear that average summer ambient temperature reductions around 2.5°C are to be expected, which may be insufficient to tackle future climate change escalation (Santamouris et al., 2017). With recent technological developments and intensified multidisciplinary approaches, many space-conditioning technologies have undergone significant breakthroughs. Within the variegated spectrum of new-generation technologies currently under investigation is solid-state cooling based on caloric materials (Garshasbi et al., 2020a; Ulpiani et al., 2019c), an environmentally friendly alternative to the conventional vapor compression-based process, which makes no use of ozone-depleting or greenhouse-gas-emitting substances. Hydrofluorocarbons, which currently sustain vapor-compression refrigeration, have a global warming potential up to a few thousand times more than carbon dioxide. Hence, the Kigali Agreement to the Montreal Protocol (Kigali Cooling Efficiency Program, n.d.), signed in 2016, already put a ban in favor of natural refrigerants (hydrocarbons, carbon dioxide, ammonia), which, however, are accompanied by a different suite of criticalities. Caloric cooling comes as a refrigerant-free solution.

The working principle typically relies on the reversible latent heat release/absorption associated with the austenitic-martensitic phase transformation in shape memory alloys (SMAs). This transformation can be attained by applying changes in magnetic/electric field or by mechanical loading/unloading, which produce changes in temperature. Elastocaloric (stress-induced) cooling exhibits greater potential for use in buildings and in the built environment (Fig. 6.2) (Schmidt et al., 2016) shows the cooling cycle: under adiabatic loading, the SMA transforms from austenite to

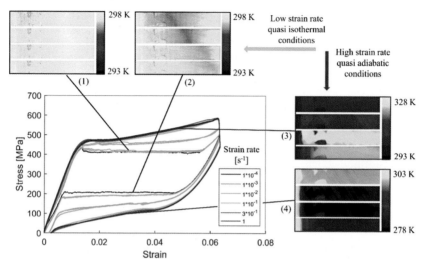

Figure 6.2 Elastocaloric cooling cycle on a stress-strain diagram. Thermograms 1–2 represent IR shots at four consequent time steps during loading/unloading in quasi isothermal conditions (low strain rates). Thermograms 3–4 represent the counterpart in quasi adiabatic conditions (high strain rates). *The data refers to NiTiFe films as in Kigali Cooling Efficiency Program [WWW Document], n.d. URL https://www.k-cep.org/.*

martensite, heats up and ceases heat to the surrounds, while upon adiabatic unloading, the SMA reverses the transformation, cools down, and absorbs heat from the surrounds. The reader is referred to Ulpiani et al. (2020a) for a detailed description of the thermo-mechanical principle and cooling cycle.

In terms of materials, SMAs based on Ni − Ti show a very high potential in comparison with other elastocaloric materials, owing to large latent heats (of up to 22 J/g) and small input work (Moya et al., 2014). A wide variety of materials has been identified, and comprehensive material reviews are available (Frenzel et al., 2018; Mañosa and Planes, 2017; Zhang et al., 2020). The search for high-performance elastocaloric materials is currently the focus of a dedicated consortium, called CaloriCool (Zarkevich et al., 2017), established in 2016 by the United States Department of Energy, which confirms the relevance of elastocalorics in the quest for sustainable cooling solutions.

In terms of devices, those based on bulk materials are comprehensively reviewed in Kirsch et al. (2018), while those operating at the miniature scale are covered in Bruederlin et al. (2018). Both reviews include basic concepts and designs, elastocaloric properties and their harmonization with device components, and extensive considerations on fabrication and characterization. Finally, a review on the state of the art of solid-state caloric cooling processes at room temperature, thus suitable for building applications, can be found in Greco et al. (2019).

This section focuses on the value of elastocalorics in shaping up a sustainable built environment in the future and reports on very recent advancements in the application of elastocalorics cooling for buildings and the built environment. Major challenges and research lines are conclusively outlined.

A variety of studies point toward elastocalorics as a promising pathway to sustainable built environments.

Elastocaloric cooling scored very high in a comparative analysis on alternatives to vapor compression performed by the US Department of Energy (Goetzler et al., 2016), thanks to a favorable combination of temperature and entropy. The analysis included aspects of energy savings potential, cost, system complexity, and nonenergy benefits.

Qian et al. proposed a coefficient of performance (COP) breakdown analysis method to compare untraditional cooling technologies, from materials (working fluids) level to the system level as a function of temperature lifts (Qian et al., 2016). Again, elastocalorics exhibited greater potential among solid-state alternatives due to higher latent heat. The potential for relatively large temperature lift ranges (as those associated with building and built environment applications) was envisioned along with significant advances in both materials and system integration.

An updated quantitative comparison of 14 air-source heat pump technologies for residential or commercial space conditioning is provided in El Fil et al. (2021) based on COP, normalized COP (exergetic efficiency), and power density. Solid-state technologies (thermoelectrics, thermionics, elastocaloric, magnetocaloric, and electrocaloric), two-phase technologies (vapor absorption, adsorption, ejector heat pump, membrane heat pump, and conventional vapor compression), and gas cycles (Stirling, Brayton, Bernoulli, vortex tube, and thermoacoustics) are compared. Elastocalorics demonstrate excellent potential, with an outstanding power density of about 344.7 kW/m^3 that exceeds that of a vapor compression cycle by 83.3% and 46.1% in heating and cooling mode, respectively. The authors concur on the need for enhanced material and system integration to tap into the flexibility of the elastocaloric cycle in terms of the range of temperature lifts required to access the building market.

With a focus on the environmental impact, Aprea et al. compared a caloric and a vapor/compression refrigerator (Aprea et al., 2018) based on the Total Environmental Warming Impact index, which combines the effect of direct refrigerant emission with the energy consumption and the related combustion of fossil fuels for the electric energy production. The use of elastocalorics ensures an environmental impact from -49% to -72% smaller than the vapor compression.

Elastocalorics are thus an intriguing pathway toward sustainable cooling. This has been recently acknowledged also by the International Energy Agency in the 2020 IEA/HPT Annex 53 on advanced cooling/refrigeration technologies (Baxter and Radermacher, 2020).

Still, minimal literature exists on the performance of elastocaloric devices for buildings and the built environment, the reason being related to major hurdles at the manufacturing level. Here we present the evidence from pioneering works on the topic.

In Luo et al. (2017), a unique air-to-air elastocaloric cooler for air conditioning applications was modeled in a fully integrated dynamic system, comprehensive of all subcomponents, including the water-to-water module as well as the related indoor and outdoor units. The model simultaneously simulates the elastocaloric effect of NiTi, the heat transfer, and the water transport in the system. The system releases 1.2 kW cooling power. Simulation for the Non-dominated Sorting Genetic Algorithm (NSGA-II) based optimization identified designs with a maximum COP of 2.7 based on 17 parameters, from SMA and tube sizes to coil rows and fin density, from the speed of fans and pumps to loading, unloading, and heat recovery times. Thousands of combinations have been found within $\pm 5\%$ of the design capacity and maximum COP, thus resulting in a large multidimensional matrix for design tradeoffs. Such an integrated approach allows investigating a variety of designs to deal with practical constraints.

Similarly, in Welsch et al. (n.d.), a system simulation tool for predicting the thermal and mechanical behavior of an elastocaloric ventilating device is developed to gain insight into the influence of SMA size, rotation frequency, airflow rate on COP, thermal power, and outlet temperature, based on the innovative concept of continuously operating elastocaloric air cooling system presented in Kirsch et al. (n.d.). The thermal output power exhibits a distinct saturation for increasing rotation frequencies (0.25–4 Hz), whose level is inversely proportional with the wire diameter (50–500 μm). The maximum hits 450 W at the highest flow rate considered in the simulations ($150\,\mathrm{m}^3/\mathrm{h}$). Higher flow rates also exhibit advantages in terms of efficiency and temperature span.

In Al-Hamed et al. (2020), the authors investigated the potential integration of elastocaloric cooling devices with solar-powered energy systems for the first time. They compared two solar energy-driven systems for dual generation of electricity and cooling, one based on traditional steam Rankine cycle and an absorption chiller (reference) and the other based on cascaded Rankine cycles and an elastocaloric cell. The elastocaloric system outperformed the traditional one in many thermodynamic aspects: (1) higher overall energy and exergy efficiencies by 4.63% and 1.10%, respectively, (2) lower exergy destruction rate by a factor of 1.7, (3) higher flexibility in tracking the cooling demand. With a COP of 4.25, the demonstrator proved its competitiveness on the market of cooling devices.

Elastocaloric cooling has all prerequisites to be a future-proof cooling strategy, with higher flexibility to meet the increasing cooling demands of the hot regions in the world compared to traditional systems. Feasible roadmaps toward the use of elastocalorics to increase energy efficiency and sustainability of buildings, as well as to counteract the UHI effect, have been delineated (Ulpiani et al., 2019c).

Major challenges are still to be tackled, both at material and device scale as reviewed in Engelbrecht (2019). These include enlarging the temperature span, increasing fatigue-resistance, perfecting the coupling to external components such as heat exchangers and fluid vectors, including work and heat recovery systems, optimizing controls and timing, and increasing cost-efficiency. The goal is to minimize the major losses, as displayed in Fig. 6.3. Material-wise, promising results are expected from additive manufacturing (Hou et al., 2019) and alternative mechanic actions, as for twistocalorics, which deliver high-intensity cooling when twisted (Wang et al., 2021, 2019). In addition, growing interest has been devoted to multicolors, namely materials that exhibit more than one type of caloric effect, thus enabling more complex cooling concepts and reduced hysteresis losses (Moya et al., 2020).

Generally speaking, active regeneration represents the boldest answer to generate the desired temperature span so far while also increasing specific cooling/heating power and COP (Emaikwu et al., n.d.; Kabirifar et al., 2019; Tušek et al., 2016). This is accomplished by establishing a temperature gradient in the working fluid direction to expand the operation range. However, the interaction with intermediate heat carriers (e.g.,

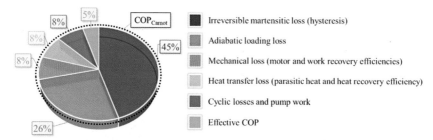

Figure 6.3 Pie chart showing how the COP is degrading from Carnot COP all the way down to the system COP on account of different sources of loss. *Indicative percent weights are taken from Qian, S., Nasuta, D., Rhoads, A., Wang, Y., Geng, Y., Hwang, Y., et al., 2016. Not-in-kind cooling technologies: a quantitative comparison of refrigerants and system performance. International Journal of Refrigeration 62, 177–192.*

fluids) is itself a limitation to the overall efficiency. Thin-walled structures with small hydraulic diameters can be used to improve the heat transfer, yet these geometries are prone to buckling instability (Wang et al., 2021). Direct, solid-to-solid transfer designs have been delivered by Saarland and Karlsruhe Universities (Bruederlin et al., 2017; Schmidt et al., 2015).

Further, alternative pathways to increase the efficiency and applicability of elastocaloric devices are currently under examination, including:
- Approaches to transfer the outstanding specific cooling capacity at the miniature scale into the macroscale by cascaded (Bruederlin et al., 2019) and parallelized setups (Schmidt et al., 2016).
- Smart thermodynamic hybridation. Elastocalorics offer the potential for implementation of different thermodynamic cycles. Combined adiabatic, isothermal, and nonadiabatic phase transformations can increase the efficiency of the process by 15% and more (Bruederlin et al., 2019). This could be achieved by variation of the strain rate, as experimentally proved in Ulpiani et al. (2020a) or by mechanical profiling (Bruederlin et al., 2018). The load of the SMA element can be adjusted by advanced stroke concepts using rotary cam tracks, whose profile can be designed to impose the desired thermodynamic cycle.
- Low-force, low-footprint actuation. Low-force designs have been proposed based on out-of-place deflection of SMA films (Schmidt et al., 2013) and bending (Ossmer et al., 2016). Actuation with magnetic SMAs, as well as rotary-to-linear transduction mechanisms (such as Scotch yoke), avoids disproportionally large sizes of actuators compared with the active material volume (Sharar et al., 2018; Slaughter

et al., 2020). Further, low-grade thermal energy can be used for actuation, such as heat-activated actuators made of high-temperature SMAs. In this case, the force-displacement characteristics of the actuator and refrigerant are intrinsically matched, resulting in footprints more than 10 times lower than with mechanical drivers (Kabirifar et al., 2019).

The interest in elastocaloric energy technology is increasing rapidly. This provides a unique window of opportunity for achieving a sustainable building stock in light of the Paris Agreement and the Kigali Amendment to the Montreal Protocol. Elastocaloric cooling outperforms vapor compression cooling as well as most not-in-kind alternatives. Their power density in cooling mode is almost twice that of vapor compression, with higher flexibility in tracking the cooling demand. System-level COPs have proved their competitiveness on the market of cooling devices. Besides, the plethora of design parameters at stake opens up a variety of strategies to better deal with practical constraints. Nonetheless, as with most nontraditional technologies, elastocalorics will require additional development before they can significantly impact the building sector. Research efforts in the direction of understanding and analyzing fully integrated systems (including all subunits for energy delivery, actuation, control, and timing) will be pivotal to usher in a new generation of efficient devices.

6.3 Using transpiration cooling to mitigate urban heat

Climate change and the UHI effect have caused a worldwide issue of urban overheating (Santamouris et al., 2015). Urban overheating will increase total energy consumption and peak electricity demand, deteriorating outdoor and indoor thermal comfort as well as public health (Santamouris, 2015a, 2020; Santamouris et al., 2020d). In order to mitigate urban overheating, researchers have proposed a variety of options: integrating cool material in the facades and roofs, like reflective materials (Santamouris et al., 2011) and radiative cooling surfaces (Feng et al., 2020); using the cool pavement on roads (Feng et al., 2020; Santamouris, 2013; Santamouris et al., 2011, 2020d), expanding green spaces in the urban environment (Santamouris, 2014a,b), etc.

Transpiration is a physiological process that happens in the stomata on the plants' leaves. Thus water lost through transpiration is usually regarded as the inevitable cost of photosynthesis.

In the related research field, the focus is usually on the transpiration rate, which describes the amount of water vapor lost to the air through transpiration per unit time. The transpiration rate is both ruled by the physiological processes of plants and the physical processes related to evaporation. Currently, the most commonly used transpiration model is the Penman − Monteith model. In 1948, Penman proposed the evaporation model, and then Monteith improved on the evaporation equation proposed by Penman and obtained the P − M model. The P − M model takes into account both physical factors and plant physiological factors. Monteith described the ability of stomata to hinder transpiration as stomatal resistance, thereby quantifying the physiological aspects of transpiration. In the Food and Agriculture Organization (FAO) manual published in 1996, the P − M equation is set as the standard method for calculating plant and ground surface evapotranspiration. The P − M equation used in FAO56 (Allen et al., 1998) is expressed as follows:

$$\lambda \text{ET} = \frac{\Delta(R_n - G) + \rho_a c_p \frac{(e_s - e_a)}{r_a}}{\Delta + \gamma\left(1 + \frac{r_s}{r_a}\right)} \tag{6.3}$$

where R_n is the net radiation, $(e_s - e_a)$ represents the vapor pressure deficit (VPD) of the air, G is the soil heat flux, ρ_a is the mean air density at constant pressure, $(e_s - e_a)$ represents the VPD of the air, c_p is the specific heat of the air, Δ represents the slope of the saturation vapor pressure − temperature relationship, r_s and r_a are the (bulk) surface and aerodynamic resistances, and γ is the psychrometric constant. In terms of each resistance term, the aerodynamic resistance is represented as:

$$r_a = \frac{\ln\left[\frac{z_m - d}{z_{om}}\right]\ln\left[\frac{z_h - d}{z_{oh}}\right]}{k^2 u_z} \tag{6.4}$$

where z_m is the height of wind measurements, z_h is the height of humidity measurements, d is the zero plane displacement height, z_{om} is the roughness length governing momentum transfer, z_{oh} is the roughness length governing the transfer of heat and vapor, k is the von Karman's constant, 0.41, and u_z is the wind speed at height z.

And the surface resistance is expressed as:

$$r_s = \frac{r_1}{\text{LAI}_{\text{active}}} \tag{6.5}$$

where r_1 is the bulk stomatal resistance of the well-illuminated leaf, and LAI_{active} is the active (sunlit) leaf area index (LAI), which equals the fraction between LAI and soil surface area.

From Eqs. (6.3) and (6.4), the radiation flux, VPD, stomatal resistance, and LAI, as well as wind speed, are the key factors that determine the transpiration rate. A strong radiation flux can intensify the transpiration activity. At the same time, when the VPD is larger, the plant's transpiration will be stronger too. In addition to directly affecting r_s, wind speed can also remove water vapor near plant leaves to increase VPD. And for LAI, larger LAI means smaller resistance, which indicates that taller plants or trees tend to have stronger transpiration capacity.

In actual measurements and calculation, in addition to the $P - M$ equation, there are many ways to calculate and measure the transpiration rate of plants. They are either a transformation of the $P - M$ equation or an indirect estimation of the transcription rate based on the side effect of the transcription on the surrounding environment. We will not elaborate on them here.

Water converts to vapor during the transpiration process, which reduces the temperature both for plants themselves and the local environment. Thus the cooling effect of transpiration does benefit to alleviating the UHI effect and urban heat waves.

Typically, transpiration can evaporate water from 0.28 to 12 L/m^2 per day (Ballinas and Barradas, 2016), which generates a cooling power ranging between 24.5 and 29.5 MJ/m^2 per day in arid environments with sufficient water supply. In temperate climates instead, this cooling power can range between 0.7 and 7.4 MJ/m^2 per day (Jones, 2013). Research from the United Kingdom (Moss et al., 2019) reported the transpiration rate of several species of urban trees between 14 and 27 kg/h/tree, contributing to cooling and humidification of air. Meanwhile, Rahman et al. (2011) found that, in Munich, the transpiration cooling power of a single tree can reach as high as 500 W per unit cubic meter of canopy volume for trees with LAI around 2, and 250W per unit cubic meter of canopy volume for trees with LAI around 1. In 2018, Rahman et al. (2018) found that the energy loss of transpiration of street trees in Munich is about 75 W m^2 in the mild summer.

At the city scale, the measurement described in Rahman et al. (2018) found that in the streets in Munich urban area, under the tree's canopy, the air temperature above the ground is 3.5°C lower than the air temperature above the canopy.

Many factors can influence the mitigation potential of plants, such as plant species (Konarska et al., 2016; Armson et al., 2013; Rahman et al., 2015) and the hydraulic architecture. Despite the difference in hydraulic architecture, the LAI is the key parameter that distinguishes the cooling capacity among species. Trees with high LAI can achieve more efficient shading and exhibit higher transpiration potential as higher LAI usually means more effective evaporative surfaces.

Broadleaf species usually have a greater transpiration rate than needle leaf species in a similar environment. Furthermore, large trees with deep roots can supply themselves with much more water resources compared with grass during drought (Rahman et al., 2015). The research from the United Kingdom shows that, in a similar urban environment, the evapotranspiration rate varies greatly among tree species (Moss et al., 2019).

Ballerinas and Barradas (2016) researched the cooling effect of *Eucalyptus camaldulensis* and *Liquidambar styraciflua*. They used the sap flow method to estimate the transpiration ratio of the two species. Then they calculated the cooling effect of the two species, and they found that 63 *E. camaldulensis* would be needed to reduce the air temperature by 1°C per hectare in the studied area. At the same time, a 2°C temperature decrease per hectare can be achieved in the studied area by only planting 24 large *L. styraciflua* trees. This case implies that tree species can be a critical factor in the mitigation potential.

Water availability is the most important factor to limit the transpiration ratio during temperature and precipitation extremes. When sufficient water is supplied, the transpiration rate can be maximized, providing a considerable cooling effect. Plenty of studies have shown that irrigating existing vegetation can generate considerable cooling effect both in normal summer and extreme heat conditions (from 0.5°C to 10°C) (Adegoke et al., 2003; Brinkmann et al., 2019; Broadbent et al., 2018; Chen et al., 2018; Qian et al., 2013; Sugimoto et al., 2019; Vahmani and Hogue, 2015).

Vahmani and Ban-Weiss (2016) conducted a simulation of Los Angeles, USA by replacing all the lawns in Los Angeles with drought-tolerant vegetation. And their results showed that the drought-tolerant vegetation led to daytime warming up to 1.9°C and a nighttime cooling of 3.2°C. They concluded that the daytime warming attributed to the decrease in irrigation and the nighttime cooling attributed to the modification in soil thermal properties. Their research both revealed the importance of plant species and water availability. Furthermore, recent studies have found that irrigation cooling is more effective in heat extremes and heatwaves compared with nonheatwave periods (Gao et al., 2020; Lam et al., 2020).

However, as irrigation brings an additional water source to evapotranspiration, it also brings extra humidity to the near-surface air, which may induce the decline of the mitigation effect in terms of thermal comfort. Their simulation of irrigation found an irrigation-induced temperature decrease of 7°C but only 2.7 reduction of heat index due to the increase of humidity. This finding should be case-sensitive as the air circulation can greatly influence the accumulation of near-surface humidity. For instance, in a basin environment, the vapor flux would be trapped within the terrain and cause greater influence in thermal discomfort. In contrast, in an environment with strong air circulation, such as the sea breeze, the vapor flux would be harder to be retained around the surface. Further, the extra latent heat flux induced by plants' transpiration can also cause nighttime heating, limiting the plants' mitigation effect. Rahman et al. (2017) also observed the nighttime warming effect of 0.5°C around the trees. Taha et al. (1991) found that in Davis, California, USA, the temperature within the studied trees' canopy can be reduce by up to 6°C and increased by about 2°C at night. If the city is not well-ventilated, the night warming effect of vegetation would impede daytime cooling and weaken the daytime cooling effect.

Also, the tree's location is a factor that influences its cooling power. A single tree that is planted on the soil can have several times larger transpiration ratio compared with the same species planted in paved streets. For instance, a five-times difference in the transpiration ratio was found by Rahman et al. (2011) when measuring the transpiration ratio of *Pyrus calleryana* in the paved street and soil. Earlier research (Souch and Souch, 1993) also observed the difference in cooling effect between trees on the sidewalk and trees on the soil. This result is quite likely caused by the difficulty of accessing water sources for the trees on sidewalks. If urban planners considered using vegetation as a heatwave and UHI mitigation strategy, the irrigation of trees on the sidewalk, car parks, and other hard surfaces should be carefully considered.

Besides, the climatic conditions of the city itself have a great influence on the cooling effect of transpiration. In coastal cities, the cooling effect of transpiration may be entirely masked by the cooling effect of the sea breeze (Broadbent et al., 2018). Similarly, in desert cities, the oasis effect of plants may boost the cooling of transpiration.

Evapotranspirative cooling is a promising cooling technique. However, a lot about this technique is unexplored. Also, evapotranspirative cooling has critical restrictions originating in water availability; hence, optimized irrigation strategies when using evapotranspirative cooling are essential.

6.4 Mist cooling

Water has always been a key cooling strategy for cities, thanks to the concerted action of sensible and latent heat transfer with ambient air. The cooling power naturally increases under hot, dry conditions, conducive to higher evaporation rates. As such, water systems automatically respond to weather extremes, especially heat waves (Broadbent et al., 2018; Nicholls and Strengers, 2018; Vanos et al., 2019). Further, extensive evidence reveals that water-based technologies exhibit the highest local impact, compared to equal coverage of other well-established heat mitigators, such as greenery, shading devices, and reflective materials (Santamouris et al., 2017).

Up until about a decade ago, most UHI-dedicated studies focused on large natural water bodies (sea, lakes, rivers), featuring high thermal mass and exhibiting high cooling potential in cooling- as well as heating-dominated climates. However, not all cities have access to the mitigating effect of large water bodies, and incessant cooling is not always beneficial. Conversely, recent studies concur that multiple, small-scale water features, especially if strategically placed and smartly controlled, achieve a much more impactful and homogeneous effect, at city scale (Gunawardena et al., 2017; Santamouris et al., 2017; Taleghani and Berardi, 2018; Ulpiani et al., 2019b).

Accordingly, artificial blue features (e.g., fountains, sprinklers, misters) are attracting increasing attention, owing to their scalability, spreadability, versatility, and controllability. They typically require no extensive or high-cost urban renewal intervention, thus preventing gentrification and other social inequities.

In particular, dry mist systems are currently regarded as the most effective water-based cooling strategy for future-proof cities. These systems pulverize water into extremely fine particles (in the order of tenths of microns) to maximize their surface-to-volume ratio and emphasize flash-evaporation. Intense locale cooling is reported in a variety of climates and for different designs while requiring a surprisingly modest withdrawal of water and electricity, as recently reviewed (Santos Nouri et al., 2018; Ulpiani, 2019). Thanks to the ease of change of state, dry misters' effectiveness does not limit to dry hot climates but finds application also in extremely humid contexts like Singapore (Wong and Chong, 2010; Zheng et al., 2019) and Chile (Desert et al., 2020). Even countries that typically benefit from cool and temperate climates like the United Kingdom and the Netherlands are faced with increasingly frequent and intense heat stress events while lacking an adequate coping infrastructure. Controlled misting systems represent an optimal, climate-adaptive mitigation strategy, targeting local, time-framed cooling actions and encouraging renewable-powered solutions (Atieh, Shariff, 2013; Barrow and Pope, 2007; Joshi et al., 2016; Montazeri et al., 2017). Technically mature and readily available on the market, these systems are the focus of a growing number of studies, especially experiment-based (see Fig. 6.4) and are expected to be incorporated in an increasing share of climate-resilient urban plans (Takebayashi, 2018).

However, the most interesting aspect is that heat mitigation is not the only effect. Water misting systems are effective against multiple climate-sensitive health threats, which makes them a holistic solution to an ever more holistic problem: the synergistic escalation of urban heat, pollution,

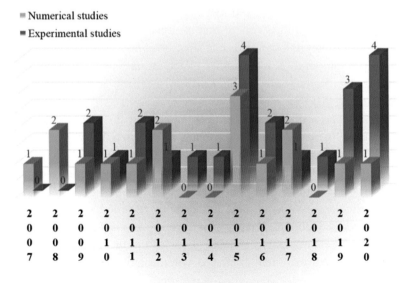

Figure 6.4 Proportion of numerical and experimental studies from 2007 to 2020.

and solar irradiation that comes as a by-product of climate change (Ulpiani, 2020). In fact, despite the limited limelight in literature, sprayed mist exhibit a variety of co-benefits, including solar attenuation (Dombrovsky et al., 2011), dust scavenging (Yu, 2014), and insect repulsion. In the following subsections, water misting is reviewed in its heat mitigation potential, its impact on local comfort, and its side benefits. Conclusive remarks are drawn to delineate future trends and research questions.

A recent review on water mist spray for outdoor cooling collects 20 years of publications from 12 countries and seven climatic zones (Yu, 2014). The findings qualify water spraying as a cost-effective, versatile, and high-impact blue mitigator to be easily integrated into existing city infrastructures and renovation projects. The potential evaporative cooling is estimated to be over 200 times the consumed power (Farnham et al., 2015) with a vertical profile that can be described by a Lorentzian distribution (Di Giuseppe et al., 2021; Ulpiani et al., 2019a). Experimental results point to local cooling ranging from few Celsius degrees (1°C–2°C) up to over 15°C at pedestrian level (Desert et al., 2020), with an average of 7°C–8°C depending on design characteristics, local microclimate, and typology of cooled space (open, semienclosed, closed area) (Desert et al., 2020). Furthermore, high-pressure systems ($>$ 3 MPa) outperform low-pressure ones (Huang et al., 2017) but reduce the chances of stable renewable-powered solutions. On the other side, humidity gains in the order of 10%–20% are expected and should be considered when hygrometric comfort is concerned (e.g., in sportive areas). As such, misting activation is suggested over about 30°C and relative humidity below 70%. Further, in the absence of wind-breaking provisions, misters should not be operated over 3 m/s wind speed, with optimal results under a dominant light breeze of 1 m/s. Notably, for every 0.8 m/s deceleration, a $+35-40\%$ boost in cooling power is expected (Di Giuseppe et al., 2021). Bespoke control logics should be contemplated to tackle site specificities and ensure optimal performances. By way of example, a comfort-oriented fuzzy logic based on real-time microclimatic parameters was found to reduce the energy cost by up to 60% (200) while closely tracking the comfort zone within a mean humidity premium of about 7% (Ulpiani et al., 2019a). Alternatively, the cooling can be adjusted by changing the flow rate or the density of active nozzles (Ulpiani et al., n.d.).

Whatever the case, the characteristics of the system are critical to generating the desired evaporation pattern. These include droplet-size distribution, the height of the nozzles from the ground, their relative distance, the nozzle type, and cone angle, while water temperature has a negligible impact (Sapit et al., 2019; Zheng et al., 2018). For instance, a validated microclimatic model in ENVI-met revealed that the cooling capacity increases by +0.2°C for any 10 l/h increment in water flow rate, while higher wind speeds reduce the local mitigation potential by about 35%—40% regardless of the water flow rate, above 2 m/s. At the same time, stronger winds expand the influence area downwind by up to tenths of meters (Ulpiani et al., 2019a). This is in line with computational fluid dynamic (CFD) analysis performed in Fluent (Montazeri et al., 2017). In the same vein, a supervised learning algorithm was used to disclose the sensitivity of the mist-induced temperature drop to the contextual microclimatic evolution (Ulpiani et al., 2020b). It was demonstrated that the cooling capacity was largely a function of the local wet bulb depression, while solar irradiation and wind speed were negatively and positively correlated, respectively. Further, mist microinstallations must be designed in view of dominant winds and synergistic evaporative processes (e.g., precipitation, evapotranspiration) so as to maximize the site-specific performance. A panoramic view of cooling potentials reported around the world is displayed in Fig. 6.5.

Heat mitigation and enhanced thermal comfort are intimately intertwined. However, when it comes to the misting system, conventional comfort metrics and physiological human models tend to fail. The reason is that the temperature drop is not descriptive of the overall impact on human thermoregulation. While people may feel cooler temperature-wise, they may also perceive a warmer sensation because of persistent high relative humidity (Oh et al., 2020a). Predicting the skin temperature in misted environments is critical as it dictates the evaporative heat loss on the skin surface and arbitrates the delicate balance between the risk of skin wetness and increase in wet feeling pleasantness (Atieh, Shariff, 2013; Barrow and Pope, 2007; Desert et al., 2020; Dombrovsky et al., 2011; Farnham et al., 2015; Joshi et al., 2016; Montazeri et al., 2017; Takebayashi, 2018; Ulpiani, 2020; Yu, 2014; Zheng et al., 2019). This is especially evident when misters work collaboratively with air blowing fans (Oh et al., 2020a).

The Standard Effective Temperature (SET*) has been frequently proposed as a suitable metric, because of its considerations of human skin, yet

Figure 6.5 Magnitude of temperature drop associated with mist coolers around the world.

it largely depends on human perspiration rather than external forcing (Gagge et al., 1986) under specific boundary conditions (Nazarian et al., 2017). The Universal Thermal Climate Index (UTCI) has been widely adopted, too (Desert et al., 2020; Dhariwal et al., 2019; Montazeri et al., 2017; Ulpiani et al., 2019a). Most recent research lines are looking into ways to adapt current models, such as ASHRAE 55−2013 (Thermal Environmental Conditions for Human Occupancy, 2013) to include mist-sensitive additional dynamics and incorporate other variables to more comprehensively quantify the cooling (Farnham et al., 2017). For instance, following the pioneering works of Farnham et al. on a dry silicone rubber skin analog (Farnham et al., 2015; Oh et al., 2020b), Oh et al. utilized both two- and three-node models to determine the skin temperature in misted conditions, demonstrating the possibility of a new environmental index considering mist wetness (Oh et al., 2021; Vanos et al., 2020). The error was in the order of 0.15°C.

However, most typically, misters improve thermal comfort under a variety of boundary conditions. By way of example, in Tempe (Arizona), the mean radiant temperature, PET and UTCI were on average 7.6°C, 6.5°C, and 4.4°C lower in misted compared to undisturbed locations ($P < .05$), with further drops when shading systems were used in combination (Remund et al., 2018). Similarly, in Ulpiani et al. (2019a), the temperature and UTCI dropped by $-8.2°C$ and $-7.9°C$, respectively. Further research will be devoted to unveiling the global comfort impact of misting systems, also considering the side-effects described below.

The hotter world depicted by most-accredited future climate scenarios is the result of ever more rising temperatures but also of higher solar radiation levels. This is advocated, for instance, in the International Panel on Climate Change AR4 A2 models for CPR 8.5 and A1B for CPR 4.5. Increasing radiation is thus expected, particularly in the near–ultraviolet part of the spectrum, where radiation can result in human tissue damage.

Mist coolers have the potential to mitigate heat and solar radiation at the same time. In Dombrovsky et al. (2011), a model was developed to quantify the hemispherical transmittance of direct and scattered solar radiation and investigate the potential of water misting systems to serve as a protection from solar irradiation, notably from harmful UV radiation. The proposed model was based on experimental data for solar irradiation, Mie theory, and radiative transfer theory. Significant attenuation was expected by finely atomized mists over the spectrum from the near–ultraviolet to the NIR spectral range. The authors concluded that "the range of application of such systems, which are inexpensive and consume modest amounts of water, can be much wider if the protection property against solar irradiation is considered." The same conclusion is drawn in Gonome and Wakabayashi (2020), where the total solar reflectance is shown to increase with increasing particle sizes until saturation.

The implications of reduced solar heat are (1) reduced air and surface temperature driving enhanced thermal and visual comfort, (2) reduced UV-related mortality and morbidity (including the risk of sunburn, skin photoaging, cutaneous malignant melanomas and nonmelanoma skin cancers, cortical cataracts, pterygium, macular degeneration, herpes, and the rare squamous cell carcinomas of the eye), (3) reduced photo-oxidative processes as those triggering the production of tropospheric ozone (Yu, 2019). Further air quality improvements may also be linked to dust scavenging, although very limited quantitative literature exists on the topic (Yu, 2014). The implications of dust removal include (1) enhanced air quality and (2) reduced respiratory illnesses.

Mist coolers are ready-to-use, versatile means to increase urban liveability at the local scale. The effect spreads over a much wider area than could be covered by a personal air conditioning unit while still at much lower costs. On-topic research studies target:

- Climate adaptation by adopting a proven strategy for decreasing temperatures and the related risk for illnesses such as heat exhaustion and stroke.
- Climate resilience by adopting a self-adjusting strategy, further controllable by smart logics.
- Health protection and promotion by reducing exposure to climate-sensitive hazards (mitigated heat hot spots, less solar gains, more local air turbulence), enhanced air quality (wet deposition of pollutants, reduced chemical potential for pollutants production, reduced UV-related illnesses including skin cancer, and ophthalmopathy).
- Health equity by promoting installations at heat island hot spots frequented by underserved individuals at risk of heat-related illness (low-income population, marginalized communities, children, the elderly, outdoor workers, individuals engaging in outdoor physical activity, and those with existing medical conditions).

Despite the proven resource-conservative, multipurpose nature, mist cooling requires further investigation toward standardization of research protocols and design criteria to avoid a cloud of incompatible results. Innovative configurations will receive further attention in the future, including public transport on-board misting systems (Kojima and Nakashima, 2012), and misting-assisted condensation for high-performance air conditioning units (Alwan et al., 2018; Heidarinejad et al., 2019; Narumi et al., 2012). Overall, the portfolio of urban water resources is broadening and applies to both semiarid and semihumid climates (Winker et al., 2019). Promoting an informed use at the industrial and private-party level will be vital in establishing mist cooling as a versatile future-proofing strategy.

6.5 Urban greenery to mitigating the urban heat

GI, also referred to as "greening," "vegetation," or "greenery," can provide a myriad of environmental, economic, and social benefits in urban areas. GI includes parks, trees, water bodies, green roofs, and vertical greenery systems (Bartesaghi-Koc et al., 2021). In particular, GI is characterized by distinctive thermal, morphological, and physiological features that can alter ambient temperatures, and thus it can potentially provide cooling effects—and sometimes warming effects under certain conditions—by: (1) lowering the penetration of short-wave solar radiation to the ground through shading, (2) blocking the upwelling long-wave radiation emitted by the ground, (3) providing evapotranspiration, (4) reducing surface run-off, (5) increasing water infiltration and storage in soils and natural surfaces (leaves), and (6) altering wind patterns, speed, and advection (Akbari et al., 2016; Bowler et al., 2010; Koc, 2018; Santamouris et al., 2018a; Santamouris, 2020). The knowledge in this area is ample, although quite fragmented and diverse, as evidence and research outputs are heterogeneous and nonstandardized, making a comparative assessment of results challenging (Santamouris, 2020).

In recent years, intensive research has been carried out to develop, evaluate, and implement innovative solutions to mitigate atmospheric overheating in urban areas (particularly dense cities) through greenery as well as their synergistic influence on air quality/pollution, thermal comfort, human health, energy, and water savings. Based on the analysis of recent research and review papers (Aflaki et al., 2017; Akbari et al., 2016; Armaghan Ahmadi et al., 2017; Charoenkit and Yiemwattana, 2016; Hunter et al., 2014; Jamei et al., 2016; Koc et al., 2018; Santamouris, 2015c, 2014a,b; Santamouris and Osmond, 2020; Wong et al., 2021; Yu et al., 2017), we have identified six high-level research foci (related to the spatio-temporal scales and extent of analysis) in which significant advancement has been achieved, namely:

1. The assessment of the mitigation potential of greenery by increasing tree canopy and vegetated surface fractions in urban areas, including the influence of greenery on ventilation and the challenges of evaluating the effects of greenery in combination with other mitigation technologies in a holistic way.

2. The influence of size, shape, composition, and spatial configuration of green spaces (parks) on air temperature of surrounding built-up areas (through evapotranspirative cooling), also referred as to "park cool islands" (PCIs); including the analysis of the intensity, extent, and spatial variability of these cooling effects.

3. The effects of plant selection, plant physiology, placement, and interactions with urban morphology (i.e., the effect of street canyon's width on vegetation's performance).

4. The evaluation of the performance of urban vegetation under extreme climatic conditions, including the effect of irrigation on cooling potential, development, and adoption of drought-tolerant species, alteration of plant physiological operation, and development of genetically engineered species able to function under very extreme conditions.

5. The effects of green roofs and vertical greenery systems on the immediate thermal environment—and behind a wall or under roof indoor
spaces—with emphasis on building energy savings.
6. Challenges, progress, and novel methods, tools, and techniques (i.e.,
on-ground, remote sensing, simulations, etc.) for monitoring and assessing vegetation's cooling potential.

Typically, most studies have analyzed the relationships between quantitative and qualitative descriptors of GI and climate-related indicators
(i.e., air temperature, mean radiant temperature, wind speed, humidity,
etc.). Table 6.1 summarizes key functional, morphological, and spatial
parameters for greenery grouped according to the above high-level
research foci. In the following sections, it is reported the main advances
and achievements from recent studies within each high-level category, as
well as potential avenues for future research.

6.5.1 Progress on atmospheric heat mitigation with green infrastructure

- Influence of vegetation fraction and abundance, and challenges for a
holistic evaluation in combination with other mitigation technologies

  Plants provide shade for buildings, pedestrian areas, and heat-
absorbing surfaces by intercepting incoming solar radiation; a crucial
aspect of reducing ambient temperatures and improving outdoor thermal comfort (OTC). However, the quality, effectiveness, and amount
of shade are determined by factors such as placement, crown shape,
size, type and angle of leaves, density, height, compactness, and conti-
nuity of canopy layers (Koc et al., 2018). These aspects are commonly
evaluated by estimating the impact of the increased tree and vegetated
surface fraction in urban areas. Whereas dimensionless indices such as
the LAI, LAD, and NDVI are commonly used as an indicator of ecological function and as a factor for controlling energy exchanges, transpiration, and photosynthetic activity, these have also emerged as
metrics for areal estimation of the shading potential of vegetation canopies (Koc et al., 2018).

  A recent review evaluating 46 studies has found that a reasonable
increment of GI fraction by 20% can result in a peak temperature drop
of up to 0.3°C, while with an increase of up to 100% it is expected a
maximum drop of average daily peak temperatures of up to 1.8°C
(Santamouris, 2020). Most studies reported higher cooling effects at
night compared to daytime conditions; so a maximum nocturnal

Table 6.1 List of key GI-related parameters of investigation grouped based on high-level research foci.

Research foci	Aspects	Key indicators	
Influence of vegetation fraction/abundance; separately and in combination with other technologies	Vegetation abundance/shading potential	– Vegetation indices: Normalized difference vegetation index (NDVI), Enhanced vegetation index (EVI)	– Leaf area index (LAI) – Leaf area density (LAD)
	Surface fractions	– Land-use/land-covers (LULC) – Tree, pervious, and water surface fractions	– Normalized difference built-up index (NDBI) – Local Climate Zones (LCZ)
Influence of size, shape, composition, and spatial configuration of parks	Landscape metrics Spatial distribution	– FRAGSTATS metrics (i.e., PLAND, PD, SHAPEAM, and ENN_AM)	– Local Moran's I indexb – Size–distance index – Distance from park/water body
	Evapotranspirative Cooling	– Evapotranspiration rate (ET₀) – Wetness Index (WI)	– Normalized difference water index (NDWI)

Effect of plant selection, physiology placement and surrounding urban morphology + Vegetation performance under extreme conditions	*Biophysical and physiological attributes*	– Type of foliage – Plant solar transmissivity – Leaf color and thickness – Geometry, size, shape, area, and height	– Leaf absorptivity – Leaf transmittance – Leaf radiation attenuation coefficient – Stomatal conductance/resistance
	Vegetated surface properties	– Emissivity (ε), albedo, reflectance – Brightness index (*BI*) – Thermal conductivity – Solar absorption capacity	– Plant permeability – Soil moisture or water content – Soil density – Irrigation rate – Normalized difference water index (*NDWI*)
	Evapotranspirative Cooling	– Evapotranspiration rate (ET_o) – Wetness Index (*WI*)	– Vegetation size, shape, and height – LAI, LAD
	Wind flow pattern	– Vegetation arrangement/configuration	
	Urban morphology influencing on greenery performance	– Aspect ratio (*H/W*) – Sky view factor (*SVF*) – Local climate zones (*LCZ*) – Volumetric density of buildings	– Altitude and elevation – Solar orientation (aspect)

(Continued)

Table 6.1 (Continued)

Cooling effects of green roofs and vertical greenery systems	*Supporting structure attributes*	–	Training system/attachment mode (for green walls/facades)
		–	Topography
		–	Type of supporting material
		–	Dimensions of structures
	Plant characteristics	–	Substrate type (green roofs)
		–	Depth of substrate (green roofs)
		–	Plant species
		–	Plant density

[a]For a full list of landscape metrics review (McGarigal and Marks, 1995).
[b]Calculated with FRAGSTATS software (McGarigal et al., 2002).
Source: Based on Table 3 in Koc, C.B., Osmond, P., Peters, A., 2018. Evaluating the cooling effects of green infrastructure: a systematic review of methods, indicators and data sources. Solar Energy 166, 486–508.

temperature decrease of up to 2.3°C may be reached by increasing GI by 80% (Santamouris, 2020). This can be attributed to (1) a lower release of sensible heat by vegetated surfaces during the night in comparison to impervious materials and (2) a reduced amount of heat stored by ground surfaces because of increased shading during the day (Bartesaghi-Koc et al., 2020; Chen et al., 2014a,b). However, tree canopy can also trap the upwelling long-wave heat reflected or emitted from horizontal and vertical surfaces, a situation that may alter the heat fluxes and inhibit convective cooling in urban areas with decreased sky view factors (SVFs). This can be perceived as a warming effect, especially at night (Arghavani et al., 2019; Bartesaghi-Koc et al., 2019; Imran et al., 2019; Morakinyo et al., 2017; Qiao et al., 2013). Despite the evidence, the expected magnitude of temperature decrease should carefully consider the influence of surrounding landscape characteristics and urban form (Koc et al., 2018). Empirical and simulation studies employing a grid approach have shown that cells (or spatial units) fully covered by greenery are cooler compared to those incorporating a mixture of vegetated and artificial surfaces (Bartesaghi-Koc et al., 2020; Fenner et al., 2017; Geletič et al., 2018; Wang et al., 2016). This is because of the presence of higher evapotranspiration and reduced heat storage during the day, which results in lower temperatures throughout the day. Conversely, the cooling capacity of vegetation may be affected by emitted and reflected heat from surrounding impervious surfaces and higher release of anthropogenic heat, particularly in dense urban areas (i.e., in overexposed street canyons) (Bartesaghi-Koc et al., 2021; Morakinyo et al., 2017; Zoulia et al., 2009).

Furthermore, the assessment of the effects of greenery should not be conducted in isolation but in a holistic manner. For instance, the implementation of additional tree cover may result in higher levels of relative humidity (due to increased evapotranspiration) that may affect OTC, a situation that can be unfavorable in tropical cities (Aflaki et al., 2017; Li and Norford, 2016; Santamouris, 2020). Like any wicked problem, heat amelioration should not only be tackled through vegetation but by implementing greenery with other mitigation technologies (Santamouris et al., 2017). A recent study has demonstrated that within an isolated street, an increment in GI in combination with reflective pavements can reduce peak ambient temperatures by an additional 0.2°C, while when combined with cool pavements, artificial shading, water spray systems, and cool roofs, the cooling effects can be raised by additional 2.9°C (Bartesaghi-Koc et al., 2021). Evidence has shown that when grass and trees are implemented separately, peak ambient temperatures are reduced by 1.2°C and 1.5°C, respectively, while when combined with increased albedo surfaces (by 0.2), it reached a reduction of 2.3°C for the same area (Santamouris et al., 2017). Nonetheless, the effects of combined technologies are not additive; thus the sum of the individual cooling capacities is 0.5°C−0.7°C lower than the maximum temperature reduction achieved together.

As the inherent complexity of overlapping multiple solutions cannot be obviated, it is of great importance to further investigate the specific contributions of each technology and the factors affecting them, including, but not limited to, differences in plant species (foliage, height, plant physiology), different arrangements of natural and artificial features/surfaces, effects from surrounding built form, aging of materials, and cofounding effects from external boundary conditions (particularly air advection) (Bartesaghi-Koc et al., 2021; Howe et al., 2017; Kotthaus and Grimmond, 2014; Kyriakodis and Santamouris, 2017; Santamouris et al., 2017; Yu et al., 2018). These aspects should be incorporated in experimental and empirical studies and better represented in climate models (i.e., urban canopy parameterizations), CFD and mesoscale simulations (Santamouris et al., 2018b).

- Influence of size, shape, composition, and spatial configuration of parks on surrounding built-up areas.

PCIs refer to the phenomena where greenspaces establish a zone of larger advective influence beyond its borders, resulting in lower temperatures than those of the immediate impervious surfaces (Chen et al., 2014a,b; Spronken-Smith and Oke, 1998). These cooling effects can be mainly attributed to shading and evapotranspiration; however, this greatly varies depending on (1) climate, (2) season, (3) plant morphology and placement, and (4) irrigation levels. Evidence suggests that air temperatures of green spaces can be 1°C−4°C cooler than those of the surrounding built-up areas (and sometimes more), with the greatest zone of influence extending downwind from parks (Doick et al., 2014; Ellis et al., 2017; Skoulika et al., 2014; Taha et al., 2018). It is not correct, however, to assume that all parks are cooler than their surrounding area at all times, as cooling intensities are influenced by diurnal and seasonal cycles (Oliveira et al., 2011; Sugawara et al.,

2016), among other factors. Accordingly, cooling effects from parks tend to be larger in summer ($>2.0°C$) compared to winter ($<0.5°C$) (Santamouris et al., 2018b; Wong et al., 2010).

The presence or absence of tree canopy and its arrangement within the green space is also a crucial factor influencing PCI formation. For instance, well-irrigated parks with dense tree canopies typically develop daytime PCIs due to the combined effects of shading and evapotranspiration, while warmer conditions at night due to the heat trapped under the canopy and reduced advection. Conversely, parks with sparse canopy cover are typically warmer during the day as most surfaces are exposed to short-wave solar radiation and are cooler at night due to long-wave radiation losses (Spronken-Smith and Oke, 1998; Taha, 2015a). Targeted configurations of tree canopies and vegetated surfaces can also significantly increase the cooling potential of parks compared to randomly sparse distributions over large areas (Taha, 2015b) as the particular arrangement of vegetation can channel cool air from parks into surrounding built-up areas (Bernard et al., 2018; Takebayashi, 2017).

In addition, the size and configuration of green spaces over a large area are also essential parameters influencing their mitigation potential (Koc et al., 2018; Yan et al., 2018). Landscape metrics, in particular FRAGSTATS indicators (McGarigal et al., 2002), have been mostly employed to analyze the relationship between the shape, extent, and spatial patterns (i.e., fragmentation, aggregation) of green–blue patches and the cooling island intensity based on land surface temperature (LST) observations (Cao et al., 2010; Masoudi et al., 2021); and, to a lesser extent, to analyze the effects on the ambient temperature gradient, with certain exceptions (Kong and Nakagoshi, 2006). On the other hand, the threshold value of efficiency for park and water body size has been proposed as an indicative value of a size to which the cooling intensity per unit area of blue-green spaces reaches a plateau or starts dropping (Yu et al., 2020). This ranges, for example, between 0.5 and 0.69 ha in temperate climates (i.e., Rome) (Yu et al., 2018); while 0.69−0.95 ha for tropical cities (Fan et al., 2019).

Generally, larger parks tend to have a more pronounced cooling effect because of the reduced anthropogenic heat and the net decrease in sensible heat, which are also greatly influenced by the presence of persistent wind (i.e., prevailing sea breezes). Contradictorily, although small green areas provide in overall a cooling potential equivalent to larger parks (Saaroni et al., 2018; Spronken–Smith and Oke, 1998), sometimes these may offer the opposite effect and behave as heat islands as microclimatic conditions tend to be more unstable throughout the day (Chen et al., 2012; Motazedian et al., 2020; Xiao et al., 2018). Generally, studies focusing on large parks are more abundant. However, reviews and studies on small green areas have increased in recent years (Armaghan Ahmadi et al., 2017; Saaroni et al., 2018). In most cases, wind velocity, wind direction, and urban geometry (i.e., SVF) have been identified as important factors influencing the intensity and extent of temperature reductions achieved by small green areas (i.e., courtyards) (Chang et al., 2007; Santamouris, 2020).

It has been found that park shape also influences the cooling capacity (Du et al., 2017). Generally, regular shapes show a higher cooling efficiency compared to complex shapes (Saaroni et al., 2018). Moreover, it has been found that the cooling efficiency of elongated (and narrow) greenspaces is lower than polygonal-shaped (i.e., circular) parks. This can be attributed to plant diversity and the lack of small trees or shrubs, so linear green spaces are more prone to be affected by heat transported from adjacent built-up areas. Another important aspect investigated is the extent or distance where the effects of PCIs are no longer present. Factors affecting cooling extent include size and structure of green areas, placement of trees, irrigation levels, plant type and placement, sky obstruction, characteristics of surrounding areas, wind conditions, and more (Koc et al., 2018). Reported distances vary considerably between studies and locations. For instance, cooling effects from 2.6°C to 4.8°C can reach 35 − 840 m away in some parks in Beijing (Lin et al., 2015), while the same cooling intensity is registered up to 330 m in greenspaces in London (Vaz Monteiro et al., 2016). In Ca et al. (1998), it is shown that a park of 0.6 km^2 can provide 1.5°C of cooling up to 1 km of distance at noon in Tokyo. Despite this, cooling intensities can be maximized if parks (of at least 1 ha) are arranged at intervals of less than 1 km.

- Effect of plant selection, physiology, placement, and surrounding urban morphology on the cooling potential of greenery.

Urban greenery is capable of modifying air and surface temperatures through the combined effects of evaporation and transpiration, a phenomenon referred to as evapotranspirative cooling. Well-watered plants can dissipate a high amount of heat by transforming water liquid present in soil and other surfaces into water vapor, keeping leaves, surrounding air, and soil substrates cool (Koc et al., 2018). Thus the selection of plant species with higher stomatal conductance (the rate at which they transpire) can help reduce temperatures and heat transfer to buildings (greenery on buildings) more rapidly (Nagase and Dunnett, 2010). Similarly, plants with high solar reflectance and thermal emissivity exhibit cooler foliage, so they are quite effective in reducing temperatures and heat flow through the canopy (Koc et al., 2018). Indeed, proper optimization of albedo values of adjacent surfaces can help increase stomatal resistance, especially in contexts of water constraint (Schweitzer and Erell, 2014). This strategy is particularly of great value when greenery is applied to buildings as this can reduce the amount of downward solar radiation transferred to and absorbed by artificial surfaces and improve the cooling performance of green roofs and vertical greenery systems (Peri et al., 2016).

Owing to the large variability in the magnitude of the thermal regulation from one tree species to another, recent studies are shifting from a generic approach into a more species- and site-specific approach (Koc et al., 2018). In the context of "right tree at right place" (Morakinyo et al., 2020), recent studies have found strong interactions between surrounding urban form, plant morphological characteristics, and thermal performance of greenery, and this deserves further study (Morakinyo and Lam, 2016; Ng et al., 2012; Norton et al., 2015). Some studies evaluated the thermal performance of generic tree forms in different urban contexts characterized by varied SVF and density; and found that the heat mitigation potential of trees decreases as urban density increases (i.e., urban canyons with low SVF values) due to competing shade from buildings (Kong et al., 2017; Tan et al., 2017; Thom et al., 2016). Thus high-density areas should have the lowest priority for tree planting, while open areas with increasing SVF values should be the top priority (Morakinyo et al., 2018). The estimation of SVF or aspect ratio (H/W height − width ratio) can help identify the best plant species (i.e., deciduous vs evergreen, or tall vs short) for a particular location and define a priority ranking for planting (Santamouris et al., 2018a).

In terms of the impact of vegetation foliage on thermal regulation, it has been found that small-leaved species are more effective in cooling the surrounding air (by regulating crown temperatures) compared to large-leaved species (Leuzinger et al., 2010). Nonetheless, total tree and trunk height also play an important role; accordingly, tall, short, crown width, high trunk, and less dense trees are recommended for high-density areas dominated by building overshadowing, while the opposite is suggested for open areas (Oh et al., 2021).

- Performance of urban vegetation under extreme climatic conditions.

Under the presence of unrestricted water, the cooling potential of plants is noticeably enhanced; however, evaporative cooling can be significantly affected by restricted water supply, poor irrigation conditions, and scarce precipitation (Gao et al., 2020). Leaf stomata may also be blocked by the presence of particulates or be closed as a response to excessive heat stress as a result of urban overheating, hot weather, heatwaves, or severe droughts. Hence, in recent years, the study of irrigation and water management in the cooling capacity of vegetation is attracting more attention from climate researchers (Coutts et al., 2013).

The frequency, duration, and intensity of heatwaves and extreme summer climatic conditions have dramatically increased in recent years (Yao et al., 2013). Several studies have found that under such conditions, the cooling performance of GI is significantly affected, and mitigation potentials are severely limited (Lu et al., 2012). Accordingly, watering of plants during heatwaves is of vital importance to maintain key physiological processes at acceptable levels, so transpiration rates (and their cooling mechanisms) are not seriously compromised (Bauweraerts et al., 2013; Teskey et al., 2015). Nonetheless, there is a large discrepancy in the way that different plant species respond to extreme climatic conditions. Therefore more research is required better to understand the cooling performance of vegetation during heatwaves, as it has been found that stomatal conductance, photosynthetic rates, and transpiration mechanism may be decoupled under such conditions (Ameye et al., 2012; Bauweraerts et al., 2013; Rogers et al., 2017). Progress in this topic also includes the development and application of a new model to predict stomatal conductance. Experimental studies analyzing the sensitivity of families of trees to extreme drought and heat indicate that plants with low height efficiency (those that grow slowly) are less susceptible to heat and water stress than families with a higher rate of height growth (Bigras, 2000). Furthermore, plants with deeper rooting systems exhibit high transpiration rates, even underwater stress conditions, as these can collect water deep in the soil profile (Drake et al., 2018).

Optimization of the location, timing, and volume of irrigation has also been identified as key aspects to consider when planning and implementing new greenspaces in urban areas (Broadbent et al., 2018). However, in cities where water supply is scarce, the implementation of drought-tolerant plants is being promoted in order to reduce water consumption required for irrigation while still increasing the vegetation cover (Koc et al., 2018). But replacing existing greenery with only drought-tolerant species without considering site-specific irrigation regimes can eventually lead to diurnal warming effects (Grossman-Clarke et al., 2010). In Vahmani and Ban-Weiss (2016), it is found that replacing lawns with drought-tolerant shrubbery can raise the daytime near-surface air temperature by up to 1.9°C, but decrease nighttime near-surface air temperature by 3.2°C on a typical summer day in Southern California. The diurnal warming effect can be attributed to the increase in leaf stomatal resistance and low evaporative cooling and latent heat fluxes due to lack of irrigation. Conversely, nocturnal cooling effects are mainly associated with decreased soil heat conductivity because of low soil moisture levels (due to lack of irrigation), which overall reduced upward ground heat fluxes. In addition, daytime cooling effects may be achieved by reducing roughness length, for example, by replacing trees with shrubs that consume less water, but this can be only achieved if the effect of increased air advection outweighs the warming effect of decreased evaporative cooling due to limited irrigation levels (Yang and Wang, 2017). Under such complex conditions, more research is needed to understand better the atmospheric processes resulting from different combinations of water regimes and drought-tolerant species to suggest better planning strategies in dry cities (Gao et al., 2020; Vahmani and Ban-Weiss, 2016).

In the context of global warming, the genetic modification of vegetation to increase plant thermo-tolerance and secure growth under extreme climatic conditions and severe drought is another aspect that should be further investigated (Santamouris et al., 2018a). Several studies have examined the impact of climate change on different tree genotypes based on the type of climate (warmer vs cooler) and found that growth performance is considerably affected by heat stress, although preconditioning can help reduce long-term damages (Savva et al., 2007). Researchers are currently developing heat-tolerant species with either traditional breeding or genetically engineered procedures.

Traditional breeding methods are considerably slow and challenging, and plant traits are typically characterized by increased membrane thermostability, higher photosynthetic rates, and heat repellent mechanisms such as early maturation, leaf rolling, and change of leaf orientation (Bita and Gerats, 2013; Wahid et al., 2007). In comparison, genetic engineering for trees is a faster alternative as specific genes for desirable traits can be introduced to improve wood properties, root formation, and increase stress tolerance (Harfouche et al., 2011). Advances to increase tree heat-tolerance capacity include the use of abiotic-stress-associated genes from other species (Vinocur and Altman, 2005); genomics, transcriptomics, and proteomics manipulation (Harfouche et al., 2011); and the implementation of the quantitative trait locus mapping method for the identification of specific chromosome segments related to heat tolerance (Bita and Gerats, 2013). However, more studies are necessary to understand the effects of these modifications over longer periods, to monitor the cycles of stress and recovery, and to define rules and regulations that can help mitigate the risk of transgene spread (i.e., creating plants with not functional flowers), and minimize public and regulatory concerns that allow the commercial use of such species in future (Koc et al., 2018).

- Effects of green roofs and vertical greenery systems.

The effects of greenery on buildings (roofs and walls) have been mostly studied in highly urbanized contexts in which spaces for greenspaces and large crown tree canopy is limited. Intensive and extensive green roofs are the two main types recognized by most authors, and multiple studies have evaluated and confirmed the cooling effect of rooftop and vertical greenery (Wong et al., 2021). Generally, green roofs are capable of reducing peak air temperatures by 1.5°C−4.1°C (an average of 3°C); however, these effects are more evident in areas located in close proximity of roofed areas (up to 1.2 m above the vegetated surfaces), and usually several meters above the ground, so the impact on pedestrians may be negligible. On the other hand, vertical greenery systems are capable of reducing peak air temperatures by 2°C−4°C (an average of 3°C) (Wong et al., 2021).

It has been found that the magnitude of the cooling capacity of rooftop and vertical greenery largely depends on prevailing boundary conditions, season, and morphological characteristics of vegetation and supporting systems. Because of their obvious insulation properties, green roofs and walls are more effective in providing cooling effects in summer because of higher evapotranspirative cooling and foliage density (more shade and higher albedo), while more effective in abating heat loss in winter (Bevilacqua et al., 2016; Sternberg et al., 2011; Vox et al., 2018). Cloudiness and rain tend to reduce the cooling intensity of green roofs and green facades, and this is attributed to reduced longwave and shortwave radiation due to changes in solar irradiance and VPD (Cascone et al., 2019; Tan et al., 2020).

Supporting systems and their placement influence the cooling capacity of green walls and green roofs. Typically, carrier systems (when plant substrate is distributed throughout the wall or roof) offer greater insulation levels than support systems (when the substrate is limited to the bottom and a mesh to support climbers). Similarly, energy savings associated with vertical greenery are larger for carrier systems. The ability of greenery on buildings also depends on morphological factors such as wall size, shape and orientation, and surrounding urban geometry. It is a fact that the cooling potential of green roofs and walls is significantly undermined by competing shade from adjacent buildings or when the lack of incoming solar radiation prevents the photosynthetic activity and thus evapotranspiration (Hohmann-Marriott and Blankenship, 2011). This problem is more obvious for green walls than for green roofs that are typically located on top of buildings. It has also been found that a higher green façade ratio can help improve thermal cooling ahead of orientation. As a very high vegetated coverage for walls is somewhat unrealistic, then at least 30% coverage ratio should be prioritized for high-density urban areas.

Plant selection also causes a significant influence on the thermal performance of green roofs and green walls as different types of foliage result in varied shading provision and evapotranspiration rates. Plant selection, however, may also depend on the type of structure and depth of soil substrate (Sailor et al., 2012). For instance, intensive green roofs have deeper substrates, so larger temperature reductions are typically expected. They exhibit larger heat absorption and have more capacity to sustain larger plant species than extensive green roofs (Jim and Tsang, 2011). In addition, plant functional attributes further contribute to temperature reductions. It has been found that plants with high leaf stomatal conductance, LAI, and thin light-colored leaves offer greater cooling effects (Monteiro et al., 2017). However, a holistic evaluation is needed as it is known that broad-leaved species can heat up more quickly than small-leaved plants, resulting in higher peak mean radiant temperatures (Tan et al., 2015). Thus factors such as albedo, height, evapotranspiration rate are also important when selecting a specific variety of traits (Cameron et al., 2014).

- Methods and tools for monitoring and assessing the cooling potential of greenery.

Various multiscale approaches have been used to assess the cooling potential of GI, including (1) on-ground field measurements, (2) remote sensing, and (3) simulated (numerical modeling) methods; and this has been extensively described and discussed in recent reviews (Nouri et al., 2014). First, on-site field measurements have been largely employed to study the impacts of greenery on air temperature and OTC conditions; these are mainly characterized by high temporal but low spatial resolution. Second, remotely sensed data have been mostly used to monitor urban and biophysical conditions of vegetation and surrounding areas, as well as to estimate well-known vegetation indices (i.e., NDVI, EVI) derived from spectral imagery and to retrieve thermal conditions of surfaces (LSTs. The integration of remotely sensed and on-ground measurements can significantly contribute to the spatio-temporal analysis of the thermal behavior of greenery (Bartesaghi-Koc et al., 2019).

In particular, considerable progress in the field is the use of active remote sensors, specifically LiDAR (light detection and ranging) point clouds, in combination with spectral/thermal imagery and field observations, as this enables an accurate retrieve of three-dimensional information on terrains and vegetation (i.e., tree heights, vegetation cover, vertical stratification). Among many parameters influencing the cooling potential of GI, evapotranspiration has been usually overlooked by climate-related remote sensing studies as its estimation is laborious, costly, and mostly conducted for agricultural purposes (Cameron et al., 2014). Indeed, few studies have focused on estimating evapotranspiration and LAI from remotely sensed data in highly heterogeneous urban contexts (Nouri et al., 2014).

Compared to empirical methods, numerical modeling or simulations have been extensively used by studies to predict thermal conditions and analyze the relationship between vegetation- and climate-related parameters. Mesoscale atmospheric models such as the Weather Research and Forecasting System (WRF) (Georgescu et al., 2011) and the coupled WRF-NOAH Land surface model (Zhou and Shepherd, 2010) are used to predict climatic impacts of GI on large areas, whole cities, or entire regions. Urban canopy models (UCM) such as the single-layer Princeton

UCM enable the parameterization of vegetation-related variables and built forms to study the physical processes and heat exchange within the urban canopy (Krayenhoff et al., 2014; Ryu et al., 2016). However, limitations of UCMs include the representation of the drag coefficient of trees and the effect of aerodynamic roughness on heat fluxes, and that they are limited in describing latent heat fluxes.

In addition, CFD models have been developed to simulate the surface − plant − air interactions to study the effect of greenery on the outdoor thermal environment, better suited at local and microscales. Common software tools or modeling environments include ENVI-met, OpenFOAM, PHOENICS, ANSYS FLUENT, TRNSYS, Rayman, and SOLWEIG. Recent studies have assessed the capacity of ENVI-met, SOLWEIG, and Rayman to estimate mean radiant temperatures in outdoor spaces; in general, these models tend to overestimate results under shade and underestimate values under prolonged periods of sunlit (Acero and Arrizabalaga, 2018; Gál and Kántor, 2020; Tsoka et al., 2018). Therefore simulation results should be carefully interpreted when assessing the impact of greenery on OTC and when identifying areas out of thermal comfort.

6.6 Conclusions

Overheating in the urban environment is a major environmental, energy, and health problem. To counterbalance the impact of the excess heat released in cities, innovative mitigation technologies, including low surface temperature coatings and materials, greenery, and evaporative systems, have been developed. The developed mitigation technologies are extensively used in cooling dominated climates in numerous large urban projects. Although high-performance mitigation techniques exhibit a considerable drop in the ambient urban temperature, important problems, including weatherization and aging problems, still have to be faced.

Intensive research in urban mitigation science has provided several new alternative techniques and structures presenting a considerably increased cooling potential. Material technologies based on super cool materials attract a high interest and most research attention. However, significant additional research has to be carried out to translate the recent scientific developments into high-performance large-scale projects.

References

Acero, J.A., Arrizabalaga, J., 2018. Evaluating the performance of ENVI-met model in diurnal cycles for different meteorological conditions. Theoretical and Applied Climatology 131, 455–469.

Adegoke, J.O., Pielke Sr, R.A., Eastman, J., Mahmood, R., Hubbard, K.G., 2003. Impact of irrigation on midsummer surface fluxes and temperature under dry synoptic conditions: a regional atmospheric model study of the US high plains. Monthly Weather Review 131, 556–564.

Aflaki, A., Mirnezhad, M., Ghaffarianhoseini, A., Ghaffarianhoseini, A., Omrany, H., Wang, Z.-H., et al., 2017. Urban heat island mitigation strategies: a state-of-the-art review on Kuala Lumpur, Singapore and Hong Kong. Cities (London, England) 62, 131–145.

Aili, A., Wei, Z.Y., Chen, Y.Z., Zhao, D.L., Yang, R.G., Yin, X.B., 2019. Selection of polymers with functional groups for daytime radiative cooling. Materials Today Physics 10, 100127.

Akbari, H., Matthews, H.D., 2012. Global cooling updates: reflective roofs and pavements. Energy and Buildings 55, 2–6.

Akbari, H., Cartalis, C., Kolokotsa, D., Muscio, A., Pisello, A.L., Rossi, F., et al., 2016. Local climate change and urban heat island mitigation techniques—the state of the art. Journal of Civil Engineering and Management 22, 1–16. Available from: https://doi.org/10.3846/13923730.2015.1111934.

Akey, A.J., Lu, C., Wu, L., Zhu, Y., Herman, I.P., 2012. Anomalous photoluminescence Stokes shift in CdSe nanoparticle and carbon nanotube hybrids. Physical Review B 85, 45404.

Al-Hamed, K.H.M., Dincer, I., Rosen, M.A., 2020. Investigation of elastocaloric cooling option in a solar energy-driven system. International Journal of Refrigeration 120, 340–356.

Allen, R.G., Pereira, L.S., Raes, D., Smith, M., 1998. Crop evapotranspiration—guidelines for computing crop water requirements—FAO Irrigation and drainage paper 56. FAO, Rome 300, D05109.

Alwan, M., Jabbar, M., Hameed, O., n.d. High efficiency A/C system development using automated water mist system. In: 2018 International Conference on Engineering Technology and Their Applications (IICETA). IEEE, pp. 1–6.

Ameye, M., Wertin, T.M., Bauweraerts, I., McGuire, M.A., Teskey, R.O., Steppe, K., 2012. The effect of induced heat waves on *Pinus taeda* and *Quercus rubra* seedlings in ambient and elevated CO_2 atmospheres. The New Phytologist 196, 448–461. Available from: https://doi.org/10.1111/j.1469-8137.2012.04267.x.

Ao, X., Hu, M., Zhao, B., Chen, N., Pei, G., Zou, C., 2019. Preliminary experimental study of a specular and a diffuse surface for daytime radiative cooling. Solar Energy Materials and Solar Cells 191, 290–296.

Aprea, C., Greco, A., Maiorino, A., Masselli, C., 2018. The environmental impact of solid-state materials working in an active caloric refrigerator compared to a vapor compression cooler. International Journal of Heat and Technology 36, 1155–1162.

Arghavani, S., Malakooti, H., Bidokhti, A.A., 2019. Numerical evaluation of urban green space scenarios effects on gaseous air pollutants in Tehran Metropolis based on WRF-Chem model. Atmospheric Environment (Oxford, England: 1994) 214, 116832.

Armaghan Ahmadi, V., Tenpierik, M., Alireza, M.H., 2017. Heat mitigation by greening the cities, a review study. Environment, Earth and Ecology 1, 5–32.

Armson, D., Rahman, M.A., Ennos, A.R., 2013. A comparison of the shading effectiveness of five different street tree species in Manchester, UK. Arboriculture & Urban Forestry 39, 157–164.

Arsenault, A.C., Puzzo, D.P., Manners, I., Ozin, G.A., 2007. Photonic-crystal full-colour displays. Nature Photonics 1, 468–472.

Atieh, A., Shariff, S., Al, 2013. Solar energy powering up aerial misting systems for cooling surroundings in Saudi Arabia. Energy Conversion and Management 65, 670–674.

Atiganyanun, S., Plumley, J.B., Han, S.J., Hsu, K., Cytrynbaum, J., Peng, T.L., et al., 2018. Effective radiative cooling by paint-format microsphere-based photonic random media. ACS Photonics 5, 1181–1187. Available from: https://doi.org/10.1021/acsphotonics.7b01492.

Babentsov, V., Sizov, F., 2008. Defects in quantum dots of IIB–VI semiconductors. Opto-Electronics Review 16, 208–225.

Ballinas, M., Barradas, V.L., 2016. The urban tree as a tool to mitigate the urban heat island in Mexico city: a simple phenomenological model. Journal of Environmental Quality 45, 157–166.

Bao, H., Yan, C., Wang, B., Fang, X., Zhao, C.Y., Ruan, X., 2017. Double-layer nano-particle-based coatings for efficient terrestrial radiative cooling. Solar Energy Materials and Solar Cells 168, 78–84.

Barrow, H., Pope, C.W., 2007. Droplet evaporation with reference to the effectiveness of water-mist cooling. Applied Energy 84, 404–412.

Bartesaghi-Koc, C., Osmond, P., Peters, A., 2019. Spatio-temporal patterns in green infra-structure as driver of land surface temperature variability: the case of Sydney. International Journal of Applied Earth Observation and Geoinformation 83, 101903. Available from: https://doi.org/10.1016/J.JAG.2019.101903.

Bartesaghi-Koc, C., Osmond, P., Peters, A., 2020. Quantifying the seasonal cooling capac-ity of 'green infrastructure types' (GITs): an approach to assess and mitigate surface urban heat island in Sydney, Australia. Landscape and Urban Planning 203, 103893. Available from: https://doi.org/10.1016/j.landurbplan.2020.103893.

Bartesaghi-Koc, C., Haddad, S., Pignatta, G., Paolini, R., Prasad, D., Santamouris, M., 2021. Can urban heat be mitigated in a single urban street? Monitoring, strategies, and performance results from a real scale redevelopment project. Solar Energy 216, 564–588. Available from: https://doi.org/10.1016/j.solener.2020.12.043.

Bauweraerts, I., Wertin, T.M., Ameye, M., Mcguire, M.A., Teskey, R.O., Steppe, K., 2013. The effect of heat waves, elevated [CO$_2$] and low soil water availability on northern red oak (*Quercus rubra* L.) seedlings. Global Change Biology 19, 517–528. Available from: https://doi.org/10.1111/gcb.12044.

Baxter, V.D., Radermacher, R., 2020. IEA/HPT Annex 53 Advanced cooling/refrigera-tion technologies development—Task 1 Report.

Berardi, U., Garai, M., Morselli, T., 2020. Preparation and assessment of the potential energy savings of thermochromic and cool coatings considering inter-building effects. Solar Energy 209, 493–504.

Berdahl, P., Chen, S.S., Destaillats, H., Kirchstetter, T.W., Levinson, R.M., Zalich, M.A., 2016. Fluorescent cooling of objects exposed to sunlight—the ruby example. Solar Energy Materials and Solar Cells 157, 312–317. Available from: https://doi.org/10.1016/j.solmat.2016.05.058.

Bernard, J., Rodler, A., Morille, B., Zhang, X., 2018. How to design a park and its sur-rounding urban morphology to optimize the spreading of cool air? Climate 6, 10.

Bevilacqua, P., Mazzeo, D., Bruno, R., Arcuri, N., 2016. Experimental investigation of the thermal performances of an extensive green roof in the Mediterranean area. Energy and Buildings 122, 63–79.

Bigras, F.J., 2000. Selection of white spruce families in the context of climate change: heat tolerance. Tree Physiology 20, 1227–1234.

Biju, V., Makita, Y., Sonoda, A., Yokoyama, H., Baba, Y., Ishikawa, M., 2005. Temperature-sensitive photoluminescence of CdSe quantum dot clusters. The Journal of Physical Chemistry. B 109, 13899−13905.

Bita, C., Gerats, T., 2013. Plant tolerance to high temperature in a changing environment: scientific fundamentals and production of heat stress-tolerant crops. Frontiers in Plant Science 4, 273.

Bowler, D.E., Buyung-Ali, L., Knight, T.M., Pullin, A.S., 2010. Urban greening to cool towns and cities: a systematic review of the empirical evidence. Landscape and Urban Planning 97, 147−155.

Brinkmann, N., Eugster, W., Buchmann, N., Kahmen, A., 2019. Species-specific differences in water uptake depth of mature temperate trees vary with water availability in the soil. Plant Biology 21, 71−81.

Broadbent, A.M., Coutts, A.M., Tapper, N.J., Demuzere, M., 2018. The cooling effect of irrigation on urban microclimate during heatwave conditions. Urban Climate 23, 309−329. Available from: https://doi.org/10.1016/j.uclim.2017.05.002.

Bruederlin, F., Ossmer, H., Wendler, F., Miyazaki, S., Kohl, M., 2017. SMA foil-based elastocaloric cooling: from material behavior to device engineering. Journal of Physics D: Applied Physics 50, 424003.

Bruederlin, F., Bumke, L., Chluba, C., Ossmer, H., Quandt, E., Kohl, M., 2018. Elastocaloric cooling on the miniature scale: a review on materials and device engineering. Energy Technology 6, 1588−1604. Available from: https://doi.org/10.1002/ente.201800137.

Bruederlin, F., Bumke, L., Quandt, E., Kohl, M., n.d. Cascaded Sma-film based elastocaloric cooling, In: 2019 20th International Conference on Solid-State Sensors, Actuators and Microsystems & Eurosensors XXXIII (TRANSDUCERS & EUROSENSORS XXXIII). IEEE, pp. 1467−1470.

Ca, V.T., Asaeda, T., Abu, E.M., 1998. Reductions in air conditioning energy caused by a nearby park. Energy and Buildings 29, 83−92.

Cameron, R.W.F., Taylor, J.E., Emmett, M.R., 2014. What's 'cool'in the world of green façades? How plant choice influences the cooling properties of green walls. Building and Environment 73, 198−207.

Cao, X., Onishi, A., Chen, J., Imura, H., 2010. Quantifying the cool island intensity of urban parks using ASTER and IKONOS data. Landscape and Urban Planning 96, 224−231.

Carotenuto, G., Nicolais, F., 2009. Reversible thermochromic nanocomposites based on thiolate-capped silver nanoparticles embedded in amorphous polystyrene. Materials (Basel) 2, 1323−1340.

Cascone, S., Coma, J., Gagliano, A., Perez, G., 2019. The evapotranspiration process in green roofs: a review. Building and Environment 147, 337−355.

Catalanotti, S., Cuomo, V., Piro, G., Ruggi, D., Silvestrini, V., Troise, G., 1975. The radiative cooling of selective surfaces. Solar Energy 17, 83−89.

Chang, C.-R., Li, M.-H., Chang, S.-D., 2007. A preliminary study on the local cool-island intensity of Taipei city parks. Landscape and Urban Planning 80, 386−395.

Charoenkit, S., Yiemwattana, S., 2016. Living walls and their contribution to improved thermal comfort and carbon emission reduction: a review. Building and Environment 105, 82−94.

Chen, X., Su, Y., Li, D., Huang, G., Chen, W., Chen, S., 2012. Study on the cooling effects of urban parks on surrounding environments using Landsat TM data: a case study in Guangzhou, southern China. International Journal of Remote Sensing 33, 5889−5914.

Chen, A., Yao, X.A., Sun, R., Chen, L., 2014a. Effect of urban green patterns on surface urban cool islands and its seasonal variations. Urban Forestry and Urban Greening 13, 646—654.

Chen, D., Wang, X., Thatcher, M., Barnett, G., Kachenko, A., Prince, R., 2014b. Urban vegetation for reducing heat related mortality. Environmental Pollution (Barking, Essex: 1987) 192, 275—284.

Chen, Z., Zhu, L., Raman, A., Fan, S., 2016. Radiative cooling to deep sub-freezing temperatures through a 24-h day-night cycle. Nature Communications 7, 1—5. Available from: https://doi.org/10.1038/ncomms13729.

Chen, F., Xu, X., Barlage, M., Rasmussen, R., Shen, S., Miao, S., et al., 2018. Memory of irrigation effects on hydroclimate and its modeling challenge. Environmental Research Letters 13, 64009.

Cho, E., Kim, T., Choi, S.M., Jang, H., Min, K., Jang, E., 2018. Optical characteristics of the surface defects in InP colloidal quantum dots for highly efficient light-emitting applications. ACS Applied Nano Materials 1, 7106—7114. Available from: https://doi.org/10.1021/acsanm.8b01947.

Coutts, A.M., Tapper, N.J., Beringer, J., Loughnan, M., Demuzere, M., 2013. Watering our cities. Progress in Physical Geography: Earth and Environment 37, 2—28. Available from: https://doi.org/10.1177/0309133312461032.

Desert, A., Naboni, E., Garcia, D., 2020. The spatial comfort and thermal delight of outdoor misting installations in hot and humid extreme environments. Energy and Buildings 224, 110202.

Dhariwal, J., Manandhar, P., Bande, L., Marpu, P., Armstrong, P., Reinhart, C.F., 2019. Evaluating the effectiveness of outdoor evaporative cooling in a hot, arid climate. Building and Environment 150, 281—288. Available from: https://doi.org/10.1016/J.BUILDENV.2019.01.016.

Di Giuseppe, E., Ulpiani, G., Cancellieri, C., Di Perna, C., D'Orazio, M., Zinzi, M., 2021. Numerical modelling and experimental validation of the microclimatic impacts of water mist cooling in urban areas. Energy and Buildings 231, 110638.

Doick, K.J., Peace, A., Hutchings, T.R., 2014. The role of one large greenspace in mitigating London's nocturnal urban heat island. The Science of the Total Environment 493, 662—671.

Dombrovsky, L.A., Solovjov, V.P., Webb, B.W., 2011. Attenuation of solar radiation by a water mist from the ultraviolet to the infrared range. Journal of Quantitative Spectroscopy & Radiative Transfer 112, 1182—1190.

Drake, J.E., Tjoelker, M.G., Vårhammar, A., Medlyn, B.E., Reich, P.B., Leigh, A., et al., 2018. Trees tolerate an extreme heatwave via sustained transpirational cooling and increased leaf thermal tolerance. Global Change Biology 24, 2390—2402. Available from: https://doi.org/10.1111/gcb.14037.

Du, H., Cai, W., Xu, Y., Wang, Z., Wang, Y., Cai, Y., 2017. Quantifying the cool island effects of urban green spaces using remote sensing data. Urban Forestry & Urban Greening 27, 24—31.

El Fil, B., Boman, D.B., Tambasco, M.J., Garimella, S., 2021. A comparative assessment of space-conditioning technologies. Applied Thermal Engineering 182, 116105.

Ellis, K.N., Hathaway, J.M., Mason, L.R., Howe, D.A., Epps, T.H., Brown, V.M., 2017. Summer temperature variability across four urban neighborhoods in Knoxville, Tennessee, USA. Theoretical and Applied Climatology 127, 701—710.

Emaikwu, N., Catalini, D., Muehlbauer, J., Hwang, Y., Takeuchi, I., Radermacher, R., n.d. Development of a cascade elastocaloric regenerator, In: Energy Sustainability. American Society of Mechanical Engineers, p. V001T16A001.

Engelbrecht, K., 2019. Future prospects for elastocaloric devices. Journal of Physics: Energy 1, 21001.

Fan, H., Yu, Z., Yang, G., Liu, T., Liu, Y., Ying, T., et al., 2019. How to cool hot-humid (Asian) cities with urban trees? An optimal landscape size perspective. Agricultural and Forest Meteorology 265, 338—348.

Fang, V., Kennedy, J.V., Futter, J., Manning, J., 2013. A review of near infrared reflectance properties of metal oxide nanostructures. GNS Science.

Farnham, C., Emura, K., Mizuno, T., 2015. Evaluation of cooling effects: outdoor water mist fan. Building Research & Information 43, 334—345. Available from: https://doi.org/10.1080/09613218.2015.1004844.

Farnham, C., Zhang, L., Yuan, J., Emura, K., Alam, A.M., Mizuno, T., 2017. Measurement of the evaporative cooling effect: oscillating misting fan. Building Research & Information 1—17. Available from: https://doi.org/10.1080/09613218.2017.1278651.

Feng, J., Gao, K., Santamouris, M., Shah, K.W., Ranzi, G., 2020. Dynamic impact of climate on the performance of daytime radiative cooling materials. Solar Energy Materials and Solar Cells 208, 110426. Available from: https://doi.org/10.1016/j.solmat.2020.110426.

Fenner, D., Meier, F., Bechtel, B., Otto, M., Scherer, D., 2017. Intra and inter local climate zone variability of air temperature as observed by crowdsourced citizen weather stations in Berlin, Germany. Meteorologische Zeitschrift 26, 525—547.

Fernandez, N., Wang, W., Alvine, K.J., Katipamula, S., 2015. Energy savings potential of radiative cooling technologies. Pacific Northwest National Laboratory, Richland, Washington, p. 99352.

Ferrari, C., Gholizadeh Touchaei, A., Sleiman, M., Libbra, A., Muscio, A., Siligardi, C., et al., 2014. Effect of aging processes on solar reflectivity of clay roof tiles. Advances in Building Energy Research 1—13. Available from: https://doi.org/10.1080/17512549.2014.890535.

Frenzel, J., Eggeler, G., Quandt, E., Seelecke, S., Kohl, M., 2018. High-performance elastocaloric materials for the engineering of bulk- and micro-cooling devices. MRS Bulletin/Materials Research Society 43, 280—284. Available from: https://doi.org/10.1557/mrs.2018.67.

Gagge, A.P., Fobelets, A.P., Berglund, L., 1986. A standard predictive Index of human reponse to thermal enviroment. Ashrae Transactions 92, 709—731.

Gál, C.V., Kántor, N., 2020. Modeling mean radiant temperature in outdoor spaces, a comparative numerical simulation and validation study. Urban Climate 32, 100571.

Gao, K., Santamouris, M., Feng, J., 2020. On the cooling potential of irrigation to mitigate urban heat island. The Science of the Total Environment 740, 139754. Available from: https://doi.org/10.1016/j.scitotenv.2020.139754.

Garshasbi, S., Santamouris, M., 2019. Using advanced thermochromic technologies in the built environment: recent development and potential to decrease the energy consumption and fight urban overheating. Solar Energy Materials and Solar Cells 191, 21—32. Available from: https://doi.org/10.1016/j.solmat.2018.10.023.

Garshasbi, S., Haddad, S., Paolini, R., Santamouris, M., Papangelis, G., Dandou, A., et al., 2020a. Urban mitigation and building adaptation to minimize the future cooling energy needs. Solar Energy 204, 708—719. Available from: https://doi.org/10.1016/j.solener.2020.04.089.

Garshasbi, S., Huang, S., Valenta, J., Santamouris, M., 2020b. Can quantum dots help to mitigate urban overheating? An experimental and modelling study. Solar Energy 206, 308—316. Available from: https://doi.org/10.1016/j.solener.2020.06.010.

Garshasbi, S., Huang, S., Valenta, J., Santamouris, M., 2020c. On the combination of quantum dots with near-infrared reflective base coats to maximize their urban overheating mitigation potential. Solar Energy 211, 111—116. Available from: https://doi.org/10.1016/j.solener.2020.09.069.

Geletič, J., Lehnert, M., Savić, S., Milošević, D., 2018. Modelled spatiotemporal variability of outdoor thermal comfort in local climate zones of the city of Brno, Czech Republic. The Science of the Total Environment 624, 385–395.

Genjima, Y., Mochizuki, H., 2002. United States Patent.

Gentle, A.R., Smith, G.B., 2010. Radiative heat pumping from the earth using surface phonon resonant nanoparticles. Nano Letters 10, 373–379.

Gentle, A.R., Smith, G.B., 2015. A subambient open roof surface under the mid-summer sun. Advancement of Science 2, n/a–n/a. Available from: https://doi.org/10.1002/advs.201500119.

Georgescu, M., Moustaoui, M., Mahalov, A., Dudhia, J., 2011. An alternative explanation of the semiarid urban area "oasis effect. Journal of Geophysical Research: Atmospheres 116, n/a–n/a. Available from: https://doi.org/10.1029/2011JD016720.

Goetzler, W., Guernsey, M., Young, J., Fujrman, J., Abdelaziz, A., 2016. The future of air conditioning for buildings. Oak Ridge National Laboratory, Oak Ridge, TN, p. 37831.

Goldstein, E.A., Raman, A.P., Fan, S., 2017. Sub-ambient non-evaporative fluid cooling with the sky. Nature Energy 2, 1–7. Available from: https://doi.org/10.1038/nenergy.2017.143.

Gonome, H., Wakabayashi, K., 2020. Solar barrier performance of water mist cooling: applications using nano-and microsized droplets and bubbles. Applied Thermal Engineering 171, 115083.

Greco, A., Aprea, C., Maiorino, A., Masselli, C., 2019. A review of the state of the art of solid-state caloric cooling processes at room-temperature before 2019. International Journal of Refrigeration 106, 66–88.

Grenier, P., 1979. Réfrigération radiative. Effet de serre inverse. Rev. Phys. Appl. 14, 87–90.

Grossman-Clarke, S., Zehnder, J.A., Loridan, T., Grimmond, C.S.B., Grossman-Clarke, S., Zehnder, J.A., et al., 2010. Contribution of land use changes to near-surface air temperatures during recent summer extreme heat events in the Phoenix metropolitan area. <https://doi.org/10.1175/2010JAMC2362.1>.

Guan, B., Ma, B., Qin, F., n.d. Application of asphalt pavement with phase change materials to mitigate urban heat island effect, In: 2011 International Symposium on Water Resource and Environmental Protection. IEEE, pp. 2389–2392.

Gunawardena, K.R., Wells, M.J., Kershaw, T., 2017. Utilising green and bluespace to mitigate urban heat island intensity. The Science of the Total Environment 584, 1040–1055.

Harfouche, A., Meilan, R., Altman, A., 2011. Tree genetic engineering and applications to sustainable forestry and biomass production. Trends in Biotechnology 29, 9–17.

Heidarinejad, G., Moghaddam, M.R.A.A., Pasdarshahri, H., 2019. Enhancing COP of an air-cooled chiller with integrating a water mist system to its condenser: investigating the effect of spray nozzle orientation. International Journal of Thermal Sciences 137, 508–525.

Heidarzadeh, H., Rostami, A., Dolatyari, M., 2020. Management of losses (thermalization-transmission) in the Si-QDs inside 3C−SiC to design an ultra-high-efficiency solar cell. Materials Science in Semiconductor Processing 109, 104936.

Hervé, A., Drévillon, J., Ezzahri, Y., Joulain, K., 2018. Radiative cooling by tailoring surfaces with microstructures: association of a grating and a multi-layer structure. Journal of Quantitative Spectroscopy & Radiative Transfer 221, 155–163. Available from: https://doi.org/10.1016/j.jqsrt.2018.09.015.

Hohmann-Marriott, M.F., Blankenship, R.E., 2011. Evolution of photosynthesis. Annual Review of Plant Biology 62, 515–548.

Hossain, M.M., Jia, B., Gu, M., 2015. A metamaterial emitter for highly efficient radiative cooling. Advanced Optical Materials 3, 1047−1051.

Hou, H., Simsek, E., Ma, T., Johnson, N.S., Qian, S., Cissé, C., et al., 2019. Fatigue-resistant high-performance elastocaloric materials made by additive manufacturing. Science 366, 1116−1121.

Howe, D.A., Hathaway, J.M., Ellis, K.N., Mason, L.R., 2017. Spatial and temporal variability of air temperature across urban neighborhoods with varying amounts of tree canopy. Urban Forestry & Urban Greening 27, 109−116. Available from: https://doi.org/10.1016/J.UFUG.2017.07.001.

Huang, Z., Ruan, X., 2017. Nanoparticle embedded double-layer coating for daytime radiative cooling. International Journal of Heat and Mass Transfer 104, 890−896. Available from: https://doi.org/10.1016/j.ijheatmasstransfer.2016.08.009.

Huang, C., Cai, J., Lin, Z., Zhang, Q., Cui, Y., 2017. Solving model of temperature and humidity profiles in spray cooling zone. Building and Environment 123, 189−199. Available from: https://doi.org/10.1016/j.buildenv.2017.06.043.

Hunter, A.M., Williams, N.S.G., Rayner, J.P., Aye, L., Hes, D., Livesley, S.J., 2014. Quantifying the thermal performance of green façades: a critical review. Ecological Engineering 63, 102−113.

Hwang, G.B., Patir, A., Allan, E., Nair, S.P., Parkin, I.P., 2017. Superhydrophobic and white light-activated bactericidal surface through a simple coating. ACS Applied Materials & Interfaces 9, 29002−29009.

Imran, H.M., Kala, J., Ng, A.W.M., Muthukumaran, S., 2019. Effectiveness of vegetated patches as green infrastructure in mitigating urban heat island effects during a heatwave event in the city of Melbourne. Weather and Climate Extremes 25, 100217.

Jamei, E., Rajagopalan, P., Seyedmahmoudian, M., Jamei, Y., 2016. Review on the impact of urban geometry and pedestrian level greening on outdoor thermal comfort. Renewable & Sustainable Energy Reviews 54, 1002−1017. Available from: https://doi.org/10.1016/j.rser.2015.10.104.

Jeevanandam, P., Mulukutla, R.S., Phillips, M., Chaudhuri, S., Erickson, L.E., Klabunde, K.J., 2007. Near infrared reflectance properties of metal oxide nanoparticles. The Journal of Physical Chemistry C 111, 1912−1918.

Jim, C.Y., Tsang, S.W., 2011. Biophysical properties and thermal performance of an intensive green roof. Building and Environment 46, 1263−1274. Available from: https://doi.org/10.1016/j.buildenv.2010.12.013.

Jing, P., Zheng, J., Ikezawa, M., Liu, X., Lv, S., Kong, X., et al., 2009. Temperature-dependent photoluminescence of CdSe-core CdS/CdZnS/ZnS-multishell quantum dots. The Journal of Physical Chemistry C 113, 13545−13550.

Jones, H.G., 2013. Plants and Microclimate: A Quantitative Approach to Environmental Plant Physiology. Cambridge Univ. Press.

Joshi, S., Gomekar, S., Joshi, S.S., Goyal, I., Ghawghawe, R., Hardas, S., et al., 2016. Design of solar powered mist cooling system for a typical semi-outdoor area in Nagpur Solar Photovoltaic Thermal system view project Liquid Spectrum Filters for Solar Photovoltaics View project Design of solar powered mist cooling system for a typical semi-outdoor area in Nagpur.

Kabirifar, P., Žerovnik, A., Ahčin, Ž., Porenta, L., Brojan, M., Tušek, J., 2019. Elastocaloric cooling: state-of-the-art and future challenges in designing regenerative elastocaloric devices. Stroj. Vestnik/Journal Mechanical Engineering (New York, N. Y.: 1919) 65.

Karlessi, T., Santamouris, M., 2015. Improving the performance of thermochromic coatings with the use of UV and optical filters tested under accelerated aging conditions. International Journal of Low-Carbon Technologies 10, 45−61. Available from: https://doi.org/10.1093/ijlct/ctt027.

Karlessi, T., Santamouris, M., Apostolakis, K., Synnefa, A., Livada, I., 2009. Development and testing of thermochromic coatings for buildings and urban structures. Solar Energy 83, 538−551. Available from: https://doi.org/10.1016/J. SOLENER.2008.10.005.

Kecebas, M.A., Menguc, M.P., Kosar, A., Sendur, K., 2017. Passive radiative cooling design with broadband optical thin-film filters. Journal of Quantitative Spectroscopy & Radiative Transfer 198, 1339−1351. Available from: https://doi.org/10.1016/j. jqsrt.2017.03.046.

Khataee, A., Moradkhannejhad, L., Heydari, V., Vahid, B., Joo, S.W., 2016. Self-cleaning acrylic water-based white paint modified with different types of TiO2 nanoparticles. Pigment & Resin Technology 198, 179−186.

Kigali Cooling Efficiency Program. [WWW Document], n.d. URL https://www.k-cep. org/.

Kim, Y.-R., Khil, B.-S., Jang, S.-J., Choi, W.-C., Yun, H.-D., 2015. Effect of barium-based phase change material (PCM) to control the heat of hydration on the mechanical properties of mass concrete. Thermochimica Acta 613, 100−107.

Kim, H.H., Im, E., Lee, S., 2020. Colloidal photonic assemblies for colorful radiative cooling. Langmuir: the ACS Journal of Surfaces and Colloids 36, 6589−6596.

Kirsch, S.-M., Welsch, F., Ehl, L., Michaelis, N., Motzki, P., Schütze, A., et al., n.d. Continuous operating elastocaloric heating and cooling device: air flow investigation and experimental parameter study, In: Smart Materials, Adaptive Structures and Intelligent Systems. American Society of Mechanical Engineers, p. V001T04A018.

Kirsch, S., Welsch, F., Michaelis, N., Schmidt, M., Wieczorek, A., Frenzel, J., et al., 2018. NiTi-based elastocaloric cooling on the macroscale: from basic concepts to realization. Energy Technology 6, 1567−1587.

Koc, C.B., 2018. Assessing the thermal performance of green infrastructure on urban microclimate. PHD thesis. The University of New South Wales, Sydney, Australia.

Koc, C.B., Osmond, P., Peters, A., 2018. Evaluating the cooling effects of green infrastructure: a systematic review of methods, indicators and data sources. Solar Energy 166, 486−508.

Kojima, M., Nakashima, K., 2012. A study of mist spraying system by urban transportation. Design for innovative value towards a sustainable society. Springer, pp. 720−723.

Konarska, J., Uddling, J., Holmer, B., Lutz, M., Lindberg, F., Pleijel, H., et al., 2016. Transpiration of urban trees and its cooling effect in a high latitude city. International Journal of Biometeorology 60, 159−172.

Kong, F., Nakagoshi, N., 2006. Spatial-temporal gradient analysis of urban green spaces in Jinan, China. Landscape and Urban Planning 78, 147−164.

Kong, L., Lau, K.K.-L., Yuan, C., Chen, Y., Xu, Y., Ren, C., et al., 2017. Regulation of outdoor thermal comfort by trees in Hong Kong. Sustainable Cities and Society 31, 12−25.

Konstantinidou, C.A., Lang, W., Papadopoulos, A.M., Santamouris, M., 2019. Life cycle and life cycle cost implications of integrated phase change materials in office buildings. International Journal of Energy Research 43, 150−166.

Kotthaus, S., Grimmond, C.S.B., 2014. Energy exchange in a dense urban environment— Part I: temporal variability of long-term observations in central London. Urban Climate 10, 261−280. Available from: https://doi.org/10.1016/j.uclim.2013.10.002.

Kou, J.L., Jurado, Z., Chen, Z., Fan, S., Minnich, A.J., 2017. Daytime radiative cooling using near-black infrared emitters. ACS Photonics 4, 626−630. Available from: https://doi.org/10.1021/acsphotonics.6b00991.

Krayenhoff, E.S., Christen, A., Martilli, A., Oke, T.R., 2014. A multi-layer radiation model for urban neighbourhoods with trees. Boundary – Layer Meteorology 151, 139–178. Available from: https://doi.org/10.1007/s10546-013-9883-1.

Kyriakodis, G.-E., Santamouris, M., 2017. Using reflective pavements to mitigate urban heat island in warm climates—results from a large scale urban mitigation project. Urban Climate . Available from: https://doi.org/10.1016/j.uclim.2017.02.002.

Lam, C.K.C., Gallant, A.J.E., Tapper, N.J., 2020. Does irrigation cooling effect intensify during heatwaves? A case study in the Melbourne botanic gardens. Urban Forestry & Urban Greening 55, 126815.

Leuzinger, S., Vogt, R., Körner, C., 2010. Tree surface temperature in an urban environment. Agriculture and Forest Meteorology 150, 56–62.

Levinson, R., Berdahl, P., Akbari, H., Miller, W., Joedicke, I., Reilly, J., et al., 2007. Methods of creating solar-reflective nonwhite surfaces and their application to residential roofing materials. Solar Energy Materials and Solar Cells 91, 304–314. Available from: https://doi.org/10.1016/j.solmat.2006.06.062.

Li, Y., Li, B.Q., 2014. Use of CdTe quantum dots for high temperature thermal sensing. RSC Advances 4, 24612–24618.

Li, X.X., Norford, L.K., 2016. Evaluation of cool roof and vegetations in mitigating urban heat island in a tropical city, Singapore. Urban Climate 16, 59–74. Available from: https://doi.org/10.1016/j.uclim.2015.12.002.

Li, B., Liu, W., Yan, L., Zhu, X., Yang, Y., Yang, Q., 2018a. Revealing mechanisms of PL properties at high and low temperature regimes in CdSe/ZnS core/shell quantum dots. Journal of Applied Physics 124, 44302.

Li, G., Huang, J., Zhu, H., Li, Y., Tang, J.-X., Jiang, Y., 2018b. Surface ligand engineering for near-unity quantum yield inorganic halide perovskite QDs and high-performance QLEDs. Chemistry of Materials: A Publication of the American Chemical Society 30, 6099–6107.

Lin, W., Yu, T., Chang, X., Wu, W., Zhang, Y., 2015. Calculating cooling extents of green parks using remote sensing: method and test. Landscape and Urban Planning 134, 66–75.

Liu, Y., Son, S., Chae, D., Jung, P.-H., Lee, H., 2020. Acrylic membrane doped with Al_2O_3 nanoparticle resonators for zero-energy consuming radiative cooling. Solar Energy Materials and Solar Cells 213, 110561.

Lu, J., Li, C., Yang, Y., Zhang, X., Jin, M., 2012. Quantitative evaluation of urban park cool island factors in mountain city. Journal of Central South University 19, 1657–1662.

Luo, D., Feng, Y., Verma, P., 2017. Modeling and analysis of an integrated solid state elastocaloric heat pumping system. Energy 130, 500–514.

Mandal, J., Fu, Y., Overvig, A.C., Jia, M., Sun, K., Shi, N.N., et al., 2018. Hierarchically porous polymer coatings for highly efficient passive daytime radiative cooling. Science 362, 315–319. Available from: https://doi.org/10.1126/science.aat9513.

Mañosa, L., Planes, A., 2017. Materials with giant mechanocaloric effects: cooling by strength. Advanced Materials 29, 1603607.

Masoudi, M., Tan, P.Y., Fadaei, M., 2021. The effects of land use on spatial pattern of urban green spaces and their cooling ability. Urban Climate 35, 100743.

Mastrapostoli, E., Santamouris, M., Kolokotsa, D., Vassilis, P., Venieri, D., Gompakis, K., 2016. On the ageing of cool roofs: measure of the optical degradation, chemical and biological analysis and assessment of the energy impact. Energy and Buildings 114, 191–199. Available from: https://doi.org/10.1016/j.enbuild.2015.05.030.

McGarigal, K., Marks, B.J., 1995. Spatial pattern analysis program for quantifying landscape structure. Gen. Tech. Rep. PNW-GTR-351. US Dep. Agric. For. Serv. Pacific Northwest Res. Stn. 1–122.

McGarigal, K., Cushman, S.A., Neel, M.C., Ene, E., 2002. FRAGSTATS: Spatial Pattern Analysis Program for Categorical Maps [WWW Document]. URL <http://www.umass.edu/landeco/research/fragstats/fragstats.html>.

Montazeri, H., Toparlar, Y., Blocken, B., Hensen, J.L.M., 2017. Simulating the cooling effects of water spray systems in urban landscapes: a computational fluid dynamics study in Rotterdam, The Netherlands. Landscape and Urban Planning 159, 85−100.

Monteiro, M.V., Blanuša, T., Verhoef, A., Richardson, M., Hadley, P., Cameron, R.W. F., 2017. Functional green roofs: importance of plant choice in maximising summer-time environmental cooling and substrate insulation potential. Energy and Buildings 141, 56−68.

Morakinyo, T.E., Lam, Y.F., 2016. Simulation study on the impact of tree-configuration, planting pattern and wind condition on street-canyon's micro-climate and thermal comfort. Building and Environment 103, 262−275. Available from: https://doi.org/10.1016/j.buildenv.2016.04.025.

Morakinyo, T.E., Kong, L., Lau, K.K.-L., Yuan, C., Ng, E., 2017. A study on the impact of shadow-cast and tree species on in-canyon and neighborhood's thermal comfort. Building and Environment 115, 1−17.

Morakinyo, T.E., Lau, K.K.-L., Ren, C., Ng, E., 2018. Performance of Hong Kong's common trees species for outdoor temperature regulation, thermal comfort and energy saving. Building and Environment 137, 157−170.

Morakinyo, T.E., Ouyang, W., Lau, K.K.-L., Ren, C., Ng, E., 2020. Right tree, right place (urban canyon): tree species selection approach for optimum urban heat mitigation-development and evaluation. The Science of the Total Environment 719, 137461.

Moss, J.L., Doick, K.J., Smith, S., Shahrestani, M., 2019. Influence of evaporative cooling by urban forests on cooling demand in cities. Urban Forestry & Urban Greening 37, 65−73.

Motazedian, A., Coutts, A.M., Tapper, N.J., 2020. The microclimatic interaction of a small urban park in central Melbourne with its surrounding urban environment during heat events. Urban Forestry & Urban Greening 52, 126688.

Moya, X., Kar-Narayan, S., Mathur, N.D., 2014. Caloric materials near ferroic phase transitions. Nature Materials 13, 439−450.

Moya, X., Phan, M.H., Srikanth, H., Albertini, F., 2020. Multicalorics. Journal of Applied Physics . Available from: https://doi.org/10.1063/5.0039106.

Nagase, A., Dunnett, N., 2010. Drought tolerance in different vegetation types for extensive green roofs: effects of watering and diversity. Landscape and Urban Planning 97, 318−327.

Narayanaswamy, A., Mayo, J., Canetta, C., 2014. Infrared selective emitters with thin films of polar materials. Applied Physics Letters 104, 183107. Available from: https://doi.org/10.1063/1.4875699.

Narumi, D., Shigematsu, K., Shimoda, Y., 2012. Effect of the evaporative cooling techniques by spraying mist water on reducing urban heat flux and saving energy in apartment house. Journal of Heat Island Institute International 7, 175−181.

Nazarian, N., Fan, J., Sin, T., Norford, L., Kleissl, J., 2017. Predicting outdoor thermal comfort in urban environments: a 3D numerical model for standard effective temperature. Urban Climate 20, 251−267.

Nazemiyan, M., Jalili, Y.S., 2013. Record low temperature Mo doped V2O5 thermochromic thin films for optoelectronic applications. AIP Advances 3, 112103.

Ng, E., Chen, L., Wang, Y., Yuan, C., 2012. A study on the cooling effects of greening in a high-density city: an experience from Hong Kong. Building and Environment 47, 256−271.

Nicholls, L., Strengers, Y., 2018. Heatwaves, cooling and young children at home: integrating energy and health objectives. Energy Research & Social Science 39, 1–9.

Norton, B.A., Coutts, A.M., Livesley, S.J., Harris, R.J., Hunter, A.M., Williams, N.S.G., 2015. Planning for cooler cities: a framework to prioritise green infrastructure to mitigate high temperatures in urban landscapes. Landscape and Urban Planning 134, 127–138.

Nouri, H., Beecham, S., Anderson, S., Nagler, P., 2014. High spatial resolution WorldView-2 imagery for mapping NDVI and its relationship to temporal urban landscape evapotranspiration factors. Remote Sensors 6, 580–602.

Oh, W., Ooka, R., Nakano, J., Kikumoto, H., Ogawa, O., 2020a. Evaluation of mist-spraying environment on thermal sensations, thermal environment, and skin temperature under different operation modes. Building and Environment 168, 106484.

Oh, W., Ooka, R., Nakano, J., Kikumoto, H., Ogawa, O., Choi, W., 2020b. Development of physiological human model considering mist wettedness for mist-spraying environments. Building and Environment 180, 106706.

Oh, W., Ooka, R., Nakano, J., Kikumoto, H., Ogawa, O., 2021. Extended standard effective temperature index for water-misting environment. Building and Environment 190, 107573.

Ohama, Y., Van Gemert, D., 2011. Application of Titanium Dioxide Photocatalysis to Construction Materials: State-of-the-Art Report of the RILEM Technical Committee 194-TDP. Springer Science & Business Media.

Oliveira, S., Andrade, H., Vaz, T., 2011. The cooling effect of green spaces as a contribution to the mitigation of urban heat: a case study in Lisbon. Building and Environment 46, 2186–2194.

Ossmer, H., Wendler, F., Gueltig, M., Lambrecht, F., Miyazaki, S., Kohl, M., 2016. Energy-efficient miniature-scale heat pumping based on shape memory alloys. Smart Materials and Structures 25, 85037.

Pal, S., Contaldi, V., Licciulli, A., Marzo, F., 2016. Self-cleaning mineral paint for application in architectural heritage. Coatings 6, 48.

Paravantis, J., Santamouris, M., Cartalis, C., Efthymiou, C., Kontoulis, N., 2017. Mortality associated with high ambient temperatures, heatwaves, and the urban heat island in Athens, Greece. Sustainability 9, 606. Available from: https://doi.org/10.3390/su9040606.

Park, Y.-S., Bae, W.K., Baker, T., Lim, J., Klimov, V.I., 2015. Effect of Auger recombination on lasing in heterostructured quantum dots with engineered core/shell interfaces. Nano Letters 15, 7319–7328.

Perez, G., Mota-Heredia, C., Sánchez-García, J.A., Guerrero, A., 2020. Compatibility between thermochromic pigments and Portland cement-based materials. Construction and Building Materials. 251, 119038. Available from: https://doi.org/10.1016/j.conbuildmat.2020.119038.

Peri, G., Rizzo, G., Scaccianoce, G., La Gennusa, M., Jones, P., 2016. Vegetation and soil-related parameters for computing solar radiation exchanges within green roofs: are the available values adequate for an easy modeling of their thermal behavior? Energy and Buildings 129, 535–548.

Pisello, A.L., Saliari, M., Vasilakopoulou, K., Hadad, S., Santamouris, M., 2018. Facing the urban overheating: recent developments. Mitigation potential and sensitivity of the main technologies. Wiley Interdisciplinary Reviews: Energy and Environment 7, e294.

Pyrgou, A., Santamouris, M., 2018. Increasing probability of heat-related mortality in a mediterranean city due to urban warming. International Journal of Environmental Research and Public Health 15. Available from: https://doi.org/10.3390/ijerph15081571.

Qian, Y., Huang, M., Yang, B., Berg, L.K., 2013. A modeling study of irrigation effects on surface fluxes and land—air—cloud interactions in the Southern Great Plains. J. Hydrometeorology 14, 700—721.

Qian, S., Nasuta, D., Rhoads, A., Wang, Y., Geng, Y., Hwang, Y., et al., 2016. Not-in-kind cooling technologies: a quantitative comparison of refrigerants and system performance. International Journal of Refrigeration 62, 177—192.

Qiao, Z., Tian, G., Xiao, L., 2013. Diurnal and seasonal impacts of urbanization on the urban thermal environment: a case study of Beijing using MODIS data. ISPRS Journal of Photogrammetry and Remote Sensing 85, 93—101.

Rahman, M.A., Smith, J.G., Stringer, P., Ennos, A.R., 2011. Effect of rooting conditions on the growth and cooling ability of *Pyrus calleryana*. Urban Forestry & Urban Greening. 10, 185—192.

Raman, A.P., Anoma, M.A., Zhu, L., Rephaeli, E., Fan, S., 2014. Passive radiative cooling below ambient air temperature under direct sunlight. Nature 515, 540—544. Available from: https://doi.org/10.1038/nature13883.

Rahman, M.A., Armson, D., Ennos, A.R., 2015. A comparison of the growth and cooling effectiveness of five commonly planted urban tree species. Urban Ecosystem 18, 371—389.

Rahman, M.A., Moser, A., Rötzer, T., Pauleit, S., 2017. Within canopy temperature differences and cooling ability of *Tilia cordata* trees grown in urban conditions. Building and Environment 114, 118—128.

Rahman, M.A., Moser, A., Gold, A., Rötzer, T., Pauleit, S., 2018. Vertical air temperature gradients under the shade of two contrasting urban tree species during different types of summer days. The Science of the Total Environment 633, 100—111.

Remund, J., Müller, S., Kunz, S., Huguenin-Landl, B., Studer, C., Cattin, R., 2018. Meteonorm Handbook part II: Theory, Global Meteorological Database Version 7 Software and Data for Engineers, Planers and Education. URL <http://www.meteonorm.com>.

Rephaeli, E., Raman, A., Fan, S., 2013. Ultrabroadband photonic structures to achieve high-performance daytime radiative cooling. Nano Letters 13, 1457—1461. Available from: https://doi.org/10.1021/nl4004283.

Rogers, A., Medlyn, B.E., Dukes, J.S., Bonan, G., Von Caemmerer, S., Dietze, M.C., et al., 2017. A roadmap for improving the representation of photosynthesis in Earth system models. The New Phytologist 213, 22—42.

Ryu, Y.-H., Bou-Zeid, E., Wang, Z.-H., Smith, J.A., 2016. Realistic representation of trees in an urban canopy model. Boundary — Layer Meteorology 159, 193—220.

Saaroni, H., Amorim, J.H., Hiemstra, J.A., Pearlmutter, D., 2018. Urban green infrastructure as a tool for urban heat mitigation: survey of research methodologies and findings across different climatic regions. Urban Climate 24, 94—110.

Sailor, D.J., Elley, T.B., Gibson, M., 2012. Exploring the building energy impacts of green roof design decisions-a modeling study of buildings in four distinct climates. Journal of Building Physics 35, 372—391. Available from: https://doi.org/10.1177/1744259111420076.

Santamouris, M., 2013. Using cool pavements as a mitigation strategy to fight urban heat island—a review of the actual developments. Renewable & Sustainable Energy Reviews 26, 224—240.

Santamouris, M., 2014a. On the energy impact of urban heat island and global warming on buildings. Energy and Buildings 82, 100—113. Available from: https://doi.org/10.1016/j.enbuild.2014.07.022.

Santamouris, M., 2014b. Cooling the cities—a review of reflective and green roof mitigation technologies to fight heat island and improve comfort in urban environments.

Solar Energy 103, 682—703. Available from: https://doi.org/10.1016/j.solener.2012.07.003.

Santamouris, M., 2015a. Analyzing the heat island magnitude and characteristics in one hundred Asian and Australian cities and regions. The Science of the Total Environment 512, 582—598. Available from: https://doi.org/10.1016/j.scitotenv.2015.01.060.

Santamouris, M., 2015b. Regulating the damaged thermostat of the cities—Status, impacts and mitigation challenges. Energy and Buildings 91, 43—56. Available from: https://doi.org/10.1016/j.enbuild.2015.01.027.

Santamouris, M., 2015c. Regulating the damaged thermostat of the cities—status, impacts and mitigation challenges. Energy and Buildings . Available from: https://doi.org/10.1016/j.enbuild.2015.01.027.

Santamouris, M., 2016. Innovating to zero the building sector in Europe: minimising the energy consumption, eradication of the energy poverty and mitigating the local climate change. Solar Energy 128, 61—94. Available from: https://doi.org/10.1016/j.solener.2016.01.021.

Santamouris, M., 2020. Recent progress on urban overheating and heat island research. Integrated assessment of the energy, environmental, vulnerability and health impact. Synergies with the global climate change. Energy and Buildings 207, 109482. Available from: https://doi.org/10.1016/j.enbuild.2019.109482.

Santamouris, M., Osmond, P., 2020. Increasing green infrastructure in cities: impact on ambient temperature, air quality and heat-related mortality and morbidity. Buildings 10, 233.

Santamouris, M., Yun, G.Y., 2020. Recent development and research priorities on cool and super cool materials to mitigate urban heat island. Renewable Energy. Available from: https://doi.org/10.1016/j.renene.2020.07.109.

Santamouris, M., Paraponiaris, K., Mihalakakou, G., 2007. Estimating the ecological footprint of the heat island effect over Athens, Greece. Climatic Change 80, 265—276. Available from: https://doi.org/10.1007/s10584-006-9128-0.

Santamouris, M., Synnefa, A., Kolokotsa, D., Dimitriou, V., Apostolakis, K., 2008. Passive cooling of the built environment—use of innovative reflective materials to fight heat islands and decrease cooling needs. International Journal of Low-Carbon Technologies 3, 71—82.

Santamouris, M., Synnefa, A., Karlessi, T., 2011. Using advanced cool materials in the urban built environment to mitigate heat islands and improve thermal comfort conditions. Solar Energy 85, 3085—3102. Available from: https://doi.org/10.1016/j.solener.2010.12.023.

Santamouris, M., Cartalis, C., Synnefa, A., Kolokotsa, D., 2015. On the impact of urban heat island and global warming on the power demand and electricity consumption of buildings—a review. Energy and Buildings 98, 119—124. Available from: https://doi.org/10.1016/j.enbuild.2014.09.052.

Santamouris, M., Ding, L., Fiorito, F., Oldfield, P., Osmond, P., Paolini, R., et al., 2017. Passive and active cooling for the outdoor built environment—analysis and assessment of the cooling potential of mitigation technologies using performance data from 220 large scale projects. Solar Energy 154, 14—33. Available from: https://doi.org/10.1016/j.solener.2016.12.006.

Santamouris, Mat, Ban-Weiss, G., Osmond, P., Paolini, R., Synnefa, A., Cartalis, C., et al., 2018a. Progress in urban greenery mitigation science—assessment methodologies advanced technologies and impact on cities. Journal of Civil Engineering and Management 24, 638—664. Available from: https://doi.org/10.3846/jcem.2018.6604.

Santamouris, Mattheos, Feng, J., Santamouris, M., Feng, J., 2018b. Recent progress in daytime radiative cooling: is it the air conditioner of the future? Buildings 8, 168. Available from: https://doi.org/10.3390/buildings8120168.

Santamouris, Mattheos, Haddad, S., Saliari, M., Vasilakopoulou, K., Synnefa, A., Paolini, R., et al., 2018c. On the energy impact of urban heat island in Sydney: climate and energy potential of mitigation technologies. Energy and Buildings 166, 154–164. Available from: https://doi.org/10.1016/J.ENBUILD.2018.02.007.

Santamouris, M., Paolini, R., Haddad, S., Synnefa, A., Garshasbi, S., Hatvani-Kovacs, G., et al., 2020d. Heat mitigation technologies can improve sustainability in cities. An holistic experimental and numerical impact assessment of urban overheating and related heat mitigation strategies on energy consumption, indoor comfort, vulnerability and heat-related mortality and morbidityin cities. Energy and Buildings 217, 110002. Available from: https://doi.org/10.1016/j.enbuild.2020.110002.

Santos Nouri, A., Costa, J.P., Santamouris, M., Matzarakis, A., 2018. Approaches to outdoor thermal comfort thresholds through public space design: a review. Atmosphere (Basel) 9, 108.

Sapit, A., Razali, M.A., Mohammed, A.N., Manshoor, B., Khalid, A., Salleh, H., et al., 2019. Study on mist nozzle spray characteristics for cooling application. International Journal of Integrated Engineering 11.

Savva, Y., Denneler, B., Koubaa, A., Tremblay, F., Bergeron, Y., Tjoelker, M.G., 2007. Seed transfer and climate change effects on radial growth of jack pine populations in a common garden in Petawawa, Ontario, Canada. Forest Ecology and Management 242, 636–647.

Schmidt, M., Schütze, A., Seelecke, S., 2013. The potential of NiTi-based solid state cooling processes. Dtsch. Kälte-und Klimatagung, Hann 1–7.

Schmidt, M., Schütze, A., Seelecke, S., 2015. Scientific test setup for investigation of shape memory alloy based elastocaloric cooling processes. International Journal of Refrigeration 54, 88–97.

Schmidt, M., Kirsch, S.-M., Seelecke, S., Schütze, A., 2016. Elastocaloric cooling: from fundamental thermodynamics to solid state air conditioning. Science and Technology for the Built Environment 22, 475–488.

Schweitzer, O., Erell, E., 2014. Evaluation of the energy performance and irrigation requirements of extensive green roofs in a water-scarce Mediterranean climate. Energy and Buildings 68, 25–32. Available from: https://doi.org/10.1016/j.enbuild.2013.09.012.

Seeboth, A., Lötzsch, D., 2008. Thermochromic Phenomena in Polymers. Smithers Rapra Technology.

Seeboth, A., Lötzsch, D., 2013. Thermochromic and Thermotropic Materials. CRC Press.

Sharar, D.J., Radice, J., Warzoha, R., Hanrahan, B., Chang, B., 2018. First demonstration of a bending-mode elastocaloric cooling 'loop,' In: 17th IEEE Intersociety Conference on Thermal and Thermomechanical Phenomena in Electronic Systems (ITherm). IEEE, pp. 218–226.

Sharifi, N.P., Mahboub, K.C., 2018. Application of a PCM-rich concrete overlay to control thermal induced curling stresses in concrete pavements. Construction and Building Materials 183, 502–512. Available from: https://doi.org/10.1016/j.conbuildmat.2018.06.179.

Shenhav, Y., Grottas, G., 2019. Cooling with anti-Stokes Fluorescence 1. United, StatesUS20190154316A1.

Skoulika, F., Santamouris, M., Kolokotsa, D., Boemi, N., 2014. On the thermal characteristics and the mitigation potential of a medium size urban park in Athens, Greece. Landscape and Urban Planning 123, 73–86. Available from: https://doi.org/10.1016/j.landurbplan.2013.11.002.

Slaughter, J., Czernuszewicz, A., Griffith, L., Pecharsky, V., 2020. Compact and efficient elastocaloric heat pumps—is there a path forward? Journal of Applied Physics 127, 194501.

Souch, C.A., Souch, C., 1993. The effect of trees on summertime below canopy urban climates: a case study Bloomington, Indiana. Journal of Arboriculture 19, 303—312.

Spronken-Smith, R.A., Oke, T.R., 1998. The thermal regime of urban parks in two cities with different summer climates. International Journal of Remote Sensing 19, 2085—2104.

Sternberg, T., Viles, H., Cathersides, A., 2011. Evaluating the role of ivy (*Hedera helix*) in moderating wall surface microclimates and contributing to the bioprotection of historic buildings. Building and Environment 46, 293—297.

Sugawara, H., Shimizu, S., Takahashi, H., Hagiwara, S., Narita, K., Mikami, T., et al., 2016. Thermal influence of a large green space on a hot urban environment. Journal of Environmental Quality 45, 125—133.

Sugimoto, S., Takahashi, H.G., Sekiyama, H., 2019. Modification of near-surface temperature over East Asia associated with local-scale paddy irrigation. Journal of Geophysical Research: Atmospheres 124, 2665—2676.

Suichi, T., Ishikawa, A., Hayashi, Y., Tsuruta, K., 2018. Performance limit of daytime radiative cooling in warm humid environment. AIP Advances 8, 55124.

Synnefa, A., Santamouris, M., Akbari, H., 2007a. Estimating the effect of using cool coatings on energy loads and thermal comfort in residential buildings in various climatic conditions. Energy and Buildings 39, 1167—1174. Available from: https://doi.org/10.1016/j.enbuild.2007.01.004.

Synnefa, A., Santamouris, M., Apostolakis, K., 2007b. On the development, optical properties and thermal performance of cool colored coatings for the urban environment. Solar Energy 81, 488—497. Available from: https://doi.org/10.1016/j.solener.2006.08.005.

Synnefa, A., Santamouris, M., Livada, I., 2006. A study of the thermal performance of reflective coatings for the urban environment. Solar Energy 80, 968—981. Available from: https://doi.org/10.1016/j.solener.2005.08.005.

Taha, H., 2015a. Cool cities: counteracting potential climate change and its health impacts. Current Climate. Change Reports 1, 163—175.

Taha, H., 2015b. Meteorological, emissions and air-quality modeling of heat-island mitigation: recent findings for California, USA. International Journal of Low-Carbon Technologies 10, 3—14.

Taha, H., Akbari, H., Rosenfeld, A., 1991. Heat island and oasis effects of vegetative canopies: Micro-meteorological field-measurements. Theoretical and Applied Climatology 44, 123—138. Available from: https://doi.org/10.1007/BF00867999.

Taha, H., Levinson, R., Mohegh, A., Gilbert, H., Ban-Weiss, G., Chen, S., 2018. Air-temperature response to neighborhood-scale variations in albedo and canopy cover in the real world: fine-resolution meteorological modeling and mobile temperature observations in the Los Angeles climate archipelago. Climate 6, 53.

Takebayashi, H., 2017. Influence of urban green area on air temperature of surrounding built-up area. Climate 5, 60.

Takebayashi, H., 2018. A simple method to evaluate adaptation measures for urban heat island. Environments 5, 70.

Takebayashi, H., Miki, K., Sakai, K., Murata, Y., Matsumoto, T., Wada, S., et al., 2016. Experimental examination of solar reflectance of high-reflectance paint in Japan with natural and accelerated aging. Energy and Buildings 114, 173—179. Available from: https://doi.org/10.1016/j.enbuild.2015.06.019.

Taleghani, M., Berardi, U., 2018. The effect of pavement characteristics on pedestrians' thermal comfort in Toronto. Urban Climate 24, 449—459.

Tan, C.L., Wong, N.H., Tan, P.Y., Jusuf, S.K., Chiam, Z.Q., 2015. Impact of plant evapotranspiration rate and shrub albedo on temperature reduction in the tropical outdoor environment. Building and Environment 94, 206−217.

Tan, Z., Lau, K.K.-L., Ng, E., 2017. Planning strategies for roadside tree planting and outdoor comfort enhancement in subtropical high-density urban areas. Building and Environment 120, 93−109.

Tan, P.Y., Wong, N.H., Tan, C.L., Jusuf, S.K., Schmiele, K., Chiam, Z.Q., 2020. Transpiration and cooling potential of tropical urban trees from different native habitats. The Science of the Total Environment 705, 135764.

Teskey, R., Wertin, T., Bauweraerts, I., Ameye, M., Mcguire, M.A., Steppe, K., 2015. Responses of tree species to heat waves and extreme heat events. Plant, Cell & Environment 38, 1699−1712. Available from: https://doi.org/10.1111/pce.12417.

Thermal Environmental Conditions for Human Occupancy, 2013. ASHRAE Standard 55.

Thom, J.K., Coutts, A.M., Broadbent, A.M., Tapper, N.J., 2016. The influence of increasing tree cover on mean radiant temperature across a mixed development suburb in Adelaide, Australia. Urban Forestry & Urban Greening 20, 233−242.

Tso, C.Y., Chan, K.C., Chao, C.Y.H., 2017. A field investigation of passive radiative cooling under Hong Kong's climate. Renewable. Energy 106, 52−61.

Tsoka, S., Tsikaloudaki, A., Theodosiou, T., 2018. Analyzing the ENVI-met microclimate model's performance and assessing cool materials and urban vegetation applications—a review. Sustainable Cities and Society. 43, 55−76.

Tušek, Jaka, Engelbrecht, K., Eriksen, D., Dall'Olio, S., Tušek, Janez, Pryds, N., 2016. A regenerative elastocaloric heat pump. Nature Energy 1, 1−6.

Ulpiani, G., 2019. Water mist spray for outdoor cooling: a systematic review of technologies, methods and impacts. Applied Energy 254, 113647. Available from: https://doi.org/10.1016/J.APENERGY.2019.113647.

Ulpiani, G., 2020. On the linkage between urban heat island and urban pollution island: three-decade literature review towards a conceptual framework. The Science of the Total Environment 141727.

Ulpiani, G., Di Giuseppe, E., Di Perna, C., D'Orazio, M., Zinzi, M., n.d. Design optimization of mist cooling for Urban Heat Island mitigation: experimental study on the role of injection density, In: IOP Conference Series: Earth and Environmental Science. IOP Publishing, p. 12025.

Ulpiani, G., Di Giuseppe, E., Di Perna, C., D'Orazio, M., Zinzi, M., 2019a. Thermal comfort improvement in urban spaces with water spray systems: field measurements and survey. Building and Environment 156, 46−61.

Ulpiani, G., di Perna, C., Zinzi, M., 2019b. Water nebulization to counteract urban overheating: development and experimental test of a smart logic to maximize energy efficiency and outdoor environmental quality. Applied Energy 239, 1091−1113.

Ulpiani, G., Ranzi, G., Bruederlin, F., Paolini, R., Fiorito, F., Haddad, S., et al., 2019c. Elastocaloric cooling: roadmap towards successful implementation in the built environment. AIMS Materials Science 6, 1135−1152.

Ulpiani, G., Bruederlin, F., Weidemann, R., Ranzi, G., Santamouris, M., Kohl, M., 2020a. Upscaling of SMA film-based elastocaloric cooling. Applied Thermal Engineering 180, 115867. Available from: https://doi.org/10.1016/j.applthermaleng.2020.115867.

Ulpiani, G., di Perna, C., Zinzi, M., 2020b. Mist cooling in urban spaces: understanding the key factors behind the mitigation potential. Applied Thermal Engineering 178, 115644.

Ulpiani, G., Ranzi, G., Shah, K.W., Feng, J., Santamouris, M., 2020c. On the energy modulation of daytime radiative coolers: a review on infrared emissivity dynamic switch against overcooling. Solar Energy . Available from: https://doi.org/10.1016/j. solener.2020.08.077.

Vahmani, P., Hogue, T.S., 2015. Urban irrigation effects on WRF-UCM summertime forecast skill over the Los Angeles metropolitan area. Journal of Geophysical Research: Atmospheres 120, 9869—9881.

Vahmani, P., Ban-Weiss, G., 2016. Climatic consequences of adopting drought-tolerant vegetation over Los Angeles as a response to California drought. Geophysical Research Letters 43, 8240—8249. Available from: https://doi.org/10.1002/2016GL069658@10.1002/(ISSN)1944-8007.2016GRLEDHIGH.

Vanos, J.K., Kosaka, E., Iida, A., Yokohari, M., Middel, A., Scott-Fleming, I., et al., 2019. Planning for spectator thermal comfort and health in the face of extreme heat: the Tokyo 2020 Olympic marathons. The Science of the Total Environment 657, 904—917.

Vanos, J.K., Wright, M.K., Kaiser, A., Middel, A., Ambrose, H., Hondula, D.M., 2020. Evaporative misters for urban cooling and comfort: effectiveness and motivations for use. International Journal of Biometeorology 1—13.

Vasudevan, D., Gaddam, R.R., Trinchi, A., Cole, I., 2015. Core-shell quantum dots: properties and applications. Journal of Alloys and Compounds . Available from: https://doi.org/10.1016/j.jallcom.2015.02.102.

Vaz Monteiro, M., Doick, K.J., Handley, P., Peace, A., 2016. The impact of greenspace size on the extent of local nocturnal air temperature cooling in London. Urban Forestry & Urban Greening 16, 160—169. Available from: https://doi.org/10.1016/j. ufug.2016.02.008.

Vinocur, B., Altman, A., 2005. Recent advances in engineering plant tolerance to abiotic stress: achievements and limitations. Current Opinion in Biotechnology 16, 123—132.

Vox, G., Blanco, I., Schettini, E., 2018. Green façades to control wall surface temperature in buildings. Building and Environment 129, 154—166.

Wahid, A., Gelani, S., Ashraf, M., Foolad, M.R., 2007. Heat tolerance in plants: an overview. Environmental and Experimental Botany 61, 199—223.

Wang, Z., Xing, W., Huang, Y., Xie, T., 2016. Studying the urban heat Island using a local climate zone scheme. Polish Journal of Environmental Studies 25, 2609—2616. Available from: https://doi.org/10.15244/pjoes/63672.

Wang, R., Fang, S., Xiao, Y., Gao, E., Jiang, N., Li, Y., et al., 2019. Torsional refrigeration by twisted, coiled, and supercoiled fibers. Science 366, 216—221.

Wang, R., Zhou, X., Wang, W., Liu, Z., 2021. Twist-based cooling of polyvinylidene difluoride for mechanothermochromic fibers. Chemical Engineering Journal 417, 128060.

Warwick, M.E.A., Binions, R., 2014. Advances in thermochromic vanadium dioxide films. Journal of Materials Chemistry A 2, 3275—3292.

Weissman, J.M., Sunkara, H.B., Albert, S.T., Asher, S.A., 1996. Thermally switchable periodicities and diffraction from mesoscopically ordered materials. Science 274, 959—963.

Welsch, F., Kirsch, S.-M., Michaelis, N., Mandolino, M., Schütze, A., Seelecke, S., et al., n.d. System simulation of an elastocaloric heating and cooling device based on SMA. In: Smart Materials, Adaptive Structures and Intelligent Systems. American Society of Mechanical Engineers, p. V001T03A005.

Winker, M., Gehrmann, S., Schramm, E., Zimmermann, M., Rudolph-Cleff, A., 2019. Greening and cooling the city using novel urban water systems: a European perspective. Approaches to water sensitive urban design. Elsevier, pp. 431—454.

Wong, N.H., Chong, A.Z.M., 2010. Performance evaluation of misting fans in hot and humid climate. Building and Environment 45, 2666—2678. Available from: https://doi.org/10.1016/J.BUILDENV.2010.05.026.

Wong, N.H., Kwang Tan, A.Y., Chen, Y., Sekar, K., Tan, P.Y., Chan, D., et al., 2010. Thermal evaluation of vertical greenery systems for building walls. Building and Environment 45, 663−672. Available from: https://doi.org/10.1016/j. buildenv.2009.08.005.

Wong, R.Y.M., Tso, C.Y., Chao, C.Y.H., Huang, B., Wan, M.P., 2018. Ultra-broadband asymmetric transmission metallic gratings for subtropical passive daytime radiative cooling. Solar Energy Materials and Solar Cells 186, 330−339.

Wong, N.H., Tan, C.L., Kolokotsa, D.D., Takebayashi, H., 2021. Greenery as a mitigation and adaptation strategy to urban heat. Nature Reviews Earth & Environment 2, 166−181.

Wu, D., Liu, C., Xu, Z., Liu, Y., Yu, Z., Yu, L., et al., 2018. The design of ultra-broadband selective near-perfect absorber based on photonic structures to achieve near-ideal daytime radiative cooling. Materials & Design 139, 104−111. Available from: https://doi.org/10.1016/j.matdes.2017.10.077.

Wuister, S.F., Van Houselt, A., de Mello Donegá, C., Vanmaekelbergh, D., Meijerink, A., 2004. Temperature antiquenching of the luminescence from capped CdSe quantum dots. Angewandte Chemie 116, 3091−3095.

Xiao, X.D., Dong, L., Yan, H., Yang, N., Xiong, Y., 2018. The influence of the spatial characteristics of urban green space on the urban heat island effect in Suzhou Industrial Park. Sustainable Cities and Society 40, 428−439.

Xu, F., Gerlein, L.F., Ma, X., Haughn, C.R., Doty, M.F., Cloutier, S.G., 2015. Impact of different surface ligands on the optical properties of PbS quantum dot solids. Materials (Basel) 8, 1858−1870.

Yalçın, R.A., Blandre, E., Joulain, K., Drévillon, J., 2020. Colored radiative cooling coatings with nanoparticles. ACS Photonics 7, 1312−1322.

Yan, H., Wu, F., Dong, L., 2018. Influence of a large urban park on the local urban thermal environment. The Science of the Total Environment 622, 882−891.

Yang, J., Wang, Z.H., 2017. Planning for a sustainable desert city: the potential water buffering capacity of urban green infrastructure. Landscape and Urban Planning 167, 339−347. Available from: https://doi.org/10.1016/j.landurbplan.2017.07.014.

Yang, Y., Taylor, S., Alshehri, H., Wang, L., 2017. Wavelength-selective and diffuse infrared thermal emission mediated by magnetic polaritons from silicon carbide metasurfaces. Applied Physics Letters 111, 051904. Available from: https://doi.org/10.1063/1.4996865.

Yang, P., Chen, C., Zhang, Z.M., 2018. A dual-layer structure with record-high solar reflectance for daytime radiative cooling. Solar Energy 169, 316−324.

Yao, Y., Luo, Y., Huang, J., Zhao, Z., 2013. Comparison of monthly temperature extremes simulated by CMIP3 and CMIP5 models. Journal of Climate 26, 7692−7707.

Ye, C., Zhou, L., Wang, X., Liang, Z., 2016. Photon upconversion: from two-photon absorption (TPA) to triplet−triplet annihilation (TTA). Physical Chemistry Chemical Physics: PCCP 18, 10818−10835.

Yu, S., 2014. Water spray geoengineering to clean air pollution for mitigating haze in China's cities. Environmental Chemistry Letters . Available from: https://doi.org/10.1007/s10311-013-0444-0.

Yu, S., 2019. Fog geoengineering to abate local ozone pollution at ground level by enhancing air moisture. Environmental Chemistry Letters 17, 565−580.

Yu, Z., Guo, X., Jørgensen, G., Vejre, H., 2017. How can urban green spaces be planned for climate adaptation in subtropical cities? Ecological Indicators 82, 152−162.

Yu, Z., Xu, S., Zhang, Y., Jørgensen, G., Vejre, H., 2018. Strong contributions of local background climate to the cooling effect of urban green vegetation. Scientific Reports 8, 6798. Available from: https://doi.org/10.1038/s41598-018-25296-w.

Yu, Z., Yang, G., Zuo, S., Jørgensen, G., Koga, M., Vejre, H., 2020. Critical review on the cooling effect of urban blue-green space: a threshold-size perspective. Urban Forestry & Urban Greening 49, 126630.

Yuxuan, Z., Yunyun, Z., Jianrong, Y., Xiaoqiang, Z., 2020. Energy saving performance of thermochromic coatings with different colors for buildings. Energy and Buildings 215, 109920.

Zapata, P.A., Rabagliati, F.M., Lieberwirth, I., Catalina, F., Corrales, T., 2014. Study of the photodegradation of nanocomposites containing TiO_2 nanoparticles dispersed in polyethylene and in poly (ethylene-co-octadecene). Polymer Degradation and Stability 109, 106–114.

Zarkevich, N.A., Johnson, D.D., Pecharsky, V.K., 2017. High-throughput search for caloric materials: the CaloriCool approach. Journal of Physics D: Applied Physics 51, 24002.

Zhai, Y., Ma, Y., David, S.N., Zhao, D., Lou, R., Tan, G., et al., 2017. Scalable-manufactured randomized glass-polymer hybrid metamaterial for daytime radiative cooling. Science (New York, N.Y.) 355, 1062–1066. Available from: https://doi.org/10.1126/science.aai7899.

Zhang, Y., Zhai, X., 2019. Preparation and testing of thermochromic coatings for buildings. Solar Energy 191, 540–548.

Zhang, J., Xu, Y., An, S., Sun, Y., Li, X., Li, Y., 2020. Giant mechanocaloric materials for solid-state cooling. Chinese Physics B 29, 76202.

Zheng, K., Ichinose, M., Wong, N.H., 2018. Parametric study on the cooling effects from dry mists in a controlled environment. Building and Environment 141, 61–70.

Zheng, K., Yuan, C., Wong, N.H., Cen, C., 2019. Dry mist systems and its impact on thermal comfort for the tropics. Sustainable Cities and Society 51, 101727.

Zhou, Y., Shepherd, J.M., 2010. Atlanta's urban heat island under extreme heat conditions and potential mitigation strategies. Natural Hazards 52, 639–668.

Zhu, X., Su, Q., Feng, W., Li, F., 2017. Anti-Stokes shift luminescent materials for bio-applications. Chemical Society Reviews 46, 1025–1039.

Zou, C., Ren, G., Hossain, M.M., Nirantar, S., Withayachumnankul, W., Ahmed, T., et al., 2017. Metal-loaded dielectric resonator metasurfaces for radiative cooling. Advanced Optical Materials 5, 1700460. Available from: https://doi.org/10.1002/adom.201700460.

Zoulia, I., Santamouris, M., Dimoudi, A., 2009. Monitoring the effect of urban green areas on the heat island in Athens. Environmental Monitoring and Assessment 156, 275–292.

CHAPTER 7

Environmental, energy, and health impact of urban mitigation technologies

Matthaios Santamouris
School of Built Environment, Faculty of Arts, Design and Architecture, University of New South Wales (UNSW), Sydney, NSW, Australia

7.1 Introduction

Several mitigation techniques and technologies are developed and implemented in numerous large-scale projects to counterbalance the serious impact of urban overheating on the environmental quality, energy, health, and sustainability in cities. The most significant of these mitigation strategies are already presented in the previous chapter of the present book.

Experimental and theoretical data collected from numerous mitigation projects worldwide have permitted to understand in concrete terms the real impact of mitigation technologies on the urban climate, energy consumption, outdoor environmental quality, heat-related mortality and morbidity, and on the global sustainability levels in cities.

Despite the limited number of existing studies, solid conclusions can be drawn, and the relative impact of the implemented and considered mitigation strategies can be assessed with sufficient accuracy. Existing data focus mainly on the impact of increased albedo technologies and additional vegetation in cities, including studies assessing the impact of a combination of both mitigation strategies.

The present chapter aims at presenting and analyzing the existing information and knowledge on the impact of the two main mitigation technologies—that is, increased albedo and vegetation—on the potential decrease of the ambient temperature in cities, the possible reduction of the energy consumption used for cooling purposes, the impact on air quality and pollutants concentration, and the reduction in the heat-related mortality and morbidity, specifically focusing on the survivability levels of the low-income population.

Urban Climate Change and Heat Islands
DOI: https://doi.org/10.1016/B978-0-12-818977-1.00001-6
297

Almost all available data have been produced with detailed simulation studies performed using advanced computational tools. It is expected that more experimental and numerical studies will become available during the next years after the first large-scale implementations of heat mitigation and may enrich the existing literature and will improve our knowledge on the topic.

7.2 The impact of increased urban albedo on urban temperature, energy consumption, and health

Urban heat islands considerably increase the energy consumption of buildings, raise pollutants concentration, and have a serious impact on health, particularly on heat-related mortality and morbidity (Santamouris, 2020). Several mitigation technologies are proposed and implemented to counteract the impact of urban overheating. Among the most efficient ones is the increase of the albedo in cities. This can be achieved by implementing highly reflective materials in roofs and pavements as well as on the skin of buildings. As presented in the previous chapters (Santamouris and Yun, 2020), several reflective technologies are proposed and commercialized. In parallel, several large-scale projects involving the use of reflective materials have been studied, and for some of them experimental data are available (Santamouris et al., 2017), revealing that implementation of reflective materials in cities can decrease in average the peak ambient temperature up to 1.5°C (Santamouris, 2014a).

Solar reflective or cool materials are designed to present a high thermal emissivity combined with high reflectance in the visible, near infrared, or in the whole solar spectrum. Several experiments and real-scale implementations demonstrated that cool materials may present much lower surface temperatures, up to 15°C, than conventional materials of the same color (Synnefa et al., 2007). Aging and optical degradation can be a serious drawback for reflective materials. Aging is mainly because of the deposition of atmospheric constituents and dust on the reflective materials affecting their mitigation potential (Mastrapostoli et al., 2016). The use of photocatalytic coatings and the implementation of self-cleaning coatings on the reflective materials are proven to reduce the optical aging caused by most of the environmental sources (Kumar et al., 2013).

The energy impact of urban overheating is well-documented and presented in the previous chapters. As already mentioned, on average, the weight of overheating on the additional peak electricity demand is close to 21 ($\pm$ 10.4) W per degree of temperature increase and per person (Santamouris et al., 2015). In parallel, the cooling energy penalty induced by urban overheating is close to 0.7 kWh per square meter of city and degree of temperature increase (Santamouris, 2014b).

Apart from the influence on energy consumption, higher ambient temperatures seriously affect human health, as human thermoregulation cannot offset extreme heat, causing a severe rise in heat-related morbidity and mortality (Johnson et al., 2005; Stafoggia et al., 2006). It is well-documented that cities present a considerably higher health risk than rural places because of the higher urban ambient temperature and the increased deprivation and vulnerability levels (Taylor et al., 2015; Jesdale et al., 2013; Schinasi et al., 2018). Recent metanalysis reviews reported that the urban population in warmer neighborhoods, on average, presents a 6% higher risk of mortality than those living in cooler precincts (Schinasi et al., 2018). Risk factors associated with demographic and socioeconomic urban factors like age and income, the availability of resources and relative institutions, and the possible impact of the urban networks highly affect heat-related health outcomes in cities (Harlan et al., 2013).

Numerous studies aiming to assess the impact of high urban albedo on reducing ambient temperature and health outcomes have been performed recently and provide helpful information on the specific potential of reflective materials in cities. The existing information is collected in many cities and is characterized by a significant nonhomogeneity and discrepancies in terms of the predicted mitigation potential and health benefits as the cooling and health potential of reflective materials are mainly determined by the specific climatic, landscape, socioeconomic, and demographic characteristics of the cities.

A significant number of studies have attempted to assess the ambient urban temperature drop caused by the potential increase of the urban albedo, using mainly mesoscale modeling techniques (Savio et al., 2006; Synnefa et al., 2008; Menon et al., 2010; Jacobson and Ten Hoeve, 2012; Sailor, 1995; Millstein and Menon, 2011). An analysis of the existing studies found that, on average, the maximum temperature drop caused by the increase of the urban albedo is between 0.14°C and 0.30°C per 0.1 increase in albedo (Santamouris, 2014a). In addition, numerous studies have assessed the impact of higher albedo values on health (Santamouris et al., 2020; Jandaghian and Akbari, 2018; Stone et al., 2014; Silva et al., 2010a; Macintyre and Heaviside, 2019; Urban Heat Island Management Study, 2017; Kalkstein et al., 2003; Vanos et al., n.d.; Jandaghian and

Akbari, 2020; Kalkstein et al., 2011; Shamila et al., 2020; Kalkstein and Sheridan, 2003). Health-related studies are based on empirical parametric associations of the local climatic conditions with the corresponding heat-related mortality (HRM) values. Unfortunately, studies on the indirect benefits of increased albedo values on urban health, like the possible decrease of the ground-level ozone, are very limited and do not provide quantitative results.

7.2.1 The impact of increased urban albedo on ambient urban temperature

Fourteen case studies assessing the potential decrease of the ambient urban temperature caused by a potential increase of the urban albedo are analyzed in Santamouris and Fiorito (2021). The study includes data for the following cities or regions: Baltimore (Vanos et al., n.d.), Detroit (Kalkstein and Sheridan, 2003), District of Columbia (Kalkstein et al., 2003), Los Angeles (Vanos et al., n.d.; Kalkstein and Sheridan, 2003), New Orleans (Kalkstein and Sheridan, 2003), New York (Vanos et al., n. d.), and Philadelphia in the United States (Kalkstein and Sheridan, 2003), Darwin (Shamila et al., 2020) and Parramatta Sydney (Santamouris et al., 2020) in Australia, Montreal (Jandaghian and Akbari, 2018, 2020) and Toronto (Jandaghian and Akbari, 2020) in Canada, and the West Midlands (Macintyre and Heaviside, 2019) in the United Kingdom. The studies in Baltimore, Detroit, District of Columbia, Los Angeles, New Orleans, New York, and Philadelphia assessed the potential ambient temperature drop caused by the modified albedo during several heat waves, while all the other studies refer to the whole summer period and finally, the research in the West Midlands refers to both heatwave and nonheatwave summer periods. Detailed information on all case studies is given in Table 7.1. As shown, the albedo value was considered to increase between 0.1 and 0.7, with most of the cases considering an increase lower than 0.5. Thirteen of the studies used mesoscale modeling techniques to assess the potential temperature decrease while the study for Darwin, Australia, was based on microscale modeling. The methodology followed to assess the impact of the modified albedo on the ambient temperature is fully reported in Santamouris and Fiorito (2021).

As expected, the ambient temperature drop was found to increase as a function of albedo. It is observed that the temperature drop per unit of albedo increase presents a remarkable variability among the considered cities. In particular, the decline in the ambient temperature per 0.1 increase

Table 7.1 Main characteristics of the 14 case studies.

No	City	Period of analysis	Characteristics climatic analysis	Resolution analysis	Reference
1	Parramatta Sydney, AU	Mortality data for summer months 2002–16	Mesoscale modeling using: Weather Research and Forecasting (WRF) Model software	500 × 500 m	Santamouris et al. (2020)
2	Darwin, AU	Wet and dry seasons in 2016	Microscale modeling using ENVImet software	6 × 6 m	Shamila et al. (2020)
3	West Midlands, UK	Summer 2006	Mesoscale modeling using WRF software	Grid of 36, 12, 3, and 1 km in the inner domain.	Macintyre and Heaviside (2019)
4	West Midlands, UK	Heat waves July 16 – 27, 2006 August 2–10, 2003	Mesoscale modeling using WRF software	Grid of 36, 12, 3, and 1 km in the inner domain	Macintyre and Heaviside (2019)
5	Baltimore, USA	Heat waves July 19 – 23, 1991 July 7 – 11, 1993 June 17 – 21, 1994 July 4 – 8, 2010	Mesoscale modeling using WRF software	1 × 1 km	Vanos et al. (n.d.)
6	New Orleans, USA	Heat waves May 31–June 2, 1998 June 16–18, 1998 July 14–18, 2000	Mesoscale modeling using WRF software	2 × 2 km	Kalkstein and Sheridan (2003)

(Continued)

Table 7.1 (Continued)

No	City	Period of analysis	Characteristics climatic analysis	Resolution analysis	Reference
7	New York, USA	Heat waves May 28 – June 1, 1991 July 31 – August 4, 1995 July 3 – July 7, 1999 July 30 – August 3, 2006	Mesoscale modeling using WRF software	2 × 2 km	Vanos et al. (n.d.)
8	Philadelphia, USA	Heat waves June 22–26, 1997 July 3–8, 1999 July 23–29, 1999	Mesoscale modeling using WRF software	2 × 2 km	Kalkstein and Sheridan (2003)
9	Detroit, USA	Heat waves July 13–16, 1995 June 6–11, 1999 August 6–9, 2001	Mesoscale modeling using WRF software	1 × 1 km	Kalkstein and Sheridan (2003)
10	Los Angeles, USA	Heat waves August 13 – 18, 1992 August 10 – 14, 1994 August 12 – 17, 1994 22 July – 6 August 2006	Mesoscale modeling using WRF software	1 × 1 km	Hajat and Kosatky (2010)

11	Los Angeles, USA	Heat waves August 10–15, 1994 October, 8 – 10, 1994 October 14–16, 1997 August 29–September 3, 1998	Mesoscale climate modeling	5 × 5 km	Kalkstein and Sheridan (2003)
12	Montreal, Canada	Heat waves in July 10–12, 2005, July 20–23, 2011, July 2–5, 2018	Mesoscale modeling using WRF software	Four two-way nested domains with a horizontal grid resolution of 333 m	Jandaghian and Akbari (2018, 2020)
13	Toronto, Canada	Heat waves in July 20–23, 2011, July 2–5, 2018	Mesoscale modeling using WRF software	Four two-way nested domains with a horizontal grid resolution of 333 m	Jandaghian and Akbari (2020)
14	District of Columbia, USA	Heat waves in July 18–23, 1991, June 17–22, 1994, June 21–25, 1997, July 21–26, 2010	Mesoscale modeling using WRF software	1000 × 1000 m	Kalkstein and Sheridan (2003)

Source: From Santamouris M., Fiorito, F., 2021. On the impact of modified urban albedo on ambient temperature and heat related mortality. Solar Energy. 216.

of the albedo is found to vary between 0.10°C and 0.38°C with a median value close to 0.2°C and a peak of the probability density function between 0.15°C and 0.20°C. Therefore it is proposed to describe and assess the decrease of the ambient temperature at 17:00 p.m., ΔT_{17} by the following relation:

$$\Delta T_{17} = \alpha \Delta(\text{Alb}) \qquad (7.1)$$

where (α) is a coefficient equal to 1.8, ΔT_{17} is the temperature drop of the daily afternoon temperature caused by the reflective materials, and $\Delta(\text{Alb})$ is the considered increase of the mean albedo. However, very significant variability of the cooling potential of reflective materials is observed under different climatic, urban land characteristics, socioeconomic, and demographic conditions.

Given the importance of the above-mentioned characteristics and considerations, it is found that the afternoon temperature drop, ΔT_{17}, can be well-described and predicted by a linear multiparameter correlation of the following form:

$$\Delta T_{17} = a_1 + \sum \left(a_j \text{AlbParam}\right) + \sum \left(a_k \text{ClimParam}\right) \\ + \sum \left(a_l \text{LandParam}\right) \qquad (7.2)$$

where AlbParam, ClimParam, and LandParam are parameters defining the albedo variability, the climate conditions, and the landscape and layout characteristics of the cities, respectively, while a_i are correlation coefficients.

Further analysis shows that the temperature drop, ΔT_{17}, can be well-predicted as a function of the albedo increase (ΔAlb) together with some landscape parameters like the percentage of green spaces (VC), the percentage of streets in the city (PCS), and also with the population density (PD).

$$\Delta T_{17} = a_1 + a_2 \Delta \text{Alb}_{in} + a_3 \text{VC} + a_4 \text{PCS} + a_5 \text{PD} \qquad (7.3)$$

The coefficients a_i and the standardized coefficients b_i are given in Table 7.2. The obtained correlation between the predicted and estimated by Eq. (7.3) data is given in Fig. 7.1. Eq. (7.3) strictly applies to the present data set as reported in Table 7.1 and should not be used out of the limits of this data range.

Table 7.2 Coefficients use in Eq. (7.3).

Equation	Coefficients
Eq. (7.3)	$a_1 = -0.261$, $a_2 = 0.935$, $a_3 = 0.01$, $a_4 = 0.013$ and $a_5 = -0.000014$
Eq. (7.3) Standardized coefficients	b_j, $b2 = 0.462$, $b3 = 0.245$, $b4 = -0.130$

Source: From Santamouris M., Fiorito, F., 2021. On the impact of modified urban albedo on ambient temperature and heat related mortality. Solar Energy 216.

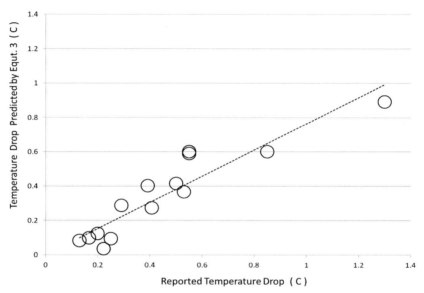

Figure 7.1 Comparison of the reported afternoon temperature drop against the corresponding values calculated by Eq. (7.3). *From Santamouris M., Fiorito, F., 2021. On the impact of modified urban albedo on ambient temperature and heat related mortality. Solar Energy 216.*

7.2.2 The impact of increased urban albedo on heat-related mortality

Fourteen studies presented in Table 7.1 are analyzed in Santamouris and Fiorito (2021). A full description of the considered studies' characteristics is given in Table 7.3. As shown, the initial HRM per day and 100,000 population, HRM_{in}, ranged between 0.46 and 42.6, with an average value close to 9.9 deaths per day. The lowest and highest mortality values corresponded to Darwin, Australia and New York, United States.

Table 7.3 Values of the main inputs in the 14 case studies.

City	Initial albedo (Alb1) [-]	Final albedo (Alb2) [-]	Albedo increase (ΔAlb) [-]	Average afternoon ambient temp. (T_{17}) [C]	Ambient temp. decrease in the afternoon (ΔT_{17}) [C]	Average initial heat-related mortality (HRM_{in}) [Death/d]	Average final heat-related mortality (HRM_{fin}) [Death/d]	Average variation of heat-related mortality [Death/d]	Pop. density (PDP) (p/km²)	Perc. of streets in the city (PCS) [%]	Poverty ratio (POV) [%]	Perc. of green spaces (VC) [%]	Total pop. (POP) [people]	Latitude (LAT) [°]
Baltimore	0.15	0.27	0.12	34.6	0.17	5.44	5.38	0.064	2931	15	21.8	9.4	603,000	39.17
Los Angeles 1	0.15	0.27	0.12	35.2	0.40	16.50	16.22	0.273	3261	26	16.5	13	3,990,000	34.05
New York	0.15	0.28	0.13	33.1	0.13	42.65	38.61	4.036	10428	13	17.3	20	8,308,999	40.73
New Orleans	0.15	0.25	0.10	34.7	0.29	1.81	1.54	0.272	888	16	20.1	26	484,674	29.95
Philadelphia	0.15	0.25	0.10	33.2	0.25	7.99	7.55	0.438	4535	15	24.5	13	1,526,000	39.95
Los Angeles 2	0.13	0.35	0.22	30.2	0.53	16.11	15.25	0.854	3261	26	16.5	13	3,990,000	34.05
Detroit	0.15	0.25	0.10	32.9	0.22	14.30	12.60	1.744	1883	13	33.4	6.1	673,000	42.33
Columbia	0.15	0.27	0.12	32.4	0.20	1.98	1.82	0.182	10146	15	16.2	22	703,000	38.89
W Mid. HW	0.20	0.70	0.50	23.5	0.85	13.00	12.14	0.904	4176	21	17.0	18	1,086,999	52.49
Montreal	0.20	0.50	0.30	31.8	0.39	13.60	12.55	1.000	911	19	16.0	15	1,780,000	45.51
Toronto	0.20	0.50	0.30	32.7	0.50	19.80	16.16	3.666	4150	25	32.0	13	2,830,000	43.65
Darwin	0.20	0.50	0.30	34.3	0.55	0.46	0.33	0.125	741	30	10.0	19	155,000	12.46
Parramatta	0.15	0.60	0.45	30.0	1.30	1.10	0.93	0.166	3070	25	9.8	45	257,000	33.86
W Mid. NHW	0.20	0.70	0.50	20.5	0.55	3.40	3.14	0.244	4176	21	17.0	18	1,086,999	52.49

Source: From Santamouris M., Fiorito, F., 2021. On the impact of modified urban albedo on ambient temperature and heat related mortality. Solar Energy 216.

It is well accepted that urban HRM is highly influenced by the local climatic, landscape, socioeconomic and demographic parameters, and housing quality (Santamouris, 2020; Hajat and Kosatky, 2010). As a result, the ambient temperature corresponding to the minimum HRM value and the temperature over which mortality starts increasing vary significantly among the various cities. Several studies have demonstrated that the threshold temperature in a city is highly dependent on the acclimatization and adaptation of the local population (Keatinge et al., 2000). In parallel, it is observed and well accepted that the threshold mortality temperature is seriously decreasing with latitude (Luo et al., 2019; Green et al., 2019). Socioeconomic and poverty levels seem to seriously influence the specific HRM levels in a city, together with parameters such as the population density of cities or the level of vegetation cover (Son et al., 2016; Klein Rosenthal et al., 2014). In parallel, demographic factors and housing quality seriously affect HRM levels (Santamouris, 2020). It is widely accepted that HRM levels increase considerably for the older population and those living in less climate-proof housing (Harlan et al., 2013).

As proposed in Santamouris and Fiorito (2021), parametric correlations between the HRM_{in} and the corresponding climatic, geographic, landscape, and socioeconomic data may be used to predict the magnitude of the initial HRM, as shown in Eq. (7.4):

$$HRM_{in} = f(LAT, \ T_{17}, \ PD, \ POV, \ VC, \ PCS, \ POP) \qquad (7.4)$$

where T_{17} is the afternoon daily ambient temperature, LAT is the latitude of the place, PD is the population density, VC is the percentage of the vegetation cover, POV is the Poverty Rate (Percentage of People in Poverty), PCS is the percentage of streets, and POP is the total population, while a_j are coefficients.

It is found that the parameters related to population and poverty present the highest significance levels with HRM_{in}. and Eq. (7.4) may be formulated accurately as below:

$$HRM_{in} = a_{00} + a_{11}POV + a_{22}POP \qquad (7.5)$$

If $HRM_{in} < 0$ then $HRM_{in} = 0$

The corresponding coefficients are given in Table 7.4. (Eq. 7.5) is valid for the specific data set and should not be used out of the limits of the input parameters.

As it concerns the final mortality, corresponding to increased albedo scenarios, it was found to vary between 38.6 and 0.93 daily deaths per

Table 7.4 Coefficients of equations.

Equation	Coefficients
Eq. (7.5) Coefficients	$a_{00} = -3{,}041$, $a_{11} = 0.236$, $a_{22} = 4.77$ e-06
Eq. (7.5) Standardized coefficients	$b_{11} = 0.14$, $b_2 = 0.94$
Eq. (7.6) Coefficients	$c_0 = -5.82$, $c_1 = 0.157$, $c_2 = -1.183$, $c_3 = 0.0395$, $c_4 = 0.098$, $c_5 = 0.043$, $c_6 = -4.13E\text{-}07$, $c_8 = 0.031$
Eq. (7.7)	$c_{00} = -3.844$, $c_{11} = 0.089$, $c_{22} = 1.59$, $c_{33} = 0.080$
Eq. (7.8)	$m_0 = -3.88$. $m_1 = 3.13$, $m_2 = 0.084$, $m_3 = 0.077$
Eq. (7.9)	$m_0 = -4.55$. $m_1 = 4.16$, $m_2 = 0.102$, $m_3 = 4.39E\text{-}7$

Source: From Santamouris M., Fiorito, F., 2021. On the impact of modified urban albedo on ambient temperature and heat related mortality. Solar Energy 216.

100,000 population. The corresponding average and median values are 9.9 and 7.5 deaths per day, while the decrease in HRM because of albedo increase is found to vary between 0.06 and 4.0 deaths per day with an average value close to 0.9 deaths per day. In parallel, the relative decrease in the initial HRM was found to vary between 1.1% and 22.6%, with an average value close to 9.3%. The decreased mortality per temperature degree of temperature drop is found to vary between 0.13 and 31.1 deaths per day and degree, with an average value close to 3.6 deaths per day and degree.

As in the case of the initial HRM, the observed reduction in HRM caused by the increase of the urban albedo is found to depend on many technical, climatic, landscape, socioeconomic, and demographic parameters. As proposed in Santamouris and Fiorito (2021), the drop of the mortality, DMort caused by the increase of the urban albedo, can be predicted accurately as a function of the initial mortality (HRM$_{in}$), the decline in the ambient temperature ΔT_{17}, the location's latitude (LAT), the poverty rate (POV), the vegetation cover (VC), the percentage of streets (PCS), and the total population of cities (POP).

$$\text{Log(DMort)} = c_0 + c_1 HRM_{in} + c_2 \Delta T_{17} + c_3 VC + c_4 PCS + c_5 PR + c_6 POP + c_7 Lat \tag{7.6}$$

The coefficients c_i are given in Table 7.4. It was observed that the drop of mortality, DMort, presents a relatively higher statistical significance association with the HRM$_{in}$, ΔT_{17}, and POV parameters. Thus the following relation can be used to accurately predict the final HRM caused by the increase of the albedo:

$$Log(DMort) = c_{00} + c_{11}HRM_{in} + c_{22}\Delta T_{17} + c_{33}POV \qquad (7.7)$$

The corresponding coefficients c_{ij} are given in Table 7.4. Using a combination of the above equations, it can be easily obtained that:

$$Log(DMort) = m_0 + m_1\Delta Alb + m_2HRM_{in} + m_3POV \qquad (7.8)$$

The coefficient m_j and the statistical characteristics of Eq. (7.8) are given in Table 7.4. When estimated values of HRM_{in}, like the ones provided by Eq. (7.5), are used as inputs to Eq. (7.8), the estimated DMort is predicted at acceptable accuracy levels, and the final equation takes the following form:

$$Log(DMort) = m_0 + m_1\Delta Alb_{in} + m_2POV + m_3POP \qquad (7.9)$$

The coefficients m_i, as well as the statistical characteristics of Eq. (7.9), are given in Table 7.4.

7.2.3 The impact of increased urban albedo on energy consumption and electricity generation

There are very few studies evaluating the impact of the potential implementation of reflective materials at the city scale on the energy consumption of the building stock. Two specific studies focusing on Darwin and Sydney, Australia, have thoroughly evaluated the energy benefits arising from the large-scale implementation of reflective materials in the cities (Shamila et al., 2020; Santamouris et al., 2020).

In Darwin, Australia, simulations using the ENVImet software model have been carried out to estimate the potential temperature decrease caused by an increase of the urban albedo from 0.2 to 0.4 and 0.6, respectively (Shamila et al., 2020). Simulations have shown that the maximum local ΔT_a decrease of the temperature was 0.7°C and 1.8°C, corresponding to albedo increases of 0.4 and 0.6, respectively, while the peak surface temperature drop was close to 15°C. Using dynamic building simulation techniques, the energy demand of representative residential and office buildings was calculated for the existing and reflective materials' scenarios. Then, the total cooling energy savings for the Darwin city were estimated based on the total floor area of residential and commercial buildings. It is found that the use of reflective materials may save 137.2 and 76.8 GWh of cooling energy for the residential and commercial buildings, respectively. The energy gains correspond to about 7% of the total cooling load of the building stock.

A similar study has been performed for the area of Western Sydney (Santamouris et al., 2020), considering an increase in the albedo of roofs 0.15 − 0.60, and from 0.08 to 0.40 for pavements. Using mesoscale simulations it was estimated that the peak ambient temperature in Western Sydney may decrease by 1°C−1.5°C. Using the existing and simulated ambient temperature data for the high albedo scenario, the energy consumption of the residential and commercial building stock was calculated. It was estimated that the decrease of the ambient temperature caused by the albedo rise could reduce the cooling load of the buildings close to 9% on an annual basis.

7.3 The impact of increased green infrastructure on urban temperature and health

7.3.1 Introduction

Green Infrastructure, GI, is a term used widely; however, its meaning may differ highly between different authors. According to the European Environmental Agency: "Green infrastructure as a term is used for a network of green features that are interconnected and therefore bring added benefits and are more resilient" (European Environmental Agency EEA, 2011). Another definition provided in Connop et al. (2016) characterizes the GI as "a network of natural and semi-natural green spaces such as forests, parks, green roofs and walls that can provide nature-based and cost-effective solutions."

Increasing the GI in a city may result in important climatic, environmental, and health benefits. Trees can reduce the magnitude of urban overheating, remove pollutants, provide carbon sequestration, retention, and detention of stormwater runoff (Connop et al., 2016; Santamouris et al., 2018a; Selmi et al., 2016; Anav et al., 2016; Pataki et al., 2011; Gascon et al., 2016). As a result, large greening projects are implemented or planned in big cities like New York, Syndey, London etc. Although existing information on the quantitative benefits of greenery is rich, there is a serious uncertainty on the magnitude of the benefits and the associated disservices, like the increased use of water, the emission of additional biogenic volatile organic compounds (BVOCs), and the additional economic cost.

7.3.2 Data and characteristics

A complete analysis of 29 articles investigating the impact of increased GI on the ambient temperature, heat-related mortality and morbidity, and air pollution is offered in Santamouris and Osmond (2020). The considered articles included 55 fully evaluated greenery scenarios for 29 cities, 19 out of which are in North America, eight in Australia, seven in Europe, three in Asia, and one in South America. Eleven studies reported the impact of the increased GI on ambient temperature and HRM, seven on ambient temperature and heat-related mortality and morbidity, five on ambient temperature and heat-related morbidity, 10 on air quality, four on ambient temperature and air quality, while 25 scenarios investigated the impact just on ambient temperature. The characteristics of all the considered scenarios are given in Table 7.5.

7.3.3 The impact of increased green infrastructure on ambient temperature—mitigation potential

Increased GI in cities may contribute to decreasing the ambient temperature because of the higher levels of evapotranspiration and shading while contributing to reducing the sensible heat emitted by the opaque massive urban elements like pavements or roads. Increased GI modifies the thermal balance of cities, and its specific contribution depends on the local climatic conditions, the availability of soil moisture, the type of vegetation, and the way it is distributed in a city.

An analysis of the specific studies given in Table 7.5, as performed in Santamouris and Osmond (2020), shows a significant correlation between the increase of the GI in a city, ΔGI, and the corresponding drop of the peak daily temperature, ΔT_{15} (Fig. 7.2).

$$\Delta T_{15} = a \ \Delta GI \tag{7.10}$$

where $a = 0.0165$. To improve the accuracy of the above expression, the available data were clustered into five clusters of ΔGI, and the corresponding average value of ΔT_{15} was correlated against the mean ΔGI value. It is found that the correlation can be better presented by a second-degree polynomial as below:

$$\Delta T_{15mean} = 0.0001 \ \Delta GI^2 + 0.0079 \ \Delta GI \tag{7.11}$$

Table 7.5 Characteristics of studies on the impact of greenery on heat related mortality and morbidity.

Part 1. Assessment of the mitigation potential, HR mortality and morbidity

City	Scenario	Simulation period	Simulation tool	Impact on temperature	Impact on mortality	Impact on morbidity	References
Darwin, Australia	Greenery increase by 20%.	Wet and dry seasons in 2016.	Envimet 6 × 6 m.	Maximum daily temperature decreased by 0.5°C.	Calculation of the HRM based on correlations between the local ambient temperature and mortality. Decrease of mortality by 19.3% against the base case.	Greenery reduces the annual excess hospital admissions of 40.14 to 27.51.	Shamila et al. (2020)
Paramatta Sydney, Australia	Addition of 2 million new trees.	Summer months 2002–2016.	WRF Resolution: 500 × 500 m.	Daily temperature decreased by 1.0°C.	Calculation of the HRM based on correlations between the local ambient temperature and mortality. Decrease of mortality by 49% against the base case.	Daily excess HR morbidity decreases from 3.66 hospital admission per day to about 2.6.	Santamouris et al. (2020)

Part 2: Assessment of the mitigation potential and HR mortality

City	Scenario	Simulation period	Simulation tool	Impact on temperature	Impact on mortality	References
New Orleans, USA	Increase of the GI from 25% to 35%.	Heat waves May 31–June 2, 1998 June 16–18, 1998 July 14–18, 2000.	WRF Resolution 2 km × 2 km	Reduction of the afternoon ambient temperature by 0.21°C.	Calculation of the HRM for the offensive air mass types. Decrease of HRM by 0.82 deaths per day and 100,000 population, or 15% of the base case.	Kalkstein and Sheridan (2003)
Philadelphia, USA	Increase of the GI from 15% to 25%.	Heat waves June 22–26, 1997 July 3–8, 1999 July 23–29, 1999.	WRF Resolution 2 km × 2 km.	Reduction of the afternoon ambient temperature by 0.32°C.	Calculation of the HRM for the offensive air mass types. Decrease of HRM by 0.67 deaths per day and 100,000 population or 5.7% of the base case.	Kalkstein and Sheridan (2003)
Detroit, USA	Increase of the GI from 15% to 25%.	Heat waves July 13–16, 1995 June 6–11, 1999 August 6–9, 2001.	WRF Resolution 1 km × 1 km.	Reduction of the afternoon ambient temperature by 0.1°C.	Calculation of the HRM for the offensive air mass types. Decrease of HRM by 0.09 deaths per day and 100,000 population or 1.5% of the base case.	Kalkstein and Sheridan (2003)
Philadelphia, USA	Increase of the GI by 6%.	Years 2020–2049.	Based on past WRF simulations.	Reduction of the maximum daily ambient temperature by 0.14°C.	Calculation of the HRM for the offensive air mass types. Decrease of the mortality by 5.6% compared to the base case.	Kalkstein and Sheridan (2003)

(Continued)

Table 7.5 (Continued)

Philadelphia, USA	Increase of the GI by 36%.	2020–2049.	Based on past WRF simulations.	Reduction of the maximum daily ambient temperature by 0.97°C.	Calculation of the HRM for the offensive air mass types. Decrease of the mortality by 26.6% compared to the base case. Reduction of deaths by 135 to 315 deaths over the period 2020 through 2049.	Kalkstein and Sheridan (2003)
Melbourne CBD area, Australia	Increase of the GI from 15% to 100%.	2009–2050.	(UCM-TAPM) Multiple one-way nesting procedure. Steps of 30, 10, 3, and 1 km.	Decrease the maximum daily temperature by 1.7°C.	Calculation of the HRM based on correlations between the local ambient temperature and mortality of elderly people. Decrease of mortality by 45% against the base case.	Chen et al. (2014)
Melbourne CBD area, Australia	Increase of the GI from 15% to 49%.	2009–2050.	(UCM-TAPM) Multiple one-way nesting procedure. Steps of 30, 10, 3, and 1 km.	Decrease the maximum daily temperature by 0.76°C.	Calculation of the HRM based on correlations between the local ambient temperature and mortality of elderly people. Decrease of mortality by 20% against the base case.	Chen et al. (2014)
Melbourne CBD area, Australia	Increase of the GI from 15% to 38%.	2009–2050.	(UCM-TAPM) Multiple one-way nesting procedure. Steps of 30,10, 3, and 1 km.	Decrease the maximum daily temperature by 0.6°C.	Calculation of the HRM based on correlations between the local ambient temperature and mortality of elderly people. Decrease of mortality by 10% against the base case.	Chen et al. (2014)

City	Scenario	Simulation period	Simulation tool	Impact on temperature	Impact on morbidity	References
Melbourne CBD area, Australia	Decrease of the GI from 15% to 5%.	2009–2050.	(UCM-TAPM) Multiple one-way nesting procedure. Steps of 30,10, and 1 km.	Increase the maximum daily temperature by 0.2°C.	Calculation of the HRM based on correlations between the local ambient temperature and mortality of elderly people. Increase of mortality by 13% against the base case.	Chen et al. (2014)
Melbourne CBD area, Australia	Increase of the GI from 15% to 33%.	2009–2050.	(UCM-TAPM) Multiple one-way nesting procedure. Steps of 30 m 10, 3, and 1 km.	Decrease the maximum daily temperature by 0.25°C.	Calculation of the HRM based on correlations between the local ambient temperature and mortality of elderly people. Decrease of mortality by 12% against the base case.	Chen et al. (2014)
Dallas, USA	Increase of the GI by 7.5%.	Summer months 2011.	WRF Resolution: 500 × 500 m.	Decrease the maximum daily temperature by 0.38°C.	Exposure – response relationship between temperature and mortality. Decrease of mortality by 5.7% against the base case.	Urban Heat Island Management Study (2017)

Part 3: Assessment of the mitigation potential and HR morbidity

City	Scenario	Simulation period	Simulation tool	Impact on temperature	Impact on morbidity	References
Phoenix, USA	Increase the GI by 5%.	2002–2006.	Zero-dimensional energy balance model.	Decrease of the average daily temperature by 1.7%.	Heat-related emergency calls decreased by 17%.	Silva et al. (2010a)
Phoenix, USA	Increase the GI by 10%.	2002–2006.	Zero-dimensional energy balance model.	Decrease of the average daily temperature by 3.6%.	Heat-related emergency calls decreased by 35%.	Silva et al. (2010a)

(Continued)

Table 7.5 (Continued)

	Scenario	Simulation period	Simulation tool	Temperature decrease	Impact on air quality	References
Phoenix, USA	Increase the GI by 15%.	2002–2006.	Zero-dimensional energy balance model.	Decrease of the average daily temperature by 5.4%.	Heat-related emergency calls decreased by 53%.	Silva et al. (2010a)
Phoenix, USA	Increase the GI by 20%.	2002–2006.	Zero-dimensional energy balance model.	Decrease of the average daily temperature by 7.2%.	Heat-related emergency calls decreased by 70%.	Silva et al. (2010a)
Oslo, Norway	Zero GI in the city.	Summer 2018.	Satellite measured surface temperature data correlated against ambient air temperature.	—	No relation between temperature and morbidity except of skin-related problem. Trees reduce the potential heat exposure for the elderly by 1.3 ± 0.1 heat risk person days.	Venter et al. (2020)

Part 4: Assessment of the air quality and mitigation potential

City	Scenario	Simulation period	Simulation tool	Temperature decrease	Impact on air quality	References
Bronx, NY, USA	Increase of GI from 24. % to 26.2%.	2010–2030.	i-tree assessment software.	The maximum temperature decrease is close to 0.09°C.	Increase of the removal PM2.5 by 9.8%, 13.7%, and 21.6% for the high, average, and low tree mortality rate.	Nyelele et al. (2019)
Northeastern USA	Increase of the urban GI from 20% to 40%.	Period of heat waves	CSUMM tool	Maximum decrease of the temperature 0.4°C	Decrease in daytime hourly ozone concentrations of 1 ppb (2.4%) with a peak decrease of 2.4 ppb (4.1%). Increases in some parts of the computational domain.	Belle Hudischewskyj Sharon and Douglas Jeffrey (2001); Nowaka et al. (2000)

City	Scenario	Simulation period	Simulation tool	Impact on temperature	Impact on air quality	References
New York, USA	Increase of GI by 30%.	Heat waves.	MMA mesoscale tool.	Decrease of the maximum temperature by 0.4°C.	Domain-wide drop of about 4 ppb of ozone (132 ppb to 128 ppb).	Calfapietra et al. (2013)
New York, USA	Increase of GI by 10%.	Heat waves	MMA mesoscale tool	Decrease of the maximum temperature by 0.15°C	Domain-wide drop of about 4 ppb of ozone (132 ppb to 128 ppb).	Calfapietra et al. (2013)

Part 5: Assessment of the air quality

City	Scenario	Simulation period	Simulation tool	Impact on air quality	References
Brooklyn, Melbourne, AU	Increase of trees from 20 to 80 trees per hectare.	Current.	i-tree assessment software.	Increase of the pollutants removal by 660% from 577 kg to 4500 kg NO_2: from 68 to 964 kg/SO_2: from 22 to 125 kg/PM_{10}: from 225 to 1474 kg. $PM_{2.5}$: from 7 to 43 kg/O_3: from 246 to 1885 kg/CO: from 9 to 10 kg.	Beckett et al. (2000)
Atlanta, USA	Reduction of urban GI by 20%.	Current.	The OZIPM4 Computer Tool.	Increase of the ozone concentration (0% − 5%).	Atkinson (2002)
Kansas City, USA	Nonquantified increase of GI.	Current.	WRF-CMAQ tool.	Increase of $PM_{2.5}$ 10% or 1.1 $\mu g\ m^{-3}$ during the night period. Decrease of the O_3 concentration by 2.0 ppbv during the daytime, and 5.2 ppbv during the night. Increase in some domains of the city.	Atkinson (2002)
West Midlands, UK	Increase GI from 3.75% to 16.5%.	—	Atmospheric FRAME model.	Reduction of the average PM_{10} concentrations by 10% from 2.3 to 2.1 $mg\ m^{-3}$ removing 110 tons per year of primary PM_{10} from the atmosphere.	McDonald et al. (2007)
West Midlands, UK	Increase GI from 3.75% to 54%.	—	Atmospheric FRAME model.	Reduction of the average PM_{10} concentration by 26%, removing 200 tons of primary PM_{10} per year.	McDonald et al. (2007)

(Continued)

Table 7.5 (Continued)

City	Scenario		Simulation tool		References
Glasgow, UK	Increase GI from 3.75% to 8.0%.	—	Atmospheric FRAME model.	Reduction of the primary PM_{10} concentrations by 2%, removing 4 ton PM_{10} per year.	McDonald et al. (2007)
Glasgow, UK	Increase GI from 3.75% to 21%.	—	Atmospheric FRAME model.	Reduction of the primary PM_{10} air concentrations by 7%, removing 13 ton of primary PM_{10} per year.	McDonald et al. (2007)
California's South Coast Air Basin, USA	Moderate increase of greenery by 6%.	—	CSUMM Tool.	If low emitting plants are used, the decrease of the population-weighted exceedance exposure to ozone above the Californian and National thresholds are up to 14% during peak afternoon hours, respectively.	Taha et al. (1997)
California's South Coast Air Basin, USA	High increase of greenery by 12%.	—	CSUMM Tool.	If low emitting plants are used, the decrease of the population-weighted exceedance exposure to ozone above the Californian and National thresholds are up to 22% during peak afternoon hours, respectively.	Taha et al. (1997)
Greater London Area, UK	Increase of GI from 20% to 30%.	—	The Urban Forest Effects Model (UFORE).	Deposition of 1109—2379 tons of PM10 (1.1%—2.6% removal) by the year 2050.	Tallis et al. (2011)

Part 6: Assessment of the mitigation potential

City	Scenario	Simulation period	Simulation tool	Temperature decrease	References
Sao Paolo, Brazil	Increase of GI by 11%. Street trees.	Summer 2014.	Envimet.	Decrease of the maximum daily temperature by 0.6°C	Duarte et al. (2015)
Sao Paolo, Brazil	Increase of GI by 11%. Pocket parks.	Summer 2014.	Envimet.	Decrease of the maximum daily temperature by 0.4°C.	Duarte et al. (2015)
Brisbane, Australia	Increase of GI from 0% to 45%.	2000—2010.	CCAM CSIRO.	Decrease of the night temperature by 1.83°C, average T by 0.99°C, and peak temperature by 0.44°C.	Santamouris et al. (2018b)
Archerfield, Australia	Increase of GI from 0% to 45%.	2000—2010.	CCAM CSIRO.	Decrease of the night temperature by 1.58°C, average T by 0.94°C, and peak temperature by 0.40°C.	Santamouris et al. (2018b)

Location	Intervention	Period	Model	Result	Reference
Logan, Australia	Increase of GI from 0% to 45%.	2000–2010.	CCAM CSIRO.	Decrease of the night temperature by 1.58°C, average T by 0.94°C, and peak temperature by 0.40°C.	Santamouris et al. (2018b)
H.K. China	Increase of the GI from 0% to 100%.	March 2000.	MM5.	Decrease of the maximum daily temperature by 1.6° C.	Carly et al. (2019)
Melbourne, Australia	Increase of the mixed forest by 20%–50%.	Heat waves January 27–30, 2009.	WRF.	Decrease of the UHI intensity from 0.5 to 5 during the night time. Nonsignificant differences during the daytime.	Stephanie et al. (1748a)
Melbourne, Australia	Increase of the GI by 5%, 10%, 40%.	12 heat waves 1990–2014.	WRF.	Decrease of the night temperature by 0.28°C, 0.38° C, and 1.08°C cooler than the control for the three scenarios.	Sarah et al. (2017)
Bochum, Germany	Increase of GI by 2%.	Heat wave of summer 2010.	WRF.	Decrease of the maximum daily temperature by 0.45°C.	Joachim et al. (2013)
Vienna	Increase of the size of urban parks by 20%.	1981–2010.	MUCLIMA 3.	Maximum decrease of the nighttime ambient temperature by 1°C.	Li, Leslie (n.d.)
New York City, USA	Increase of trees by 8%.	Heat waves 2002.	MM5.	Average decrease at 03:00 p.m. close to 0.22°C.	Luley and Bond (2002)
Mid-Manhattan West, NY, USA	Increase of trees by 8.8%.	Heat waves 2002.	MM5.	Average decrease at 03:00 p.m. close to 0.275°C.	Luley and Bond (2002)
Lower Manhattan East, NY, USA	Increase of trees by 9.9%.	Heat waves 2002.	MM5.	Average decrease at 0300p m close to 0.275°C.	Luley and Bond (2002)
Fordham Bronx, NY USA	Increase of trees by 6.2%.	Heat waves 2002.	MM5.	Average decrease at 03:00 p.m. close to 0.33°C.	Luley and Bond (2002)
Maspeth Queens, NY USA	Increase of trees by 14.4%.	Heat waves 2002.	MM5.	Average decrease at 03:00 p.m. close to 0.22°C.	Luley and Bond (2002)
Crown Heights, NY USA	Increase of trees by 13.4%.	Heat waves 2002.	MM5.	Average decrease at 03:00 p.m. close to 0.495°C.	Luley and Bond (2002)
Ocean Parkway, NY, USA	Increase of trees by 8%.	Heat waves 2002.	MM5.	Average decrease at 03:00 p.m. close to 0.495°C.	Luley and Bond (2002)

(Continued)

Table 7.5 (Continued)

Singapore	Increase of trees by 45%.	April and May period.	WRF.	Decrease of the maximum daily temperature close to 0.3°C, and close to 1.5°C during the night.	Arghavani et al. (2019)
Stuttgart, Germany	Increase of trees by 12%.	Heat waves August 2003.	WRF.	Decrease of the average peak temperature by 0.13°C. Maximum decrease up to 2°C.	Žuvela-Aloise et al. (2016)
Tehran, Iran	Increase of trees by 20%.	June 2016.	WRF.	Decrease of the maximum daily temperature up to 0.6°C. Increase of the nighttime temperature up to 1.5°C.	Tong et al. (2005)
Brampton, Toronto, Canada	Increase of trees by 50%.	Heat waves 2018.	WRF.	Decrease of the peak daily temperature by 0.39°C.	Kalkstein and Sheridan (2003)
Brampton, Toronto, Canada	Increase of trees by 80%.	Heat waves 2018.	WRF.	Decrease of the peak daily temperature by 1.61°C.	Kalkstein and Sheridan (2003)
Caledon, Toronto, Canada	Increase of trees by 50%.	Heat waves 2018.	WRF.	Decrease of the peak daily temperature by 0.59°C.	Kalkstein and Sheridan (2003)
Caledon, Toronto, Canada	Increase of trees by 80%.	Heat waves 2018.	WRF.	Decrease of the peak daily temperature by 1.29°C.	Kalkstein and Sheridan (2003)
Sydney, Australia	Increase of trees by 55%.	Whole summer period.	Envimet.	Decrease of the peak daily temperature by 1.2°C.	Santamouris et al. (2020)

Source: From Santamouris M., Osmond, P., 2020. Increasing green infrastructure in cities—impact on ambient temperature, air quality and heat related mortality and morbidity. Buildings 10, 233; doi:10.3390/buildings10120233.

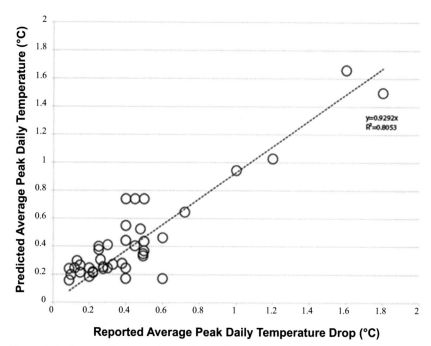

Reported Average Peak Daily Temperature Drop (°C)

Figure 7.2. Reported against the predicted from Eq. (7.1), values of the average peak daily ambient temperature drop. *From Santamouris M., Osmond, P., 2020. Increasing green infrastructure in cities—impact on ambient temperature, air quality and heat related mortality and morbidity. Buildings 10, 233; doi:10.3390/buildings10120233.*

As it concerns the temperature decrease caused by the increased GI during the night time it is found that it can be described by a linear correlation as below (Fig. 7.3):

$$\Delta T_{NGH} = b \ \Delta GI \qquad (7.12)$$

where $b = 0.0277$. The whole analysis of data performed in Santamouris and Osmond (2020) resulted in the following conclusions:

1. The potential maximum decrease of the average daily peak temperature caused by the increased urban GI may not exceed 1.8°C even if the GI increases up to 100%. For a reasonable increase of the GI by 20%, the expected peak temperature drop may not exceed 0.3°C.

2. The maximum decrease of the ambient temperature during the night time may not exceed 2.3°C, while an increase of the GI by 20% may decrease the ambient temperature up to 0.5°C.

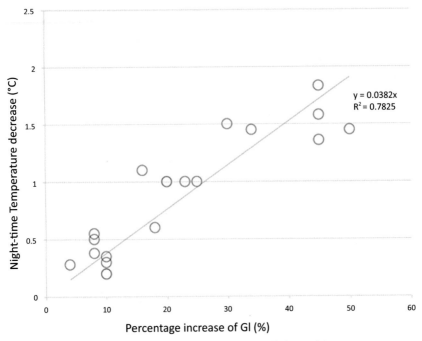

Figure 7.3 Correlation between the maximum drop of the ambient temperature at night against the corresponding increase of the tree covering. *From Santamouris M., Osmond, P., 2020. Increasing green infrastructure in cities—impact on ambient temperature, air quality and heat related mortality and morbidity. Buildings 10, 233; doi:10.3390/buildings10120233.*

3. It is found that 19 out of the 22 considered studies reported higher temperature decrease during the night than the daytime period. This is attributed to two main reasons: (1) higher levels of GI may seriously reduce the stored heat in the ground during the daytime and the corresponding release of sensible heat by the ground during the night, and (2) the sensible heat by the trees released during the night is substantially lower than the heat released by the impervious urban surfaces, In parallel, it is important to mention that the daytime evapotranspiration is the main cooling mechanism; however, its contribution on the surface energy budget may not compensate the heat fluxes because of the advection and sensible heat by the impervious surfaces.

4. Several studies report warming effects up to 2°C during the night time associated with an increased GI (Stephanie et al., 1748b; Tong et al.,

2005). This is due to the decreased sky view factor in the canyons that limits the escape of longwave radiation.

5. An increase in the urban GI may result in higher ambient-relative humidity levels and lower levels of thermal comfort. This is mainly observed in tropical cities where the relative humidity is considerably high.

7.3.4 Impact of increased green infrastructure on heat-related mortality

An analysis of 13 case studies evaluating the impact of increased GI on HRM is offered in Table 7.5, as studied in Santamouris and Osmond (2020). The studies present 13 different scenarios for seven cities: New Orleans, Dallas, Philadelphia, and Detroit in the United States, and Melbourne, Parramatta, Sydney, and Darwin in Australia. Correlations were investigated between the HRM historical data and climatic parameters to assess the mortality variability. All the considered studies estimated the potential decrease of the daily temperature and the mortality levels before and after the increase of the urban GI.

As shown in Santamouris and Osmond (2020), an increase of the urban GI may reduce HRM levels between 1.5% and 49%. A linear correlation is observed between the drop in the maximum daily temperature and the corresponding percentage of HRM decrease (Fig. 7.4). It is observed that a decrease of the maximum ambient temperature by 1°C reduces the HRM by about 29.5%, as in the following relation:

$$\text{Percentage Decrease of HRM} = 29.44 \times \text{DTmax} \qquad (7.12)$$

where DTmax, is the decrease of the maximum daily temperature. There was no important association between the decrease of the heat-related mortality caused by the increase of the urban GI with local poverty rates and population size. Although there is a clear positive association between the increased tree cover, GI, and the corresponding mortality, several issues may be considered as discussed below.

1. Given that more than 20 years are needed to achieve the full benefits of the increased GI, and the corresponding change of the demographic conditions and the expected adaptation of human societies during the same period, the use of the actual correlations between the ambient conditions and the HRM based on past health data incorporate a very significant uncertainty regarding the future mortality assessments.

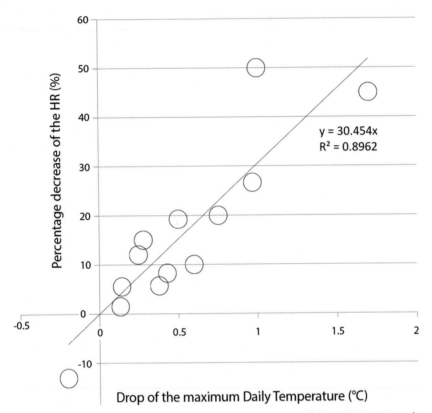

Figure 7.4 Correlation between the percentage decrease of the HRM against the corresponding decrease of the peak daily temperature at 1500 p.m. *From Santamouris M., Osmond, P., 2020. Increasing green infrastructure in cities—impact on ambient temperature, air quality and heat related mortality and morbidity. Buildings 10, 233; doi:10.3390/buildings10120233.*

2. Given that the duration and the amplitude of heat waves are increasing, the synergy between the global and local climate change that increase further the ambient temperature and the lack of cooling capacity of greenery above a threshold temperature, the current relations between the impact of increased GI on the ambient temperature may be seriously modified altering the capacity of urban greenery to reduce HRM.
3. Given that additional greenery may increase the concentration of pollutants in cities because of the emission of BVOCs and the important impact of these pollutants on health, holistic studies considering the whole spectrum of the benefits and drawbacks of increased GI have to be carried out.

7.3.5 Impact of green infrastructure on heat-related morbidity

Contrary to the high availability of studies on the impact of GI on HRM, very few studies are available on the potential effect on heat-related morbidity (Shamila et al., 2020; Maurice Middendorp, 2013; Silva et al., 2010b; Venter et al., 2020). The characteristics of all the studies are summarized in Table 7.5. In particular, it is reported that when in Phoenix, Arizona, in increasing the baseline fraction of vegetated area from 30.4% by 5%, 10%, 15%, and 20%, the average 24 hours temperature is lowered by 1.7%, 3.6%, 5.4%, and 7.2%, which corresponds to a decrease in the number of heat-related emergency calls by 17%, 35%, 53%, 70%, respectively.

A second study analyzing the impact of GI on heat-related morbidity in Oslo, Norway, found that if the most common nontree cover replaces trees, the threshold of health risk may increase from 23% to 29%, in all neighborhoods of Oslo exceeding 30°C.

A third study carried out for Darwin, Australia, found that a possible increase of the GI by 20% may reduce the peak ambient temperature by 0.5°C, and decrease the annual excess hospital admissions of 40.1 to 27.5 per 100,000 population. A similar study for Sydney, Australia, found that rise of the maximum ambient temperature by 1°C, increases the heat-related hospital admissions by 1.1%, or 4.6% for the days with maximum temperature above 27°C. It is also found that when 2 million new trees are planted the daily excess hospital admission decreases from 3.66 to about 2.6 per 100,000 people.

7.3.6 Impact of increased green infrastructure on urban pollution levels

Trees cause a significant and dynamic change of the climatic and chemical conditions in the lower atmosphere. Through the stomata of the leaves, urban vegetation absorbs gaseous pollutants such as NOx and O_3 while it accumulates particulate matter by impaction, sedimentation, and interception (Fantozzi et al., 2015). Tree species preset different absorption characteristics and potential. In parallel, several parameters like the tree species traits, the concentration of the pollutants, the stickiness, porosity and roughness as well as the leaf area, the length of the leaf season, the size of particles, meteorological condition, the canopy structure, and possible precipitation, determine the pollutants' removal capacity of trees (Manes et al., 2007; Beckett et al., 2000). On the other hand, the emission of BVOCs by trees, like isoprenoids, monoterpenes, and sesquiterpenes, is a serious concern for the built environment. BVOCs are several times more reactive than the emissions from petrol combustion and also much more reactive than the anthropogenic VOC emissions (Atkinson, 2002; Carter, 1994; Calfapietra et al., 2013).

An increase in the urban GI has positive and negative impacts on urban atmospheric pollution levels. A decrease in the ambient temperature caused by trees reduces the BVOCs' emission rates while decreasing the rate of photochemical reactions. However, it causes a reduction of the buoyancy and turbulence mixing resulting in lower Planetary Boundary Layer Heights (PBLH). Lower PBLH can increase the concentration of gaseous pollutants, like NOx and primary airborne particles (Long et al., 2018). Lower surface temperatures in coastal areas may limit the circulation of sea breeze and the transport of precursors and pollutants in the city.

An analysis of 11 studies around the impact of additional GI in cities is provided in Santamouris and Osmond (2020). The characteristics of the case studies are given in Table 7.5. As it concerns the impact of additional greenery on the concentration of the atmospheric particulate matter, the study has concluded the following:

1. More urban trees increase the atmospheric removal of particles considerably mainly through deposition processes. Several parameters like the trees species, the type of tree covering, the concentration of the particles in the atmosphere, and the specific transport and climatic conditions affect the magnitude and the rate of the removal.

2. The deposition of the pollutants depends mainly on the size of particles as small-diameter particles with a diameter between 1 mm and 10 mm, which are deposited because trees force the air flow to bend particles, while large particles (> 10 mm in diameter) fall on the soil below the trees by sedimentation Finally, ultrafine particles (below 1 mm) are transported by diffusion (Sara, 2015; Hinds, 1999).

3. It is found that trees in urban canyons change the roughness properties resulting in a significant increase in the concentration of particulate matter (Buccolieri et al., 2011; Salmond et al., 2016; Wania et al., 2012).

4. An increase of the urban GI may increase the concentration of the atmospheric particles as lower surface temperature reduces the height of the PBL and blocks the particles in the lower atmosphere.

As it concerns the impact of additional urban greenery on the concentration of ground-level ozone, the following conclusions are drawn:

1. Given that the atmospheric dynamics and the chemistry determining the ground-level ozone concentration are quite complex, the relationship between increased tree cover and ozone concentration is not simple. Ozone is generated because of the photolysis of NO_2 under the presence of VOCs. More urban trees result in increased BVOCs emissions, a decrease of the emission rate, a higher dry deposition and absorption of ozone and NOx, and a slow down of the photochemical reactions. In parallel, a decrease in surface temperature results in a lower PBL blocking NOx and other pollutants in the lower atmosphere, resulting in decreased ozone concentrations because of titration processes.

2. Dry deposition of the ozone determines its concentration in the atmosphere. According to (Nowak et al., 2018), during 2010, urban trees in 55 US cities removed almost 523 ktons/year of ozone and 12.87 ktons/year in 87 cities in Canada.

3. Most of the existing studies agree that urban trees decrease the concentration of ground-level ozone, although several experimental studies found similar ozone concentration in tree canopies compared to adjacent nonvegetative open zones.

4. It is important to use low BVOC emission trees, especially in urban areas. Low emitting trees under 2 μg/g/h of isoprene and 1 μg/g/h of monoterpenes usually result in a net decrease of the ground-level ozone. Highly or moderately emitting vegetation may result in increased ozone concentrations.

7.4 Conclusion

To fight urban overheating as well as the impact of the increased frequency and magnitude of extreme heat phenomena, high-performance heat mitigation and adaptation strategies based on the use of natural and man-made cooling technologies have to be developed and implemented.

An increase of the urban albedo and additional trees offer undeniable benefits to the urban climate and urban environment and improve the quality of life. In particular, a decrease in the ambient temperature during the warm period reduces the cooling energy consumption and peak electricity demand, the concentration of harmful pollutants, and decreases heat-related mortality and morbidity.

The existing literature on the benefits and potential drawbacks of increased urban albedo and additional urban greenery is quite limited, fragmented, and characterized by serious heterogeneity in the research questions asked and the methodology followed. Therefore the provided solutions, results, and conclusions cover a vast spectrum, can be conflicting and, in some cases, confusing and may be a significant source of misconception leading to the adoption and implementation of nonoptimum mitigation strategies.

A successful increase of the urban albedo and plantation of additional trees requires complete knowledge and assessment of the potential benefits and drawbacks. It is evident that scientific knowledge on the topic should improve considerably, mainly through detailed and precise experimental studies.

References

Anav, A., De Marco, A., Proietti, C., Alessandri, A., Dell'Aquila, A., Cionni, I., et al., 2016. Comparing concentration-based [AOT40] and stomatal uptake [PODY] metrics for ozone risk assessment to European forests. Global Change Biology 22, 1608–1627.

Arghavani, Somayeh, Malakooti, Hossein, Bidokhti, Abbasali Aliakbari, 2019. Numerical evaluation of urban green space scenarios effects on gaseous air pollutants in Tehran Metropolis based on WRF-Chem model. Atmospheric Environment 214.

Atkinson, R., 2002. Atmospheric chemistry of VOCs and NOx. Atmospheric Environment (Oxford, England: 1994) 34, 2063–2101.

Beckett, P.K., Freer-Smith, P., Taylor, G., 2000. The capture of particulate pollution by trees at five contrasting urban sites. Arboricultural Journal 24, 209–230.

Belle Hudischewskyj Sharon G., Douglas Jeffrey R. Lundgren, 31 January 2001. Meteorological and air quality modeling to further examine the effects of urban heat island mitigation measures on several cities in the northeastern U.S. SYSAPP-01-001.

Buccolieri, R., Salim, S.M., Leo, L.S., Di Sabatino, S., Chan, A., Ielpo, P., et al., 2011. Analysis of local scale tree-atmosphere interaction on pollutant concentration in idealised street canyons and application to a real urban junction. Atmospheric Environment (Oxford, England: 1994) 45 (9).

Calfapietra, C., Fares, S., Manes, F., Morani, A., Sgrigna, G., Loreto, F., 2013. Role of biogenic volatile organic compounds (BVOC) emitted by urban trees on ozone concentration in cities: a review. Environmental Pollution (Barking, Essex: 1987) 183, 71–80.

Ziter, D.C., Pedersen, E.J., Kucharik, C.J., Turner, M.G., 2019. Scale-dependent interactions between tree canopy cover and impervious surfaces reduce daytime urban heat during summer. PNAS 116 (15), 7575–7580.

Carter, W.P.L., 1994. Development of ozone reactivity scales for volatile organic compounds. Air & Waste: Journal of the Air & Waste Management Association 44, 881–899.

Chen, Dong, Wang, Xiaoming, Thatcher, Marcus, Barnett, Guy, Kachenko, Anthony, Prince, Robert, 2014. Urban vegetation for reducing heat related mortality. Environmental Pollution 192, 275–284.

Connop, S., Vandergert, P., Eisenberg, B., Collier, M.J., Nash, C., Clough, J., et al., 2016. Renaturing cities using a regionally-focused biodiversity-led multifunctional benefits approach to urban green infrastructure. Environmental Science and Pollution Research 62, 99—111.

Duarte, Denise H.S., Shinzato, Paula, dos Santos Gusson, Carolina, Abrahão Alves, Carolina, 2015. The impact of vegetation on urban microclimate to counterbalance built density in a subtropical changing climate. Urban Climate 14, 224—239.

European Environmental Agency [EEA], 2011. Green infrastructure and territorial cohesion; European Environmental Agency [EEA]: Copenhagen, Denmark.

Fantozzi, F., Monaci, F., Blanusa, T., Bargagli, R., 2015. Spatio-temporal variations of ozone and nitrogen dioxide concentrations under urban trees and in a nearby open area. Urban Climate 12, 119—127.

Gascon, M., Triguero-Mas, M., Martínez, D., Dadvand, P., Rojas-Rueda, D., Plasència, A., et al., 2016. Residential green spaces and mortality: a systematic review. Environment International 86, 60—67.

Green, H., Bailey, J., Schwarz, L., Vanos, J., Ebi, K., Benmarhnia, T., 2019. Impact of heat on mortality and morbidity in low and middle income countries: a review of the epidemiological evidence and considerations for future research. Environmental Research 171, 80—91. Available from: https://doi.org/10.1016/j.envres.2019.01.010.

Hajat, S., Kosatky, T., 2010. Heat-related mortality: a review and exploration of heterogeneity. Journal of Epidemiology and Community Health 64 (9), 753—760. Available from: https://doi.org/10.1136/jech.2009.087999.

Harlan, S.L., Declet-Barreto, J.H., Stefanov, W.L., Petitti, D.B., 2013. Neighborhood effects on heat deaths: social and environmental predictors of vulnerability in Maricopa County, Arizona. Environmental Health Perspectives 121 (2), 197—204. Available from: https://doi.org/10.1289/ehp.1104625.

Hinds, W.C., 1999. Aerosol Technology: Properties, Behavior, and Measurement of Airborne Particles, second edp. 504.

Jacobson, M.Z., Ten Hoeve, J.E., 2012. Effects of urban surfaces and white roofs on global and regional climate. Journal of Climate 25 (3), 1028—1044. Available from: https://doi.org/10.1175/JCLI-D-11-00032.1.

Jandaghian, Z., Akbari, H., 2018. The effects of increasing surface reflectivity on heat-related mortality in Greater Montreal Area, Canada. Urban Climate. 25, 135—151. Available from: https://doi.org/10.1016/j.uclim.2018.06.002.

Jandaghian, Z., Akbari, H., 2020. Heat mitigation strategy to reduce heat death in Toronto and Montreal, Canada. Energy and Buildings .

Jesdale, B.M., Morello-Frosch, R., Cushing, L., 2013. The racial/ethnic distribution of heat risk-related land cover in relation to residential segregation. Environmental Health Perspectives 121 (7), 811—817. Available from: https://doi.org/10.1289/ehp.1205919.

Joachim, F., Stefan, E., Peter, S., 2013. Mitigation of urban heat stress—a modelling case study for the area of Stuttgart. DIE ERDE Journal of the Geographical Society of Berlin 144 (3-4).

Johnson, H., Kovats, R.S., McGregor, G., Stedman, J., Gibbs, M., Walton, H., et al., 2005. The impact of the 2003 heat wave on mortality and hospital admissions in England. Health Statistics Quarterly/Office for National Statistics 25.

Kalkstein, L.S., Sheridan, S., 2003. The impact of heat island reduction strategies on health-debilitating oppressive air massive in urban areas. U.S. EPA Heat Island Reduction Initiative.

Kalkstein, L.S., Sailor, D., Shickman, K., Sheridan, S., Vanos, J., 2003. Assessing the health impacts of urban heat island reduction strategies in the District of Columbia. Global Cool Cities Alliance, Contract No.: DDOE ID#2013-10-OPS.

Kalkstein, L.S., Greene, S., Mills, D.M., Samenow, J., 2011. An evaluation of the progress in reducing heat-related human mortality in major U.S. cities. Natural Hazards 56 (1), 113−129. Available from: https://doi.org/10.1007/s11069-010-9552-3.

Keatinge, W.R., Donaldson, G.C., Cordioli, E., Martinelli, M., Kunst, A.E., Mackenbach, J.P., et al., 2000. Heat related mortality in warm and cold regions of Europe: observational study. British Medical Journal 321 (7262), 670−673. Available from: https://doi.org/10.1136/bmj.321.7262.670.

Klein Rosenthal, J., Kinney, P.L., Metzger, K.B., 2014. Intra-urban vulnerability to heat-related mortality in New York City, 1997-2006. Health and Place 30, 45−60. Available from: https://doi.org/10.1016/j.healthplace.2014.07.014.

Kumar, S., Verma, N.K., Singla, M.L., 2013. Study on reflectivity and photostability of Al-doped TiO_2 nanoparticles and their reflectors. Journal of Materials Research 28 (3), 521−528. Available from: https://doi.org/10.1557/jmr.2012.361.

Li, Xian-Xiang, Leslie, K.Norford, 2016. Evaluation of cool roof and vegetations in mitigating urban heat island in a tropical city, Singapore. Urban Climate. 16, 59−74.

Long, X., Wu, J., Li, X., Feng, T., Xing, L., Zhao, S., et al., 2018. Does afforestation deteriorate haze pollution in Beijing − Tianjin − Hebei (BTH), China? Atmospheric Chemistry and Physics 18, 10869−10879. Available from: https://doi.org/10.5194/acp-18-10869-2018.

Luley, C.J., Bond, J., 2002. A report to North East State Foresters Association. A plan to integrate management of urban trees into air quality planning. Davey Resource Group, New York State Department of Environmental Conservation, Albany, NY Division of Lands and Forests and Division of Air Resources USDA Forest Service, Northeastern Research Station, Syracuse, NY March.

Luo, Q., Li, S., Guo, Y., Han, X., Jaakkola, J.J.K., 2019. A systematic review and meta-analysis of the association between daily mean temperature and mortality in China. Environmental Research 173, 281−299. Available from: https://doi.org/10.1016/j.envres.2019.03.044.

Macintyre, H.L., Heaviside, C., 2019. Potential benefits of cool roofs in reducing heat-related mortality during heatwaves in a European city. Environment International 127, 430−441. Available from: https://doi.org/10.1016/j.envint.2019.02.065.

Manes, F., Vitale, M., Fabi, M.A., De Santis, F., Zona, D., 2007. Estimates of potential ozone stomatal uptake in mature trees of Quercus ilex in a Mediterranean climate. Environmental and Experimental Botany 59, 235−241.

Mastrapostoli, E., Santamouris, M., Kolokotsa, D., Vassilis, P., Venieri, D., Gompakis, K., 2016. On the ageing of cool roofs: measure of the optical degradation, chemical and biological analysis and assessment of the energy impact. Energy and Buildings 114, 191−199. Available from: https://doi.org/10.1016/j.enbuild.2015.05.030.

Middendorp, M., 2013. Mesoscale Modelling the Influence of Urban Vegetation on the Urban Temperatures and Human Thermal Comfort (Msc. Thesis) Meteorology and Air Quality Wageningen University, Wageningen, the Netherlands. Supervisors: G.J. Steeneveld; N.E. Theeuwes October 2013.

McDonald, W.J., Bealey, D., Fowler, U., Dragosits, U., Skiba, R.I., Smith, R.G., et al., 2007. Quantifying the effect of urban tree planting on concentrations and depositions of PM_{10} in two U.K. conurbations. Atmospheric Environment 41, 8455−8467.

Menon, S., Akbari, H., Mahanama, S., Sednev, I., Levinson, R., 2010. Radiative forcing and temperature response to changes in urban albedos and associated CO_2 offsets. Environmental Research Letters 5 (1). Available from: https://doi.org/10.1088/1748-9326/5/1/014005.

Millstein, D., Menon, S., 2011. Regional climate consequences of large-scale cool roof and photovoltaic array deployment. Environmental Research Letters 6 (3). Available from: https://doi.org/10.1088/1748-9326/6/3/034001.

Nowak, D.J., Hirabayashi, S., Doyle, M., McGovern, M., Pasher, J., 2018. Air pollution removal by urban forests in Canada and its effect on air quality and human health. Urban Forestry and Urban Greening 29, 40−48.

Nowaka, David J., Civerolo, Kevin L., Trivikrama Rao, S., Sistla, Gopal, Luley, Christopher J., Crane, Daniel E., 2000. A modeling study of the impact of urban trees on ozone. Atmospheric Environment 34, 1601−1613.

Nyelele, P.C., Kroll, C.N., Nowak, D., J., 2019. Present and future ecosystem services of trees in the Bronx, NY. Urban Forestry and Urban Greening 42.

Pataki, D.E., Carreiro, M.M., Cherrier, J., Grulke, N.E., Jennings, V., Pincetl, S., et al., 2011. Coupling biogeochemical cycles in urban environments: ecosystem services, green solutions, and mis-conceptions. Frontiers in Ecology And the Environment 9, e27−e36.

Sailor, D.J., 1995. Simulated urban climate response to modifications in surface albedo and vegetative cover. Journal of Applied Meteorology 34 (7), 1694−1704. Available from: https://doi.org/10.1175/1520-0450-34.7.1694.

Salmond, J.A., Tadaki, M., Vardoulakis, S., Arbuthnott, K., Coutts, A., Demuzere, M., et al., 2016. Health and climate related ecosystem services provided by street trees in the urban environment. Environmental Health: A Global Access Science Source 15 (1), 95.

Santamouris, M., 2014a. Cooling the cities—a review of reflective and green roof mitigation technologies to fight heat island and improve comfort in urban environments. Solar Energy 103, 682−703. Available from: https://doi.org/10.1016/j.solener.2012.07.003.

Santamouris, M., 2014b. On the energy impact of urban heat island and global warming on buildings. Energy and Buildings 82, 100−113.

Santamouris, M., Cartalis, C., Synnefa, A., Kolokotsa, D., 2015. On the impact of urban heat island and global warming on the power demand and electricity consumption of buildings—a review. Energy and Buildings 98, 119−124. Available from: https://doi.org/10.1016/j.enbuild.2014.09.052.

Santamouris, M., Ding, L., Fiorito, F., Oldfield, P., Paul Osmond, R., Paolini, D., et al., 2017. Passive and active cooling for the outdoor built environment—analysis and assessment of the cooling potential of mitigation technologies using performance data from 220 large scale projects. Solar Energy 154, 14−33.

Santamouris, M., Ban-Weiss, G., Cartalis, C., Crank, C., Kolokotsa, D., Morakinyo, T.E., et al., 2018a. Progress in urban greenery mitigation science—assessment methodologies advanced technologies and impact on cities. Journal of Civil Engineering and Management 24, 638−671.

Santamouris, Mattheos, Haddad, Shamila, Saliari, Maria, Vasilakopoulou, Konstantina, Synnefa, Afroditi, Paolini, Riccardo, et al., 2018b. On the energy impact of urban heat island in Sydney: climate and energy potential of mitigation technologies. Energy & Buildings 166, 154−164.

Santamouris, M., 2020. Recent progress on urban overheating and heat island research. Integrated assessment of the energy, environmental, vulnerability and health impact synergies with the global climate change. Energy and Buildings 207, 109482.

Santamouris, M., Yun, G.Y., 2020. Recent development and research priorities on cool and super cool materials to mitigate urban heat island. Renewable Energy 792−807.

Santamouris, M., Osmond, P., 2020. Increasing green infrastructure in cities—impact on ambient temperature, air quality and heat related mortality and morbidity. Buildings 10, 233. Available from: https://doi.org/10.3390/buildings10120233.

Santamouris, M., Paolini, R., Haddad, S., Synnefa, A., Garshasbi, S., Hatvani-Kovacs, G., et al., 2020. Heat mitigation technologies can improve sustainability in cities. An holistic experimental and numerical impact assessment of urban overheating and related heat mitigation strategies on energy consumption, indoor comfort, vulnerability and heat-related mortality and morbidity in cities. Energy and Buildings 217. Available from: https://doi.org/10.1016/j.enbuild.2020.110002.

Santamouris, M., Fiorito, F., 2021. On the impact of modified urban albedo on ambient temperature and heat related mortality. Solar Energy .

Janhäll, S., 2015. Review on urban vegetation and particle air pollution—deposition and dispersion. Atmospheric Environment 105.

Sarah, C., Thatcher, M., Alazar, A., Watson, J.E.M., Mcalpine, C.A., 2017. The effect of urban density and vegetation cover on the heat iIsland of a subtropical city. Journal of Applied Climatology and Meteorology 57, 2532.

Savio, P., Rosenzweig, C., Solecki, W.D., Slosberg, R.B., 2006. Mitigating New York City's heat island with urban forestry, living roofs, and light surfaces. New York City Regional Heat Island Initiative. New York State Energy Research and Development Authority (NYSERDA).

Schinasi, L.H., Benmarhnia, T., De Roos, A.J., 2018. Modification of the association between high ambient temperature and health by urban microclimate indicators: a systematic review and *meta*-analysis. Environmental Research 161, 168–180. Available from: https://doi.org/10.1016/j.envres.2017.11.004.

Selmi, W., Weber, C., Rivie're, E., Blond, N., Mehdi, L., Nowak, D.J., 2016. Air pollution removal by trees in public green spaces in Strasbourg city, France. Urban Forestry and Urban Greening 17, 192e201.

Shamila, Haddad, Paolini, Riccardo, Ulpiani, Giulia, Synnefa, Afroditi, Hatvani-Kovacs, Gertrud, Garshasbi, Samira, et al., 2020. Holistic approach towards urban sustainability: co-benefits of urban heat mitigation in a hot humid region of Australia. Scientific Reports 10, 14216.

Silva, H.R., Phelan, P.E., Golden, J.S., 2010a. Modeling effects of urban heat island mitigation strategies on heat-related morbidity: a case study for Phoenix, Arizona, USA. International Journal of Biometeorology 54 (1), 13–22. Available from: https://doi.org/10.1007/s00484-009-0247-y.

Silva, Humberto R., Patrick, E.Phelan, Golden, Jay S., 2010b. Modeling effects of urban heat island mitigation strategies on heat-related morbidity: a case study for Phoenix, Arizona, USA. International Journal Biometeorology 54, 13–22.

Son, J.Y., Lane, K.J., Lee, J.T., Bell, M.L., 2016. Urban vegetation and heat-related mortality in Seoul, Korea. Environmental Research 151, 728–733. Available from: https://doi.org/10.1016/j.envres.2016.09.001.

Stafoggia, M., Forastiere, F., Agostini, D., Biggeri, A., Bisanti, L., Cadum, E., et al., 2006. Vulnerability to heat-related mortality: a multicity, population-based, case-crossover analysis. Epidemiology (Cambridge, Mass.) 17 (3), 315–323. Available from: https://doi.org/10.1097/01.ede.0000208477.36665.34.

Stephanie, J.Jacobs, Gallant, Ailie J.E., Tapper, Nigel J., Li, Dan, 1748a. Use of cool roofs and vegetation to mitigate urban heat and improve human thermal stress in Melbourne, Australia. Journal of Applied Meteorology and Climatology 77.

Jacobs, S.J., Gallant, A.J.E., Tapper, N.J., Li, D., 1748b. Use of cool roofs and vegetation to mitigate urban heat and improve human thermal stress in Melbourne, Australia. Journal of Applied Meteorology and Climatology 77.

Stone Jr, B., Vargo, J., Liu, P., Habeeb, D., DeLucia, A., Trail, M., et al., 2014. Avoided heat-related mortality through climate adaptation strategies in three U.S. cities. PLoS One 9 (6). Available from: https://doi.org/10.1371/journal.pone.0100852.

Synnefa, A., Santamouris, M., Apostolakis, K., 2007. On the development, optical properties and thermal performance of cool colored coatings for the urban environment. Solar Energy 81 (4), 488–497. Available from: https://doi.org/10.1016/j.solener.2006.08.005.

Synnefa, A., Dandou, A., Santamouris, M., Tombrou, M., Soulakellis, N., 2008. On the use of cool materials as a heat island mitigation strategy. Journal of Applied Meteorology and Climatology 47 (11), 2846−2856. Available from: https://doi.org/10.1175/2008JAMC1830.1.

Taha, H., Douglas, Sharon, Haney, Jay, 1997. Mesoscale meteorological and air quality impacts of increased urban albedo and vegetation. Energy and Buildings 25, 169−177.

Tallis, Matthew, Taylor, Gail, Sinnett, Danielle, 2011. Peter Freer-Smith: estimating the removal of atmospheric particulate pollution by the urban tree canopy of London, under current and future environments. Landscape and Urban Planning 103, 129−138.

Taylor, J., Wilkinson, P., Davies, M., Armstrong, B., Chalabi, Z., Mavrogianni, A., et al., 2015. Mapping the effects of urban heat island, housing, and age on excess heat-related mortality in London. Urban Climate 14, 517−528. Available from: https://doi.org/10.1016/j.uclim.2015.08.001.

Tong, Hua, Walton, Andrew, Sang, Jianguo, Johnny, C.L.Chan, 2005. Numerical simulation of the urban boundary layer over the complex terrain of Hong Kong. Atmospheric Environment 39, 3549−3563.

Urban Heat Island Management Study. Dallas 2017. Texas Trees Foundation, 2017.

Vanos J, Kalkstein LS, Sailor D, Shickman K, Sheridan S. (n.d.) Assessing the health impacts of urban heat island reduction strategies in the cities of Baltimore, Los Angeles, and New York. Available from: https://www.adaptationclearinghouse.org/resources/assessing-the-health

Venter, Zander S., Hjertager Krog, Norun, David, N.Barton, 2020. Linking green infrastructure to urban heat and human health risk mitigation in Oslo, Norway. Science of the Total Environment 709, 136193.

Wania, A., Bruse, M., Blond, N., Weber, C., 2012. Analysing the influence of different street vegetation on traffic-induced particle dispersion using microscale simulations. Journal of Environmental Management 94 (1), 91.

Žuvela-Aloise, M., Koch, R., Buchholz, S., Früh, B., 2016. Modelling the potential of green and blue infrastructure to reduce urban heat load in the city of Vienna. Climate Change . Available from: https://doi.org/10.1007/s10584-016-1596-2.

Index

Note: Page numbers followed by "*f*" and "*t*" refer to figures and tables, respectively.

Printed in the United States
by Baker & Taylor Publisher Services